Elementary Chemical Reactor Analysis

Rutherford Aris

DOVER PUBLICATIONS, INC.
Mineola, New York

Copyright

Copyright © 1989 by Rutherford Aris.
All rights reserved under Pan American and International Copyright Conventions.

Published in Canada by General Publishing Company, Ltd., 30 Lesmill Road, Don Mills, Toronto, Ontario.

Bibliographical Note

This Dover edition, first published in 1999, is an unabridged republication of the work of the same title published in 1989 by Butterworths Publishers, Stoneham, Mass. Several minor errors have been corrected.

Library of Congress Cataloging-in-Publication Data

Aris, Rutherford.
    Elementary chemical reactor analysis / Rutherford Aris.
        p. cm.
    Originally published: Boston : Butterworths, c1989.
    Includes bibliographical references and index.
    ISBN 0-486-40928-7 (pbk.)
        1. Chemical reactors. I. Title.
TP157 .A72 1999
660'.2832—dc21   99-047132

Manufactured in the United States of America
Dover Publications, Inc., 31 East 2nd Street, Mineola, N.Y. 11501

**AD
Patrem Meum**

*doctorem meum de rebus magnis
hunc librum de rebus parvis dedico.*

Preface

This book is a more thorough rewriting of my *Introduction to the Analysis of Chemical Reactors* than might appear at first sight, for though I have retained the structure and most of the headings of the former volume there are few sections that have not been largely rewritten. I am still convinced of the soundness of the general plan and of the necessity of making the undergraduate grasp the fundamental ideas and become aware of the more advanced topics and difficult problems. What I have endeavored to do here is to soften the mathematical aspect which some found rather forbidding in the *Introduction* and to answer the criticism that too few practical examples were given. This has been done by introducing each topic with a fairly extensive discussion of the simplest example, by treating the mathematics descriptively as well as analytically, and by omitting certain topics. I hope that the result is not too much "like a missionary talking to cannibals"—as Littlewood said of a much better and more famous book—or that, if it is, then it at least escapes the venial sin of condescension, in the unpleasant sense which that word has lately acquired. I have added a great number of illustrative examples taken where possible from actual reactions. At the same time I feel that no apology is needed for using examples such as $A \longrightarrow B$, $A_1 - A_2 - A_3 = 0$, $A \longrightarrow B \longrightarrow C$, or $\sum \alpha_{ij} A_j = 0$; they are entirely suitable to the task of illustrating the principles of the subject, and what they lack in the charm of Titchmarshian "picturesqueness" they make up for by the virtues of Boudartian "asceticism."

A partial list of examples used in the text and in the exercises is given after the table of contents, for, though all these items could be found from the index, it is useful to know what there is to look for. The teacher's manual contains a discussion of some of the pedagogical problems of a course in reactor analysis, as well as a number of quiz questions and the solutions of the 200 problems of the text. A few of these problems have been taken from the Cambridge University Tripos and Qualifying Examinations and I am indebted to the Syndics of the Cambridge University Press for permission to use them. They are denoted by the initials C.U. A few of the figures are reproduced from papers in the journal literature and I am most grateful to the several editors and authors for allowing this. The specific acknowledgements are made in the legends to these figures.

In writing this book I have incurred even more numerous obligations than before and I must start by asking the pardon of any whom I inadvertently overlook. I owe to Michel Boudart the original suggestion that it would be possible, not to say desirable, to present the scheme of the *Introduction* in a more elementary form. Many correspondents took up points of difficulty, supplied corrigenda for the previous book, or commented on their use of it in the classroom. Among these my colleagues A. G. Fredrickson, R. W. Carr, and L. D. Schmidt should have first mention, but D. W. Condiff, J. M. Douglas, L. C. Eagleton, G. C. Frazier, G. R. Gavalas, J. M. Prausnitz, T. W. F. Russel, G. D. Shilling, J. C. R. Turner, and J. Wei have also made valuable contributions. Among the students who have assisted in the course, looked over manuscript or proof and have made useful comment K. J. Anselmo, I. Copelowitz, D. W. Drott, M. Gackstetter, A. P. Jackman, J. G. Jouven, R. H. Knapp, D. J. Kudish, R. D. Megee, R. Palas, T. M. Pell, S. Rester, P. A. Rouyer, D. V. Rudd, D. R. Schneider, J. M. Wheeler deserve mention. As usual my colleagues, notably N. R. Amundson, H. T. Davis, J. S. Dahler, K. H. Keller and L. E. Scriven, have helped to clarify various points. The business of deciphering my holograph and producing an excellent typescript was performed by Mrs. John Grose, Mrs. Jay Poissant, Miss Linda Anderson, and Miss Sharon Wellmann. As always Messrs John Davis and John McCanna of Prentice-Hall have been most helpful and considerate, and it has been a pleasure to work with the production editor Mr. John S. Covell on the book and Mr. Nicholas Romanelli on the teacher's manual.

To all these I am most grateful for their several favors and most particularly to my wife for her patience with my preoccupation and perverseness—even if at times it has driven her to the philosophy of Charlie Brown, who learned to "dread only one day at a time"!

RUTHERFORD ARIS

Contents

1. **What Is Chemical Reactor Analysis?**   1

    1.1 A general look at the subject, 1.   *1.2* A note on presentation, problems, and prerequisites, 5.   References, 5.

2. **Stoichiometry**   8

    2.1 What it is and why we need it, 8.   *2.2* Entire reactions and reaction mechanisms, 9.   *2.3* Independence of reactions, 11.   *2.4* Measurement of quantity and its change by reaction, 15.   *2.5* Measures of concentration, 19.   *2.6* Concentration changes with a single reaction, 21.   *2.7* Concentration changes with several reactions, 23.   *2.8* Rate of reaction, 25.   Notation, 27.

3. **Thermochemistry and Chemical Equilibrium**   29

    3.1 Introduction, 29.   *3.2* Heat of formation, 30.   *3.3* Heat of reaction, 31.   *3.4* Variation of the heat of reaction, 35.   *3.5* Rate of generation of heat by reaction, 36.   *3.6* Chemical equilibrium, 37.   *3.7* The calculation of homogeneous equilibrium compositions, 40.   *3.8* Equilibrium of simultaneous and heterogeneous reactions, 47.   References, 51.   Notation, 51.

4. Reaction Rates — 53

4.1 What is needed and what can be obtained, 53. 4.2 Homogeneous reaction rates, 54. 4.3 Variation of the reaction rate with extent and temperature, 62. 4.4 A portrait of the second order reversible reaction, 69. 4.5 Reaction rates near equilibrium, 72. 4.6 Reaction mechanisms, 74. 4.7 Heterogeneous reaction rate expressions, 77. 4.8 Reaction rates in other concentration measures, 78. 4.9 Classification and orders of magnitude of reaction rates, 79. References, 82. Notation, 82.

5. The Progress of the Reaction in Time — 84

5.1 Introduction, 84. 5.2 The first order reaction, 88. 5.3 The general irreversible reaction, 92. 5.4 The general homogeneous reaction, 94. 5.5 Concurrent reactions of Low order, 98. 5.6 Consecutive first order reactions, 100. 5.7 Systems of first order reactions, 104. 5.8 Numerical methods, 107. References, 111. Notation, 112.

6. The Interaction of Chemical and Physical Rate Processes — 113

6.1 Effective reaction rate expressions, 114. 6.2 The concept of the rate determining step, 118. 6.3 External mass transfer, 127. 6.4 Diffusion within the catalyst pellet, 128. 6.5 The combination of external mass transfer and internal diffusion, 140. 6.6 The effect of temperature variations, 141. 6.7 Applications, 147. References, 150. Notation, 153.

7. The Continuous Flow Stirred Tank Reactor — 156

7.1 The basic mass balances, 158. 7.2 The energy balance, 166. 7.3 The design of a single reactor, 168. 7.4 Some considerations in detailed design, 181. 7.5 Stability of the steady state, 188. 7.6 Control of the steady state, 198. 7.7 Sequences of stirred tank reactors, 201. 7.8 Optimal sequences of stirred tank reactors, 208. 7.9 Mixing in the reactor, 215. References, 223. Notation, 227.

8. Adiabatic Reactors — 229

8.1 General principles, 229. 8.2 The adiabatic stirred tank, 230. 8.3 Sequences of adiabatic stirred tanks, 234. 8.4 The adiabatic tubular or batch reactor, 238. 8.5 Multistage adiabatic reactors, 244. 8.6 Combined types of adiabatic reactor, 249. 8.7 Stability of adiabatic reactors, 252. References, 255. Notation, 256.

Contents ix

9. The Tubular Reactor 259

9.1 Types of tubular reactor, 259. 9.2 The mass balance, 262. 9.3 The pressure drop through the reactor, 268. 9.4 The energy balance, 269. 9.5 The basic design problems, 271. 9.6 Optimal designs for tubular reactors, 275. 9.7 Tubular reactors cooled or heated from the wall, 283. 9.8 Sensitivity and stability, 301. 9.9 The effect of flow profile, 306. 9.10 Axial dispersion in tubular reactors, 309. References, 314. Notation, 320.

10. The Batch Reactor 322

10.1 The equations for the batch reactor, 323. 10.2 Intermittent operation, 327. 10.3 Optimal control, 330. References, 334. Notation, 335.

Appendix: Some Less Well-known Functions of Use in Reactor Analysis 337

Modified Bessel functions, 337. The exponential integral, 341. References, 343.

Index 345

Illustrative Examples Used in the Text and Exercises

Stoichiometry of methanol synthesis reaction (2.3).
Dependence of reactions in hydrogen bromide mechanism (2.3).
Stoichiometry of ammonia synthesis (2.6.1), with side reaction (2.7.2).
Heat of reaction forming lead sulfate (3.3).
Heat of reaction for methanol synthesis (3.3.2).
Heat of formation of sulfur trioxide (3.3.3).
Equilibrium of ammonia synthesis (3.7.5), hexane dehydrogenation (3.7.6), and formation of alcohol from ethylene (3.7.7).
Equilibrium for the reduction of carbon dioxide (3.8; 3.8.3).
Equilibrium for the combustion of producer gas (3.8.2) and decomposition of calcium carbonate (3.8.4).
Hydrolysis of ethyl acetate (4.2.4).
Autocatalytic reaction $A \longrightarrow B + C$ (4.3).
Mechanism of phosgene formation (4.6).
Michaelis-Menten kinetics (4.6; 5.3.3).
Decomposition of acetaldehyde (5.3.5; 9.5.5).
Decomposition and formation of hydrogen iodide (5.4.3, 4).
Trypsin-trypsinogen reaction (5.3.5).
Decomposition of an alcohol (5.5).
Denbigh's reaction (5.6.5; 7.3.7; 7.7.6).
Triangular reaction $A \rightleftarrows B \rightleftarrows C \rightleftarrows A$ (5.7).
Hydrogenation of ethylene (6.2).
Hydrogenation catalyst (6.7).
Hydrolysis of acetic anhydride (7.1; 7.7).
Economic analysis of stirred tank with $A_1 \longrightarrow A_3$, $A_1 + A_2 \longrightarrow A_4$ (7.4).
Saponification (7.7.5).
Decomposition of phosphine (8.4.4).
Irreversible decomposition with change in the total number of moles (9.2).
Oxidation of sulfur dioxide (9.2).
Dehydration of alcohol (9.5.4).
Cracking of acetone (9.7).
Ammonia synthesis (9.7).
Inversion of sugar (9.9.5).
Nitration of benzene sulphonic acid (10.1).
First order decomposition $A \longrightarrow 2B + 3C$ (10.1).

Elementary Chemical Reactor Analysis

What is Chemical Reactor Analysis? 1

1.1 A General Look at the Subject

The analysis and synthesis of chemical reactors is par excellence the domain of the chemical engineer. For, while other engineers share his interest in fluid mechanics and transport phenomena and the chemist his concern with the kinetics and mechanisms of reactions, it is his business to combine a knowledge of these subjects for the better understanding, design, and control of the reactor. This book discusses the basic principles of reactor design and if the word "analysis" figures in the title it is not there to exclude synthesis, but to emphasize that the parts can only be put together when the functioning of each one and its relation to others have first been understood. Thus in the first half of the book we shall be examining the parts severally, with our principal concern the chemical reaction; in the second half, our attention will center on the reactor (see Fig. 1.1).

In practice the chemical engineer will meet a situation much less clearly defined than that studied in the classroom. Information about reaction kinetics will always be incomplete and usually inaccurate, and he must hazard a design on such a basis. Moreover, there are economic considerations in

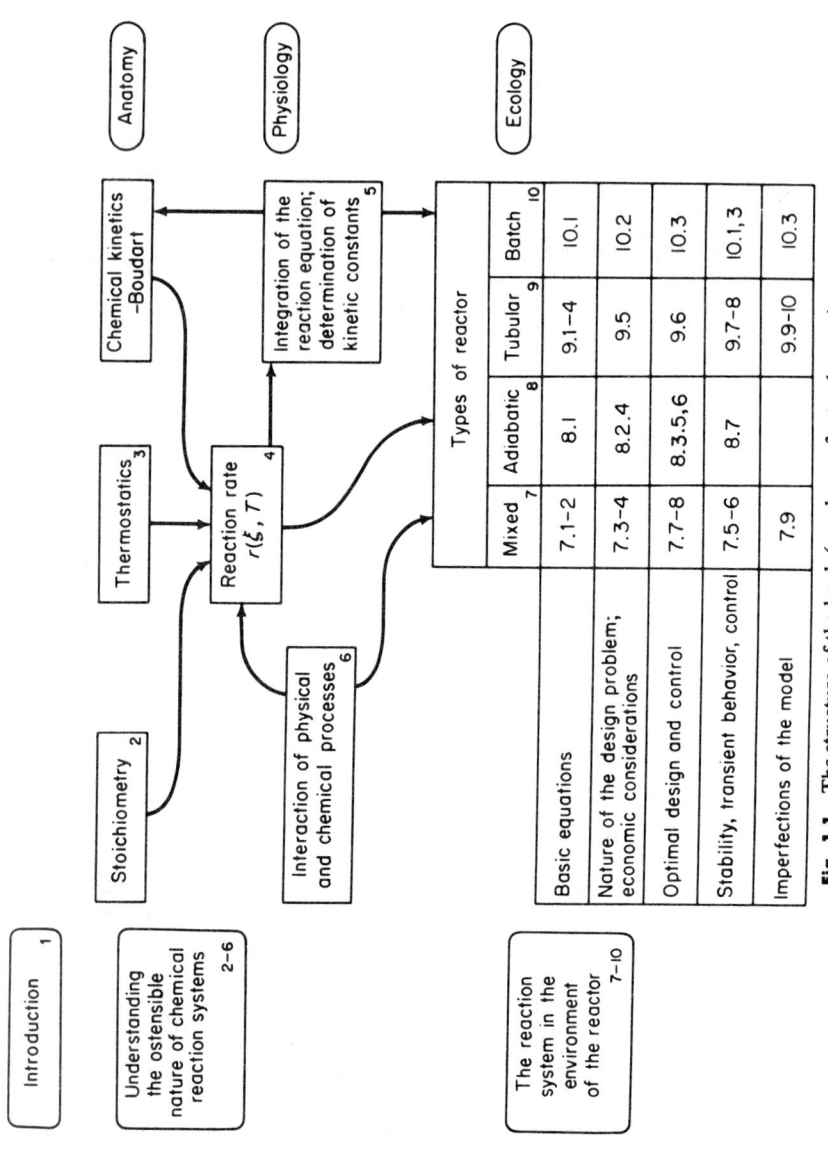

Fig. 1.1 The structure of the book (numbers refer to chapters).

design which fall quite outside the scope of this book. If then the art of the engineer lies in making intelligent guesses in the presence of uncertainty we may wonder if general analytical considerations are not rather beside the point. On the contrary, a general "feel" for the form of the subject becomes all the more important for this will help to fill out, as well as bound, his partial knowledge. For example, it is shown in Chap. 7 that if the maximum reaction rate of an exothermic reaction rate is known for several conversions, then an optimal design of several stirred tanks in series can be calculated. Although this result is established from equations that contain a precise reaction rate expression for all temperatures and pressures, its final form shows that much less complete information would be usable. Such considerations may allow the engineer to design with greater confidence from available data or may show him in what direction further data are needed. Provided that no essential feature is ignored, the quantitative study of a simplified model may give qualitatively valid results in the sense that it reveals trends that would otherwise be difficult to perceive and shows their directions even when it cannot accurately predict their magnitudes.

With the assurance that analysis and synthesis, like theory and experiment, are complementary rather than competitive, let us take a general look at the territory we wish to explore. The subject is also referred to as "chemical reaction engineering" or "applied chemical kinetics," and the latter suggests that we should first make the distinction between "pure" and "applied" kinetics. Pure chemical kinetics is concerned with elucidating the chemical features of the reaction and is interested in the reaction rate as one of the clues in this search. The notion of the reaction rate is also central to applied kinetics but in the sense that it bears on the behavior of the reactor rather than on the mechanism of the reaction. Thus the applied chemical kineticist can be content with a comparatively gross description of the reaction provided that it yields a usable expression for the reaction rate. In keeping with this, we shall be quite content to draw on the results of chemical kinetics, recognizing its importance but not having occasion here to delve into it. Fortunately, an excellent introduction is to be found in Boudart's text in this series.

The three principal ingredients of the reaction rate expression are shown in the top boxes of Fig. 1.1. Stoichiometry deals with the changes of composition that may take place by reaction. From thermostatics we can learn much about the heat effects of reactions and the nature of equilibrium, and from chemical kinetics we shall take any result that is available and useful. These three factors form the anatomy of our subject, whose physiology emerges when we ask just what they mean in terms of the behavior of the reaction rate expression. In Chap. 4 we will examine the reaction rate expression and in Chap. 5 we will see what this implies for the course of the reaction in time. But we must also look at the interaction of physical and chemical processes for it may be that the step that controls the rate of the process is one of mass transfer rather than of reaction. The ecological aspect of the subject comes into prominence when we study the reaction in its habitat, the reactor. The final chapters are therefore devoted to four types of reactor: the perfectly mixed reactor, the adiabatic, the tubular, and the batch reactor. The simplest model of the steady state of the tubular reactor is described by the same equations as for the batch reactor. The term *adiabatic* really refers to the mode of operation of the reactor rather than to its construction. We therefore really have two main types of reactor, the tubular (in which it is generally desirable to have as little mixing as possible as the reactants flow through it) and the stirred tank (in which good mixing is deliberately cultivated). For each type we first set up the basic equations and from these we see what elements enter into the design problem. This brings up the question of how good the design is and what methods are available for finding an optimal design. The full analysis of the control of a reactor is generally very difficult, but a surprising amount of information can be obtained from a careful examination of the basic equations. The same remark applies to questions of stability and sensitivity which, though very deep, can be usefully elucidated in quite simple terms. Finally, we turn and look again at the basic model to see how it can be made more realistic.

The process of mathematical modeling is an iterative one in which careful study of a simple model leads to a more realistic, but more complex, one, which is itself the precursor of a better model. What is needed is a good feel for the right kind of simplification to be made in building a model and the ability to see the shape of the results that come from analyzing it. The word "shape" is used deliberately for it is only safe to start detailed computations when the form of the solution is understood. To do otherwise is to invite an expensive output of meaningless numbers and to sacrifice one's birthright of rationality for a mess of digits.

Two further remarks may be made about the boundaries of the territory we wish to explore. No reference is made to fluidized beds or to noncatalytic reactions with solid particles, not because they are unimportant, but because a line had to be drawn somewhere. An excellent introduction to these topics will be found in Levenspiel's book. Neither is any real attention paid to the hardware and technology of reactors. Again this is not due to any lack of importance of these topics—they are simply inappropriate to our approach. It is a disputed question whether real technological know-how can properly be taught in a university. It is important that the student should not despise technology in any way, but experience suggests that technology is best learned in its own setting.

To sum up, the viewpoint is structural and the methods are analytic. But structure is to be understood in the broadest sense, not as a static concept, but as including both function and interaction. Analysis is not to be thought of as the antithesis of synthesis, but rather as the ground on which sound synthetic methods can be based.

1.2 A Note on Presentation, Problems, and Prerequisites

I have endeavored to introduce each topic by taking the simplest example that brings out the characteristic features. At times, the analysis may seem painfully long and heavy-handed and the student who is able to do so should jump ahead. I hope that this will temper the wind of generality for those who find an algebraic notation obscure. At the same time, I make no apology for using the reactions $\sum \alpha_{ij} A_j = 0$, nor even the time-honored $A \longrightarrow B$ and $A \longrightarrow B \longrightarrow C$. Often little is added to the understanding of principles by giving particular instances, but they do make for a useful familiarity and where possible I have introduced an example.

Some of the problems extend the text by opening up the way into more general cases. Most of them are rather easy, for it is hard to find a problem in this area which is difficult without being unusually lengthy. I have added some problems which involve estimation of the importance of different effects and some which contain reference to actual processes. One or two of the problems do not have very definite answers and these are the type that will be met most of the time in practice. No attempt has been made to separate the problems into classes of various degrees of difficulty—life isn't like that.

A working knowledge of elementary calculus is presumed as is some acquaintance with elementary differential equations. Section 5.1 is a thumbnail sketch of some particularly important equations. A thorough course in thermodynamics is one of the staples of a chemical engineer's diet and should precede a course on reactors. Chapter 3 is therefore a bare outline of familiar thermochemistry in a notation conformable to the rest of the book. It is impossible to avoid duplications in notation and a list has been provided at the end of each chapter.

REFERENCES

The literature of chemical reactor analysis is vast and awaits an industrious bibliographer. We give here a partially annotated list of texts and monographs of general use in the field of applied chemical kinetics. The undergraduate should familiarize himself with the chemical engineering journals; those of particular value are as follows.

Chemical Engineering Science
Industrial and Engineering Chemistry
  Fundamentals
  Process Design and Development
A. I. Ch. E. Journal
Chemie-Ingenieur-Technik
Canadian Journal of Chemical Engineering

Three particularly valuable supplements to *Chemical Engineering Science* have been issued. They are the papers given at the first three European Symposia on Chemical Reaction Engineering in 1957, 1960, and 1964. The first was issued as a monograph and as Vol. 8; the second as Vol. 14. The third symposium is issued as a supplement by Pergamon Press. A fourth symposium took place in 1968.

One of the first texts in the field was:

O. A. Hougen and K. M. Watson, *Chemical Process Principles*, Vol. III. New York: John Wiley & Sons, Inc., 1947.

(The earlier volumes of this work, to which reference is made later, have been revised with the collaboration of R. A. Ragatz.)

See also:

J. M. Smith, *Chemical Engineering Kinetics*. New York: McGraw-Hill Book Co., 1956.

S. M. Walas, *Reaction Kinetics for Chemical Engineers*. New York: McGraw-Hill Book Co., 1959.

W. Brötz, *Grundriss der Chemischen Reaktionstechnik*. Weinheim: Verlag Chemie, 1958. Translated by D. A. Diener and J. A. Weaver as *Fundamentals of Chemical Reaction Engineering*. Reading, Mass.: Addison-Wesley Publishing Co., 1965.

K. Dialer, F. Horn, and L. Küchler, "Chemische Reaktionstechnik" in *Chemische Technologie*, Vol. I. München: Carl Hanser Verlag, 1958.

J. C. Jungers et al., *Cinétique Chimique Appliquée*. Paris: Technip, 1958.

O. Levenspiel, *Chemical Reaction Engineering*. New York: John Wiley & Sons, Inc., 1962.

H. Kramers and K. R. Westerterp, *Elements of Chemical Reactor Design and Operation*. New York: Academic Press Inc., 1963.

K. G. Denbigh, *Chemical Reactor Theory*. New York: Cambridge University Press, 1965.

E. E. Petersen, *Chemical Reaction Analysis*. Englewood Cliffs, N. J.: Prentice-Hall, Inc., 1965.

The volumes of Academic Press' series *Advances in Chemical Engineering* should also be consulted. For example, Vol. 3 (1962) contains an article:

J. Beek, "Design of Packed Catalytic Reactors";

and in Vol. 4 (1963) is found:

O. Levenspiel and K. B. Bischoff, "Patterns of Flow in Chemical Process Vessels."

An introduction to fluidized beds is given in Levenspiel's book quoted above and in his forthcoming:

O. Levenspiel and D. Kunii, *Fluidization Engineering*. New York: John Wiley & Sons, Inc., 1968.

See also:

J. F. Davidson and D. Harrison, *Fluidized Particles*. New York: Cambridge University Press, 1963.

On the pure kinetics side, the *Advances in Catalysis* series, also published by Academic Press, is of particular interest; Walas has a convenient listing of their contents at the end of his book. Of texts in this cognate area we will only mention:

M. Boudart, *The Kinetics of Chemical Processes*. Englewood Cliffs, N. J.: Prentice-Hall, Inc., 1968.

This contains an excellent exposition of chemical kinetics and an entry into its vast literature.

Stoichiometry 2

2.1 What It Is and Why We Need It

The strict meaning of the word "stoichiometry" is measurement of the elements, but it is commonly used to refer to all manner of calculations regarding the composition of a chemical system. We shall give it a broad meaning here for we are concerned as much with the algebra of the relationship between the chemical species as with the arithmetical calculations that are implied. Stoichiometry is essentially the bookkeeping of the material components of the chemical system.

Its importance lies in the fact that the changes in composition in a reactor are not haphazard, but are of two distinct kinds. First, there are the advective changes due to material brought into the system or removed from it; this may be by forced flow, convection, or diffusion. Second, there is the internal change of composition by reaction; this change would be seen in a well-stirred batch reactor, for example, where advection has been deliberately eliminated. The advection will have to be expressed by certain terms in the material balance for a particular reactor, but the reactive changes are common to all types and deserve study first.

To see the nature of the restrictions on possible changes in composition let us recall a universally known reaction

$$2 H_2 + O_2 = 2 H_2O. \tag{2.1.1}$$

An equation such as this can have two meanings. It may be a kinetic description of the reaction and imply that two molecules of hydrogen combine directly with one of oxygen to form two molecules of water. In this particular case, Eq. (2.1.1) is not true as a kinetic description. On the other hand, it may be a stoichiometric description of the reaction, and the equation will then mean that the numbers of hydrogen and oxygen molecules combining to form water are in the ratio 2 : 1. It is clear that if a reaction expression is true in a kinetic sense it is also true in a stoichiometric sense, but the converse statement is false. We shall only be concerned in passing with the kinetic meaning of the reaction expression. The proportions 2 : 1 in the stoichiometric description are absolutely definite and an equation such as $3\,H_2 + O_2 = 3\,H_2O$ is nonsense. It follows that the changes in composition will be definite, for every time a given number of molecules of water are formed, the same number of hydrogen and half that number of oxygen molecules will disappear.

In keeping with our need to understand the structure of these changes, we shall not want to talk about particular reactions except by way of illustration, and our first task is to devise a suitably general notation.

2.2 Entire Reactions and Reaction Mechanisms

From a chemical engineer's point of view, the important thing about a reaction like the one above is that three chemical species (here hydrogen, oxygen, and water) are observed to react in certain proportions. If he can account for the rate at which this takes place in terms of the concentrations of the observed chemical species, he has all he needs to know about the reactions so far as designing a reactor is concerned. Such a reaction and its reaction rate expression will be called *entire*. It is the chemist who is interested in the fact that the entire reaction does not provide a kinetic description of what is taking place, and in devising a *mechanism* for a reaction he looks for elementary steps such that the stoichiometric equation of each step does correspond to the kinetics. In doing so he may be led to hypothesize certain intermediate species or radicals which are present in only trace quantities. This activity of the chemical kineticist is most important if valid and intelligible reaction rate expressions are to be built up, but it is part of pure chemical kinetics and we shall only have occasion to refer to it in passing.

We want our notation to reflect these facts and to be suitably general. Let us start with a set of S chemical species and denote them by $A_1, A_2, \ldots, A_S$ or A_j, where $j = 1, 2, \ldots, S$. Thus if $A_1 \equiv H_2O$, $A_2 \equiv H_2$, and $A_3 \equiv O_2$, the reaction given by Eq. (2.1.1) would be $2A_2 + A_3 = 2A_1$. It is convenient to write all the chemical species on one side of the equation and to give a positive sign to the species which are regarded as the products of the reaction. Thus if water is being formed we would write

$$2\,H_2O - 2\,H_2 - O_2 = 0 \quad \text{or} \quad 2A_1 - 2A_2 - A_3 = 0. \quad (2.2.1)$$

This is purely a convention, but, since in most engineering situations the intention is to make a definite product, it is a useful one. If the object of the reaction were to form hydrogen and oxygen we would write

$$-2\,H_2O + 2\,H_2 + O_2 = 0 \quad \text{or} \quad -2A_1 + 2A_2 + A_3 = 0. \tag{2.2.2}$$

This convention should not be given undue importance but it is usually possible to observe it, at least for the main reaction.

The numbers 2, -2, and -1 in Eq. (2.2.1) are called the *stoichiometric coefficients* and a natural way to write the general reaction is

$$\sum_{j=1}^{S} \alpha_j A_j \equiv \alpha_1 A_1 + \cdots + \alpha_S A_S = 0. \tag{2.2.3}$$

Thus A_j denotes the jth chemical species and α_j is its stoichiometric coefficient in the given reaction. Subject to the above convention it is convenient to call the species with positive stoichiometric coefficients the *products* of the reaction and those with negative coefficients the *reactants*. The products are formed from the reactants by the *forward reaction*, while the reverse reaction converting the products into the reactants will be called the *back reaction*. Both forward and back reactions are usually going on simultaneously and equilibrium is reached when they go at equal and opposite rates. It is sometimes useful to include an inert chemical species in the set A_j, and since it does not take part in the reaction it is given a stoichiometric coefficient of zero. An *entire* reaction is thus one whose behavior can be fully described in terms of the concentrations of the species $A_1, \ldots, A_S$.

The important thing about the stoichiometric coefficients of a reaction is their ratio rather than their absolute magnitude. Thus the reaction for the formation of water could just as well be written $H_2O - H_2 - \tfrac{1}{2} O_2 = 0$, and in fact often is. We can say therefore that the stoichiometric coefficients of a reaction are given up to a constant multiplier, for the equation $\sum \lambda \alpha_j A_j = 0$ has exactly the same meaning as $\sum \alpha_j A_j = 0$, provided that λ is not zero.

Exercise 2.2.1. If the molecular species A_j is a combination of elements

$$E_1, \ldots, E_T, \qquad A_j = \sum_{k=1}^{T} \epsilon_{jk} E_k,$$

show that

$$\sum_{j=1}^{S} \alpha_j \epsilon_{jk} = 0, \qquad k = 1, 2, \ldots, T$$

[e.g., in Eq. (2.2.1), $A_1 = H_2O = 2\,H + O = 2E_1 + E_2$].

Exercise 2.2.2. If m_j is the molecular weight of A_j show that $\sum \alpha_j m_j = 0$. What physical law is used here?

Exercise 2.2.3. If two reactions

$$A_1 + A_2 - A_3 - A_4 = 0$$
$$- A_2 + 2A_3 - A_4 = 0$$

take place between the four species $A_1, \ldots, A_4$ show that
$$\tfrac{1}{2}m_4 < m_3 < 2m_4.$$

Exercise 2.2.4. Show that the reaction scheme
$$A_1 + A_2 - A_3 + A_4 = 0$$
$$A_1 - A_2 + A_3 - A_4 = 0$$
is impossible.

2.3 Independence of Reactions

Our way of writing a chemical reaction can easily be extended to simultaneous reactions. Thus if there are R simultaneous reactions between S species we can write

$$\sum_{j=1}^{S} \alpha_{ij} A_j = 0, \qquad i = 1, 2, \ldots, R. \tag{2.3.1}$$

Here α_{ij} is the stoichiometric coefficient of A_j in the ith reaction. The summation is over the range of the repeated suffix j and it is often unambiguous to write $\sum \alpha_{ij} A_j = 0$. As far as possible we shall reserve the suffix j for the species and the suffix i for the reaction; thus j is an integer between 1 and S and i between 1 and R.

As an example of this notation consider the methanol synthesis reaction
$$CO + 2 H_2 \rightleftharpoons CH_3OH$$
with the side reaction
$$CO_2 + H_2 \rightleftharpoons H_2O + CO.$$
If we write
$$A_1 \equiv CH_3OH, \qquad A_2 \equiv CO, \qquad A_3 \equiv H_2, \qquad A_4 \equiv CO_2, \qquad A_5 \equiv H_2O,$$
the reactions can be written
$$A_1 - A_2 - 2A_3 \qquad\qquad = 0,$$
$$A_2 - A_3 - A_4 + A_5 = 0.$$
Here $R = 2$, $S = 5$, and $\alpha_{11} = \alpha_{22} = \alpha_{25} = 1$, $\alpha_{21} = \alpha_{14} = \alpha_{15} = 0$, $\alpha_{12} = \alpha_{23} = \alpha_{24} = -1$, and $\alpha_{13} = -2$. Obviously these two equations represent quite different reactions. On the other hand the conceivable reaction
$$CO_2 + 3 H_2 \rightleftharpoons CH_3OH + H_2O$$
does not really tell us any more about the system than is contained in the two previous reactions. It would be written
$$A_1 - 3A_3 - A_4 + A_5 = 0$$
and is clearly the sum of the other two reactions. It is important to know just how many independent reactions there are in a given system even though

it is not always necessary to eliminate the dependent ones and work with the independent set.

We have seen that the reactions $\sum \alpha_j A_j = 0$ and $\sum \lambda \alpha_j A_j = 0$ are really the same provided that λ is not zero (which would make the second reaction trivial) or infinite (which would make it meaningless). Another way of saying this is that the reactions $\sum \alpha_{1j} A_j = 0$ and $\sum \alpha_{2j} A_j = 0$ are the same if the ratio α_{1j}/α_{2j} is constant and neither zero nor infinite. The usual way to express this mathematically is to say that the reactions are linearly dependent if we can find two numbers λ_1 and λ_2 (not both zero) for which

$$\lambda_1 \alpha_{1j} + \lambda_2 \alpha_{2j} = 0, \qquad j = 1, 2, \ldots, S. \tag{2.3.2}$$

If such numbers cannot be found the reactions are independent.

With three reactions we have not only the possibility that one will be a multiple of one of the others but also that it may be a linear combination of the other two. For example, if the first two reactions above are $\sum \alpha_{1j} A_j = 0$ and $\sum \alpha_{2j} A_j = 0$ and the third reaction, $A_1 - 3A_3 - A_4 + A_5 = 0$, is $\sum \alpha_{3j} A_j = 0$, then the fact that it is the sum of the other two is expressed by $\alpha_{3j} = \alpha_{1j} + \alpha_{2j}$, or

$$\alpha_{1j} + \alpha_{2j} - \alpha_{3j} = 0, \qquad j = 1, 2, \ldots, S.$$

In general we shall say that the reactions are dependent if we can find three numbers λ_1, λ_2, and λ_3 (not all zero) such that

$$\lambda_1 \alpha_{1j} + \lambda_2 \alpha_{2j} + \lambda_3 \alpha_{3j} = 0, \qquad j = 1, 2, \ldots, S. \tag{2.3.3}$$

In the above case $\lambda_1 = \lambda_2 = -\lambda_3 = 1$. If no such multipliers can be found the three reactions are independent. Notice that this formulation of the criterion also covers the case of two reactions, for Eq. (2.3.2) can be obtained from Eq. (2.3.3) by setting $\lambda_3 = 0$.

The general criterion for independence of R simultaneous reactions can be stated as follows. If no set of multipliers λ_i, where $i = 1, 2, \ldots, R$, can be found (other than the trivial set $\lambda_i = 0$, where $i = 1, 2, \ldots, R$) such that

$$\sum_{i=1}^{R} \lambda_i \alpha_{ij} = 0, \qquad j = 1, 2, \ldots, S, \tag{2.3.4}$$

then the reactions $\sum \alpha_{ij} A_j = 0$ are said to be independent, or, more particularly, stoichiometrically independent. We see that this implies that none of the reactions can be expressed as a linear combination of the others. Such a criterion is quite satisfactory from a mathematician's point of view, but in practice the failure to find a nontrivial set of multipliers might be due to our own stupidity rather than the independence of the reactions. It would therefore be useful to have a constructive test for the number of independent reactions. We give such a test (without proof) by means of an example.

Consider for example the set of reactions that is said to be the kinetic description of the formation of hydrogen bromide. These are

Sec. 2.3 Independence of Reactions

$$Br_2 \longrightarrow 2\, Br,$$
$$Br + H_2 \longrightarrow HBr + H,$$
$$H + Br_2 \longrightarrow HBr + Br,$$
$$H + HBr \longrightarrow H_2 + Br,$$
$$2Br \longrightarrow Br_2.$$

A moment's inspection shows that the first and fifth and the second and fourth are pairs of identical reactions; but it is not safe to conclude from this that there are three independent reactions without further investigation. Let us write

$$A_1 = Br_2, \quad A_2 = Br, \quad A_3 = H_2, \quad A_4 = H, \quad A_5 = HBr;$$

then the reactions are

$$-A_1 + 2A_2 = 0,$$
$$-A_2 - A_3 + A_4 + A_5 = 0,$$
$$-A_1 + A_2 - A_4 + A_5 = 0,$$
$$A_2 + A_3 - A_4 - A_5 = 0,$$
$$A_1 - 2A_2 = 0.$$

It is convenient to detach the stoichiometric coefficients and write them in matrix form with α_{ij} in the ith row and jth column:

$$\begin{matrix} -1 & 2 & 0 & 0 & 0 \\ 0 & -1 & -1 & 1 & 1 \\ -1 & 1 & 0 & -1 & 1 \\ 0 & 1 & 1 & -1 & -1 \\ 1 & -2 & 0 & 0 & 0. \end{matrix}$$

Now take the first row with a nonzero element in the first column (in this case it is the first row) and divide through by its leading element (here -1) to make this $+1$. In general, if $\alpha_{11} \neq 0$ this gives a new first row

$$1, \quad \frac{\alpha_{12}}{\alpha_{11}}, \quad \ldots, \quad \frac{\alpha_{1S}}{\alpha_{11}}$$

and here we have

$$1, \quad -2, \quad 0, \quad 0, \quad 0.$$

Now use this row to make all the elements in the first column zero. Thus from the ith row we would subtract α_{i1} times this new first row to give for $i = 2, 3, \ldots, R$,

$$0, \quad \alpha_{i2} - \frac{\alpha_{i1}\alpha_{12}}{\alpha_{11}}, \quad \ldots, \quad \alpha_{iS} - \frac{\alpha_{i1}\alpha_{1S}}{\alpha_{11}}.$$

In the present case we obtain

$$\begin{matrix} 1 & -2 & 0 & 0 & 0 \\ 0 & -1 & -1 & 1 & 1 \\ 0 & -1 & 0 & -1 & 1 \\ 0 & 1 & 1 & -1 & -1 \\ 0 & 0 & 0 & 0 & 0. \end{matrix}$$

Now ignore the first row and first column and repeat the procedure with the remaining $R - 1$ rows. Here this would give

$$\begin{matrix} 1 & -2 & 0 & 0 & 0 \\ 0 & 1 & 1 & -1 & -1 \\ 0 & 0 & 1 & -2 & 0 \\ 0 & 0 & 0 & 0 & 0 \\ 0 & 0 & 0 & 0 & 0, \end{matrix}$$

and it is not necessary to continue the process any farther for we have 1's down the diagonal as far as they will go and zeros beneath them. If this had not been so, we would have ignored the first two rows and columns and repeated the process on the remaining $R - 2$. By this means we can always get an array with 1's on the diagonal as far as possible and zeros beneath them. If the R elements on the diagonal are all 1 then the R reactions are all independent; in fact the number of independent reactions is the number of 1's in the diagonal. Alternatively, it is the number of reactions minus the number of rows in the final array that are all zero.

We can always choose to work with only an independent subset of reactions and for most chemical engineering purposes it is convenient to do so. It is in the study of reaction mechanisms that systems of reactions that are not independent most naturally arise. An independent subset of the above five reactions is

$$A_1 - 2A_2 \qquad\qquad = 0,$$
$$A_2 + A_3 - A_4 - A_5 = 0,$$
$$A_3 - 2A_4 \qquad = 0.$$

The overall reaction $-A_1 - A_3 + 2A_5 = 0$, which this mechanism is said to explain, is not independent of these three; it is (-1) times the first plus (-2) times the second plus the third.

Exercise 2.3.1. Consider the steps of the test given in this section and show that this is equivalent to the criterion Eq. (2.3.4).

Exercise 2.3.2. If to each i there corresponds a j (different from that corresponding to any other i) such that $\alpha_{ij} \neq 0$, but $\alpha_{i'j} = 0$, where $i' = 1, 2, \ldots, R$, and $i' \neq i$, show that the reactions are independent.

Sec. 2.4 Measurement of Quantity and Its Change by Reaction 15

Exercise 2.3.3. How many of the following reactions are independent?

(a) $2\,C_2H_4 + O_2 = 2\,C_2H_4O$
$C_2H_4 + 3\,O_2 = 2\,CO_2 + 2\,H_2O$
$2\,C_2H_4O + 5\,O_2 = 4\,CO_2 + 4\,H_2O$

(b) $4\,NH_3 + 5\,O_2 = 4\,NO + 6\,H_2O$
$4\,NH_3 + 3\,O_2 = 2\,N_2 + 6\,H_2O$
$4\,NH_3 + 6\,NO = 5\,N_2 + 6\,H_2O$
$2\,NO + O_2 = 2\,NO_2$
$2\,NO = N_2 + O_2$
$N_2 + 2\,O_2 = 2\,NO_2$

(c) $2\,NaCl + H_2SO_4 = Na_2SO_4 + 2\,HCl$
$Na_2SO_4 + 4\,C = Na_2S + 4\,CO$
$Na_2S + CaCO_3 = Na_2CO_3 + CaS$

2.4 Measurement of Quantity and Its Change by Reaction

The amount of any chemical species present in a system can be measured by its mass or number of moles. If m_j is the molecular weight of A_j, M_j the mass, and N_j the number of moles present, then

$$M_j = m_j N_j. \qquad (2.4.1)$$

Where a unit is given for the mass, say kg or lb, this equation allows it to be expressed as a certain number of kg- or lb-moles. Thus 1 kg-mole of iron is 56 kg and 1 kg of H_2 is half a kg-mole. The mole is a convenient unit of quantity since by Avogadro's hypothesis it contains the same number of molecules (approximately 6.023×10^{23}) for all substances. It follows from what we have said about the meaning of the stoichiometric equation $\sum \alpha_j A_j = 0$, that whenever α_j moles of A_j are formed $\alpha_{j'}$ moles of $A_{j'}$ are also formed. If α_j is negative we interpret this to mean that α_j moles of A_j are destroyed.

If N_j is the number of moles of A_j present at any time and N_{j0} the number present at some arbitrary time and only the reaction is changing the composition, it follows that

$$\frac{N_j - N_{j0}}{\alpha_j}$$

is the same for $j = 1, 2, \ldots, S$. Let the common value be X; then

$$N_j = N_{j0} + \alpha_j X \qquad (2.4.2)$$

is an equation expressing the variation of the amount of A_j present during the course of the reaction. X is called the *molar extent* or *degree of advancement* of the reaction and is an extensive variable measured in moles. Its virtue lies in the fact that it is a variable linked with the reaction and not with any particular A_j, the choice of which would be arbitrary. The change in number of moles of one species can, of course, be expressed in terms of any other by eliminating X between the two expressions obtained from Eq. (2.4.2) for j and j':

$$N_{j'} = N_{j'0} + \frac{\alpha_{j'}}{\alpha_j}(N_j - N_{j0}). \tag{2.4.3}$$

The change in mass can also be obtained by combining Eqs. (2.4.1) and (2.4.2),

$$M_j = M_{j0} + \alpha_j m_j X, \tag{2.4.4}$$

where M_{j0} is the mass of A_j originally present.

If there are several reactions, an extent can be defined for each of them. If X_i is the extent of the ith reaction, it has contributed $\alpha_{ij} X_i$ moles to the total change in the number of moles of A_j. Thus,

$$N_j = N_{j0} + \sum_{i=1}^{R} \alpha_{ij} X_i. \tag{2.4.5}$$

We see why it is unnecessary to consider reactions that are dependent on others. Suppose the Rth reaction were dependent on the first $(R-1)$,

$$\alpha_{Rj} = \sum_{i=1}^{R-1} \nu_i \alpha_{ij};$$

then we could write

$$N_j - N_{j0} = \sum_{i=1}^{R-1} \alpha_{ij} X_i + \alpha_{Rj} X_R = \sum_{i=1}^{R-1} \alpha_{ij}(X_i + \nu_i X_R).$$

By taking $(X_i + \nu_i X_R) = X'_i$, where $i = 1, \ldots, R-1$, to be the extent of the ith reaction, we have expressed the composition changes in terms of only $(R-1)$ quantities, namely extents that can be associated with the first $(R-1)$ reactions.

As an illustration of this, consider the reaction in which a substance A_1 is formed from two reactants A_2 and A_3 according to the equation

$$A_1 - A_2 - A_3 = 0.$$

If there is initially one mole each of A_2 and A_3 present then $N_{10} = 0$, $N_{20} = N_{30} = 1$, and

$$N_1 = X, \quad N_2 = 1 - X, \quad N_3 = 1 - X. \tag{2.4.6}$$

We could represent the state of the system by a point in three dimensions as is shown in Fig. 2.1. The point P represents the initial state and, as the molar extent increases from zero to 1, the representative point moves down the line PQ. Notice that the direction of this line is fixed by the stoichiometric

Sec. 2.4   Measurement of Quantity and Its Change by Reaction   17

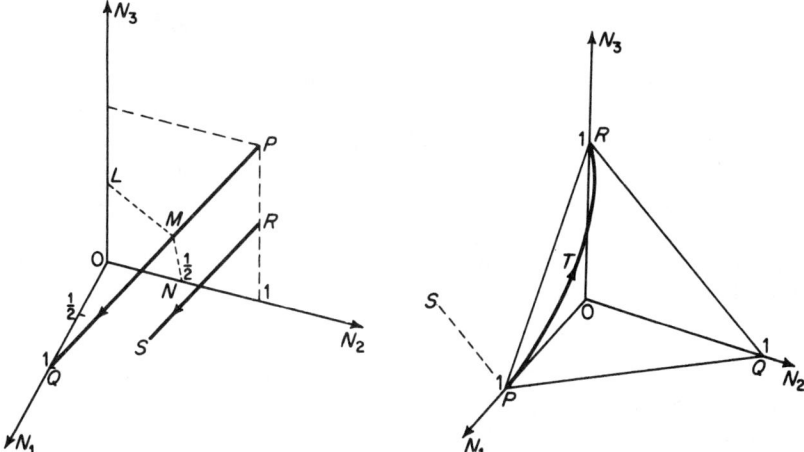

Fig. 2.1 Reaction paths for a single reaction $A_1 - A_2 - A_3 = 0$.

Fig. 2.2 The reaction plane for the two reactions $A_1 \rightarrow A_2 \rightarrow A_3$.

coefficients but its position in space depends on the reference composition N_{j0}. However, this reference composition could be any point on the line PQ, for example $N_{10} = N_{20} = N_{30} = \frac{1}{2}$. Thus the reference composition is a composition that could occur by reaction but not necessarily one which does occur. If the initial composition had been half a mole of each species then it would still have been possible to use $N_{10} = 0$, $N_{20} = N_{30} = 1$ as the reference composition with an initial value for the extent of $X = \frac{1}{2}$. Since N_j cannot be negative, only the part of the reaction line that lies in the positive octant means anything physically. Thus with $N_{10} = 0, N_{20} = N_{30} = 1$ we must have $0 \leq X \leq 1$; with $N_{10} = N_{20} = N_{30} = \frac{1}{2}$, $-\frac{1}{2} \leq X \leq \frac{1}{2}$. If the initial composition had been $N_{10} = 0$, $N_{20} = 1$, and $N_{30} = \frac{1}{2}$, the reaction would have been represented by the line RS in Fig. 2.1. We see that it is impossible to get from a composition on line RS to one on line PQ by reaction alone; half a mole of A_3 must be added to the system to get from one line to the other.

To illustrate the case of two simultaneous reactions we may take the pair

$$-A_1 + A_2 = 0, \quad -A_2 + A_3 = 0,$$

which is the standard way of writing $A_1 \longrightarrow A_2 \longrightarrow A_3$. In this case, Eq. (2.4.5) gives

$$N_1 = N_{10} - X_1, \quad N_2 = N_{20} + X_1 - X_2, \quad N_3 = N_{30} + X_2, \quad (2.4.7)$$

and the points (N_1, N_2, N_3) representing the composition lie in a plane through the points (N_{10}, N_{20}, N_{30}). In Fig. 2.2, point P would represent $N_{10} = 1$, $N_{20} = N_{30} = 0$. Line PQ represents the first reaction when the second reaction does not take place. Line QR would represent the second reaction when there is no A_1, i.e., $N_{20} = 1$ and $N_{30} = 0$. Only points in the plane PQR are accessible from an initial composition lying in that plane, such as $N_{10} = 1$ and $N_{20} = N_{30} = 0$. The actual course of reactions $A_1 \longrightarrow A_2 \longrightarrow A_3$ might be represented by such a curve as PTR lying in the plane. Again we notice that the orientation of the plane is fixed by the stoichiometry of the reactions and its position is determined by the reference composition.

We also observe that by adding the parts of Eq. (2.4.7) we have the equation of the plane PQR, namely

$$N_1 + N_2 + N_3 = N_{10} + N_{20} + N_{30}.$$

This says that a certain linear combination of the N_j, $\sum \beta_j N_j$, does not change during the reaction; in this case $\beta_1 = \beta_2 = \beta_3 = 1$. The direction given by $(\beta_1, \ldots, \beta_S)$ is always at right angles to the path of the reaction. Here the direction (1, 1, 1), shown by PS in Fig. 2.2, is normal to the plane and so orthogonal to any path in it. In the case of a single reaction between three species there are two independent linear combinations of the N_j that are invariant. In the first example, these are

$$N_1 + N_2 = N_{10} + N_{20}, \qquad N_1 + N_3 = N_{10} + N_{30},$$

and the two directions (1, 1, 0) and (1, 0, 1), shown by LM and MN in Fig. 2.1, are at right angles to the reaction path.

In general, when there are R independent reactions between S species, there are $(S - R)$ independent linear combinations of the N_j which are constant during reaction. The full exploitation of these ideas carries us rather rapidly into deeper algebraic waters and is outside our scope here.

Exercise 2.4.1. Show that the greatest possible value of X is the least positive value of $N_{j0}/(-\alpha_j)$.

Exercise 2.4.2. What is the nature of the bounds on X_1 and X_2 for two reactions? Use the example of $A_1 + A_2 - A_3 - A_4 = 0$, $A_1 - 2A_2 - A_3 = 0$ to illustrate this.

Exercise 2.4.3. If the reaction $\sum \alpha_j A_j = 0$ is replaced by $\sum \alpha'_j A_j = 0$ where $\alpha'_j = \lambda \alpha_j$, what is the relation between the extents X and X'?

Exercise 2.4.4. A mixture of original composition N_{j0} reacts to an extent X, and another of original composition N'_{j0} goes to an extent X'. They are then mixed together. Show that the composition is the same as if they had been mixed before reaction and reacted to an extent $(X + X')$.

Exercise 2.4.5. Interpret in the light of Fig. 2.1 the operation of taking 1 mole each of A_2 and A_3, reacting them until $\frac{1}{2}$ mole of A_1 is formed, then adding ½ mole of A_3 and 1 mole of A_2 and carrying out the reaction to completion.

Exercise 2.4.6. The reactions $A_1 \longrightarrow A_2 \longrightarrow A_3$ proceed from $N_{10} = 1$, $N_{20} = N_{30} = 0$ to extents X_1 and X_2. Then n_1 and n_2 moles of A_1 and A_2 are added and the reactions proceed further by extents X_1' and X_2'. Show that X_1' cannot exceed $(1 + n_1 - X_1)$ nor can X_2' exceed $(1 + n_1 + n_2 - X_2)$. In terms of Fig. 2.2 show that the subsequent reaction is represented by a curve in the plane parallel to PQR through the points $N_1 = 1 + n_1 + n_2$, $N_2 = N_3 = 0$, and hence might have been achieved by starting with this mixture and reacting to extents $(n_2 + X_1 + X_1')$ and $(X_2 + X_2')$, respectively.

Exercise 2.4.7. The reactions

$$-A_1 - 2A_2 + A_3 = 0$$
$$2A_1 - A_2 + A_4 = 0$$

can take place simultaneously. If there are initially 2 moles of A_1, 8 moles of A_2, and 1 mole of A_3, draw a careful diagram in the plane of the two extents X_1 and X_2 to show the region in which physically possible values of X_1 and X_2 lie. Find the values of X_1 and X_2 for which the greatest number of moles of A_3 would be formed. How many moles of A_3 are present when the reactions proceed to these extents?

Exercise 2.4.8. Copper and sulfuric acid react to give sulfur dioxide:

$$\mathrm{Cu} + 2\,\mathrm{H_2SO_4} = \mathrm{CuSO_4} + 2\,\mathrm{H_2O} + \mathrm{SO_2}.$$

How much copper and how much 94% acid are needed to make 1 kg of sulfur dioxide? If 1 kg of copper dissolves in 15 kg of 94% acid and all of the sulfur dioxide leaves the solution, what is the strength of the acid in which the copper sulfate is dissolved? (All acid compositions are in mass %.)

2.5 Measures of Concentration

Up to this point we have used the extensive composition variable N_j, the number of moles of A_j present, and the extensive molar extent X. By this we mean, as in thermodynamics, that doubling the system in every way would double the value of these variables. Intensive composition variables are measures of concentration, and it is the purpose of this section to define several commonly used measures and to relate them to one another.

(i) *Molar concentration.* If there are N_j moles of A_j uniformly dispersed in a volume V, the molar concentration is

$$c_j = \frac{N_j}{V}. \tag{2.5.1}$$

(ii) *Mass concentration.* If the mass of A_j present in the volume V is M_j, then the density or mass concentration is

$$\rho_j = \frac{M_j}{V}. \tag{2.5.2}$$

(iii) *Mole fraction.* If the total number of moles present is

$$N = \sum_{j=1}^{S} N_j \tag{2.5.3}$$

then the mole fraction is

$$x_j = \frac{N_j}{N}. \tag{2.5.4}$$

(iv) *Mass fraction.* If the total mass present is

$$M = \sum_{j=1}^{S} M_j \tag{2.5.5}$$

then the mass fraction is

$$g_j = \frac{M_j}{M}. \tag{2.5.6}$$

There are two corresponding measures of total concentration. The *total molar concentration* is

$$c = \sum_{j=1}^{S} c_j = \frac{N}{V} \tag{2.5.7}$$

and the *total mass concentration* or *density*

$$\rho = \sum_{j=1}^{S} \rho_j = \frac{M}{V}. \tag{2.5.8}$$

The relation between these is best shown in a table, in which the diagonal boxes give the definition and in any section the row variable is defined in terms of the column variable.

	c_j	x_j	c	ρ_j	g_j	ρ
c_j	$\frac{N_j}{V}$	cx_j	cx_j	$\frac{\rho_j}{m_j}$	$\frac{\rho g_j}{m_j}$	$\frac{\rho g_j}{m_j}$
x_j	$\frac{c_j}{c}$	$\frac{N_j}{N}$	$\frac{c_j}{c}$	$\frac{\rho_j}{m_j} / \sum \frac{\rho_j}{m_j}$	$\frac{g_j}{m_j} / \sum \frac{g_j}{m_j}$	...
c	$\sum c_j$	...	$\frac{N}{V}$	$\sum \frac{\rho_j}{m_j}$	$\rho \sum \frac{g_j}{m_j}$	$\rho \sum \frac{g_j}{m_j}$
ρ_j	$m_j c_j$	$m_j c x_j$	$m_j x_j c$	$\frac{M_j}{V}$	ρg_j	ρg_j
g_j	$\frac{m_j c_j}{\sum m_j c_j}$	$\frac{m_j x_j}{\sum m_j x_j}$	...	$\frac{\rho_j}{\rho}$	$\frac{M_j}{M}$	$\frac{\rho_j}{\rho}$
ρ	$\sum m_j c_j$	$c \sum m_j x_j$	$c \sum m_j x_j$	$\sum \rho_j$	...	$\frac{M}{V}$
Dimensions	$\frac{\text{moles}}{\text{volume}}$	dimensionless	$\frac{\text{moles}}{\text{volume}}$	$\frac{\text{mass}}{\text{volume}}$	dimensionless	$\frac{\text{mass}}{\text{volume}}$

In an ideal gas mixture the partial pressure is a useful measure of concentration. By Dalton's law, the partial pressure P_j of A_j is related to the total pressure by

$$P_j = x_j P. \qquad (2.5.9)$$

Again, for an ideal gas

$$P_j V = N_j RT$$

or

$$P_j = c_j RT, \qquad (2.5.10)$$

whence

$$c_j = \frac{x_j P}{RT}. \qquad (2.5.11)$$

For nonideal gases this equation must be modified by the introduction of a compressibility factor Z_j to be determined from the available tables. Consult, for example, Hougen, Watson, and Ragatz or Hall and Ibele among many sources (see references to Chap. 3). In this case

$$P_j = c_j Z_j RT \qquad (2.5.12)$$

and

$$c_j = \frac{x_j P}{Z_j RT}. \qquad (2.5.13)$$

2.6 Concentration Changes with a Single Reaction

Formulae for the change of concentration with extent of reaction can be derived by combining the various definitions of concentration with the basic relation Eq. (2.4.2). Thus, if the volume V is constant

$$\begin{aligned} c_j &= \frac{N_j}{V} = \frac{N_{j0}}{V} + \alpha_j \frac{X}{V} \\ &= c_{j0} + \alpha_j \xi, \end{aligned} \qquad (2.6.1)$$

where ξ is the intensive extent of reaction in moles per unit volume. In the reference composition, c_{j0} is the molar concentration of A_j.

If c_0 denotes $\sum_{j=1}^{S} c_{j0}$ and $\bar{\alpha} = \sum_{j=1}^{S} \alpha_j$

then

$$c = c_0 + \bar{\alpha}\xi, \qquad (2.6.2)$$

and hence

$$x_j = \frac{c_j}{c} = \frac{c_{j0} + \alpha_j \xi}{c_0 + \bar{\alpha}\xi} = \frac{x_{j0} + \alpha_j \xi'}{1 + \bar{\alpha}\xi'}, \qquad (2.6.3)$$

where $\xi' = \xi/c_0$ is a dimensionless measure of the extent of reaction. If $\bar{\alpha} = 0$, then the total number of moles is unchanged by reaction and this equation becomes linear in ξ'. If $\bar{\alpha} \neq 0$, then the mole fraction of an inert species ($\alpha_j = 0$) will change during reaction.

Since $\rho_j = m_j c_j$, were m_j is the molecular weight of A_j,

$$\rho_j = \rho_{j0} + \alpha_j m_j \xi. \qquad (2.6.4)$$

This formula is valid when there is no volume change by reaction and since $\sum \alpha_j m_j$ is always zero (see Exercise 2.2.2) we observe that the density is constant, $\rho = \rho_0$. Dividing through by it we have

$$g_j = \frac{\rho_j}{\rho} = g_{j0} + \alpha_j m_j \xi'', \qquad (2.6.5)$$

where

$$\xi'' = \frac{\xi}{\rho} = \frac{X}{M} \qquad (2.6.6)$$

is the extent in moles per unit mass. This last equation can be used even when the volume is changing because the total mass does not change and Eq. (2.6.5) can be obtained by dividing Eq. (2.4.4) by M.

The intensive extent variables ξ, ξ', and ξ'' are associated with the reaction and not with an arbitrarily selected key species. It is always possible to express concentrations in terms of one of them, say the concentration of A_1, by eliminating the extent between the general formula and that for $j = 1$. Thus

$$\xi = \frac{c_j - c_{j0}}{\alpha_j} = \frac{c_0(x_j - x_{j0})}{\alpha_j - \bar{\alpha} x_j} = \frac{\rho_j - \rho_{j0}}{\alpha_j m_j} \qquad (2.6.7)$$

so that

$$c_j = c_{j0} + \frac{\alpha_j}{\alpha_1}(c_1 - c_{10}), \qquad (2.6.8)$$

$$x_j = x_{j0} + \frac{\alpha_j - \bar{\alpha} x_{j0}}{\alpha_1 - \bar{\alpha} x_{10}}(x_1 - x_{10}), \qquad (2.6.9)$$

$$\rho_j = \rho_{j0} + \frac{\alpha_j m_j}{\alpha_1 m_1}(\rho_1 - \rho_{10}). \qquad (2.6.10)$$

The notion behind Eq. (2.6.1) should by this time be rooted in the student's mind; however, none of these formulae should be applied by rote. The extent variables or the concentration variables should be used with versatility so that the equations for any particular problem assume the simplest possible form.

Exercise 2.6.1. In a mixture of N_2, H_2, and inert gases the ratio of the mole fractions of N_2 to H_2 is $1:3\gamma$, and δ is the mole fraction of inert gases. NH_3 is formed by reaction, and when its mole fraction is z, the mole fractions of N_2 and H_2 have changed by the ratios $(1 - b_1 z)$ and $(1 - b_2 z)$, respectively. Show that

$$b_1 = \frac{\frac{1}{2}(3\gamma - 1) + \delta}{1 - \delta}, \qquad b_2 = \frac{\frac{1}{2}(\gamma + 1) + \gamma\delta}{\gamma(1 - \delta)}.$$

Exercise 2.6.2. Show that the mole fraction ratio of N_2 to H_2 remains constant only if $\gamma = 1$.

Exercise 2.6.3. A unit volume of a mixture of initial composition c_{j0} is reacted to an extent ξ without any change of volume and a volume v of composition c'_{j0} is added. Show that the resulting composition can be related to a reference composition

$$\frac{c_{j0} + v c'_{j0}}{1 + v}.$$

Exercise 2.6.4. A unit volume of initial composition c_{j0} is reacted to an extent ξ_1 and a volume v_1 of the original mixture is added. The reaction proceeds to a further extent ξ_2 and a volume v_2 of the original mixture is added. Show that after n such operations, $c_j = c_{j0} + \alpha_j \xi$, where

$$\xi = \frac{\xi_1 + (1 + v_1)\xi_2 + (1 + v_1 + v_2)\xi_3 + \cdots + (1 + v_1 + v_2 + \cdots + v_{n-1})\xi_n}{1 + v_1 + v_2 + \cdots + v_n}.$$

Exercise 2.6.5. Show that if after each cycle of reaction and dilution the apparent extent of reaction is to be a constant ξ, then

$$\xi_r = \frac{\xi v_r}{1 + v_1 + v_2 + \cdots + v_{r-1}}.$$

2.7 Concentration Changes with Several Reactions

The concentration changes with several reactions follow the same pattern that was seen in the extensive formulation of Sec. 2.4. An intensive extent variable can be defined for each reaction

$$\xi_i = \frac{X_i}{V}, \qquad \xi'_i = \frac{X_i}{N_0}, \qquad \xi''_i = \frac{X_i}{M}, \tag{2.7.1}$$

where V, N_0, and M are the volume, original number of moles, and mass, respectively. Thus, from Eq. (2.4.5) we have

$$c_j = c_{j0} + \sum_{i=1}^{R} \alpha_{ij} \xi_i. \tag{2.7.2}$$

If

$$\bar{\alpha}_i = \sum_{j=1}^{S} \alpha_{ij}, \tag{2.7.3}$$

then

$$c = c_0 + \sum_{i=1}^{R} \bar{a}_i \xi_i \qquad (2.7.4)$$

and

$$x_j = \frac{x_{j0} + \sum_{i=1}^{R} \alpha_{ij}\xi'_i}{1 + \sum_{i=1}^{R} \bar{a}_i \xi'_i}. \qquad (2.7.5)$$

Similarly,

$$\rho_j = \rho_{j0} + \sum_{i=1}^{R} \alpha_{ij} m_j \xi_i \qquad (2.7.6)$$

and

$$g_j = g_{j0} + \sum_{i=1}^{R} \alpha_{ij} m_j \xi''_i. \qquad (2.7.7)$$

Again it is possible to select certain key species and to express the concentrations of the others in terms of their concentrations. Without loss of generality we may take these to be $A_1, \ldots, A_R$ and express c_j, where $j = R + 1, \ldots, S$, in terms of $c_1, \ldots, c_R$ by eliminating $\xi_1, \ldots, \xi_R$ between the $(R + 1)$ equations:

$$c_j - c_{j0} = \alpha_{1j}\xi_1 + \cdots + \alpha_{Rj}\xi_R$$
$$c_1 - c_{10} = \alpha_{11}\xi_1 + \cdots + \alpha_{R1}\xi_R$$
$$\overline{}$$
$$c_R - c_{R0} = \alpha_{1R}\xi_1 + \cdots + \alpha_{RR}\xi_R.$$

When the R reactions are independent the eliminant is

$$\begin{vmatrix} c_j - c_{j0} & \alpha_{1j} & \cdots & \alpha_{Rj} \\ c_1 - c_{10} & \alpha_{11} & \cdots & \alpha_{R1} \\ \hline c_R - c_{R0} & \alpha_{1R} & \cdots & \alpha_{RR} \end{vmatrix} = 0,$$

and expanding on the first column we have

$$c_j = c_{j0} + \sum_{i=1}^{R} \frac{\Delta_{ij}}{\Delta}(c_i - c_{i0}), \qquad (2.7.8)$$

where $\Delta = \det |\alpha_{11}\alpha_{22} \cdots \alpha_{RR}|$, and Δ_{ij} is the determinant in which the ith row of Δ is replaced by $\alpha_{1j}, \alpha_{2j}, \ldots, \alpha_{Rj}$.

Exercise 2.7.1. Find a formula similar to Eq. (2.7.8) for the mole fraction x_j.

Exercise 2.7.2. Write the simultaneous reactions $3 H_2 + N_2 = 2 NH_3$, $H_2 + CO_2 = CO + H_2O$ in standard form treating NH_3 as the principal product. A mixture of 68.4% H_2, 22.6% N_2, and 9% CO_2 reacts until

15% NH_3 and 5% H_2O are formed. (These are mole percentages.) What are then the percentages of H_2 and N_2?

2.8 Rate of Reaction

In an isolated system where changes of composition are only due to chemical reaction, the extensive rate of the reaction $\sum \alpha_j A_j = 0$ may be defined as

$$R^* = \frac{dX}{dt}. \tag{2.8.1}$$

Since the reference composition and stoichiometric coefficients are independent of time, the rate of change of the number of moles of A_j is

$$\frac{dN_j}{dt} = \alpha_j \frac{dX}{dt} = \alpha_j R^*. \tag{2.8.2}$$

The units of R^* are moles per unit time.

The intensive rate of reaction in moles per unit time per unit volume is given by

$$r = \frac{R^*}{V}. \tag{2.8.3}$$

If the volume is constant, then clearly

$$r = \frac{d\xi}{dt} \tag{2.8.4}$$

and

$$\frac{dc_j}{dt} = \alpha_j r. \tag{2.8.5}$$

More generally,

$$r = \frac{1}{V}\frac{d}{dt}(V\xi) = \frac{d\xi}{dt} + \xi\frac{d \ln V}{dt} \tag{2.8.6}$$

and

$$\frac{dc_j}{dt} = \frac{d}{dt}\frac{N_j}{V} = \frac{\alpha_j}{V}\frac{dX}{dt} - \frac{N_j}{V^2}\frac{dV}{dt} = \alpha_j r - c_j \frac{d \ln V}{dt}. \tag{2.8.7}$$

The rate of change of the total concentration, $c = \sum c_j$, is

$$\frac{dc}{dt} = \bar{\alpha} r - c\frac{d \ln V}{dt}, \tag{2.8.8}$$

and hence

$$\frac{dx_j}{dt} = \frac{d}{dt}\left(\frac{c_j}{c}\right) = \frac{1}{c}\frac{dc_j}{dt} - \frac{c_j}{c^2}\frac{dc}{dt} = (\alpha_j - \bar{\alpha} x_j)\frac{r}{c}. \tag{2.8.9}$$

The mass does not change by reaction so that the rate of reaction per unit mass is

$$r'' = \frac{R^*}{M} = \frac{d}{dt}\left(\frac{X}{M}\right) = \frac{d\xi''}{dt} = \frac{r}{\rho}. \tag{2.8.10}$$

The mass concentration is $\rho_j = m_j c_j$ and multiplying Eq. (2.8.7) by m_j we have

$$\frac{d\rho_j}{dt} = \alpha_j m_j r - \rho_j \frac{d \ln V}{dt}. \tag{2.8.11}$$

The rate of change of the mass fraction is

$$\frac{dg_j}{dt} = \frac{\alpha_j m_j r}{\rho} = \alpha_j m_j r''. \tag{2.8.12}$$

For simultaneous reactions, we can define a set of extensive reaction rates

$$R_i^* = \frac{dX_i}{dt} \quad i = 1, 2, \ldots, R \tag{2.8.13}$$

and intensive rates by

$$r_i = \frac{R_i^*}{V}. \tag{2.8.14}$$

This group of reaction rates belongs to the set of reactions as a whole. From the formulae of the last section we see that

$$\frac{dN_j}{dt} = \sum_{i=1}^{R} \alpha_{ij} R_i^*, \tag{2.8.15}$$

$$\frac{dc_j}{dt} = \sum_{i=1}^{R} \alpha_{ij} r_i - c_j \frac{d \ln V}{dt}, \tag{2.8.16}$$

$$\frac{dx_j}{dt} = \sum_{i=1}^{R} (\alpha_{ij} - \bar{\alpha}_i x_j) \frac{r_i}{c}, \tag{2.8.17}$$

$$\frac{dM_j}{dt} = \sum_{i=1}^{R} \alpha_{ij} m_j R_i^*, \tag{2.8.18}$$

$$\frac{d\rho_j}{dt} = \sum_{i=1}^{R} \alpha_{ij} m_j r_i - \rho_j \frac{d \ln V}{dt}, \tag{2.8.19}$$

$$\frac{dg_i}{dt} = \sum_{i=1}^{R} \frac{\alpha_{ij} m_j r_i}{\rho} = \sum_{i=1}^{R} \alpha_{ij} m_j r_i''. \tag{2.8.20}$$

Rates R^* and r are functions of composition and of the thermodynamic variables about which we shall have much more to say later.

Exercise 2.8.1. Show that at constant pressure and temperature, the volume change of an ideal gas mixture is related to the reaction rate by

$$\frac{1}{V}\frac{dV}{dt} = \frac{\bar{\alpha}}{c_0} r,$$

where c_0 is the original total molar concentration. Hence, or otherwise,

show that
$$\frac{dc_j}{dt} = \left(\alpha_j - \frac{\bar{\alpha}c_j}{c_0}\right)r.$$

Exercise 2.8.2. At constant pressure, a reacting ideal gas mixture changes volume both by a change in the total number of moles and by a change in temperature proportional to the extent. If the latter is linear, $T = T_0 + JX$, show that
$$\frac{d \ln V}{dt} = \left(\frac{\bar{\alpha}}{c} + J\frac{N\mathsf{R}}{P}\right)r.$$

(Here R is the gas constant.)

Exercise 2.8.3. Consider the change in partial pressure when the reaction takes place at constant volume along the lines of the two preceding problems.

NOTATION

A_j	jth chemical species
c	total molar concentration (moles/unit vol.)
c_j	molar concentration of A_j (moles/unit vol.)
c_0	total molar concentration in reference composition (moles/unit vol.)
c_{j0}	molar concentration of A_j in reference composition (moles/unit vol.)
g_j	mass fraction of A_j
g_{j0}	mass fraction of A_j in reference composition
M	total mass present (mass)
M_j	mass of A_j present (mass)
m_j	molecular weight of A_j (mass/mole)
N	total number of moles present
N_j	number of moles of A_j present
N_{j0}	number of moles of A_j in reference composition
P	total pressure (pressure)
P_j	partial pressure of A_j (pressure)
R	gas constant
R	number of reactions
R^*	rate of reaction (moles/unit time)
R_i^*	rate of reaction i (moles/unit time)
r	rate of reaction (moles/unit vol. /unit time)
r_i	rate of reaction i (moles/unit vol./unit time)
r''	rate of reaction (moles/unit mass/unit time)

S	number of chemical species present
T	temperature
V	volume (vol.)
X	extent of reaction (moles)
X_i	extent of reaction i (moles)
x_j	mole fraction of A_j
x_{j0}	mole fraction in reference composition
Z_j	compressibility factor of A_j
α_j	stoichiometric coefficient of A_j
α_{ij}	stoichiometric coefficient of A_j in reaction i
$\bar{\alpha}$	$\sum_1^S \alpha_j$
$\bar{\alpha}_i$	$\sum_1^S \alpha_{ij}$
Δ	determinant of the first R rows and columns of the stoichiometric matrix
Δ_{ij}	Δ with the ith row replaced by $(\alpha_{1j}, \ldots, \alpha_{Rj})$
λ	multipliers
ξ	extent of reaction (moles/unit vol.)
ξ'	ξ/c_0
ξ''	ξ/ρ (moles/unit mass)
ξ_i, ξ_i', ξ_i''	extents for reaction i
ρ	density or total mass concentration (mass/unit vol.)
ρ_j	mass concentration of A_j (mass/unit vol.)
ρ_{j0}	mass concentration of A_j in reference composition (mass/unit vol.)

SUFFIXES

i	denotes reaction $i (i = 1, 2, \ldots, R)$
j	denotes species $j (j = 1, 2, \ldots, S)$

Thermochemistry and Chemical Equilibrium 3

3.1 Introduction

Any attempt at a profound and extensive treatment of the thermostatics and thermodynamics of a chemical reaction would be out of place here for two reasons. First, this is an elementary text in which we wish to present the principles governing chemical reactions with a minimum of complication. Although it is true that in practice intricate thermodynamic calculations of nonideal behavior often have to be carried out, the foundation must be securely laid on a careful consideration of the simpler situations where the underlying principles show up most clearly. There is a relative abundance of thermodynamic data but care must be exercised so that its use does not lead to a false sophistication quite out of keeping with the character of the kinetic data with which it must be used. The second reason for not going into greater detail is the wealth, not to say the plethora, of texts on thermodynamics.

Thermochemistry has to do with the heat effects that accompany a chemical reaction and leads to a discussion of the nature of the equilibrium state. Figure 3.1 shows the structure of the chapter; the numbers in the boxes refer to the sections. The material of this chapter should be quite familiar, from any elementary thermodynamics course, but it will be helpful to restate it within the framework of the notation we are using in this book.

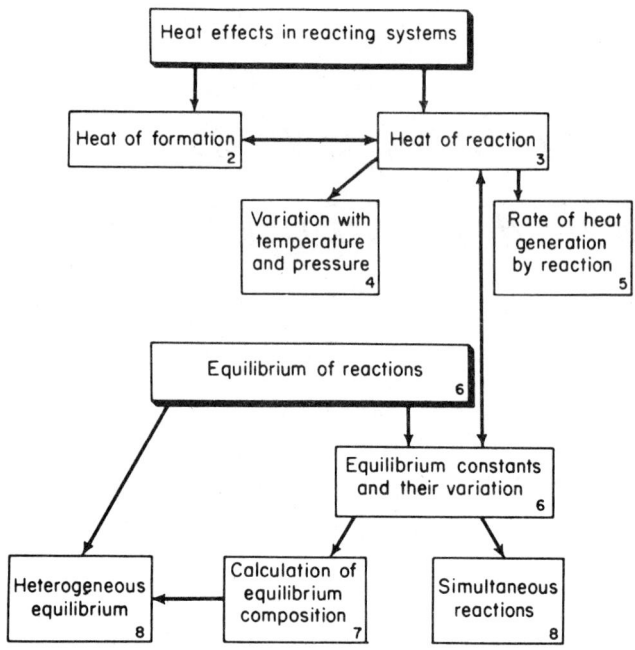

Fig. 3.1 The structure of the chapter.

3.2 Heat of Formation

The difference between the enthalpy of one mole of a pure compound and the total enthalpy of the elements from which it is composed is called the *heat of formation* of the compound. It represents the change in energy due to rearrangement of the atoms, and to be made precise it must be specified under certain standard conditions and the state of aggregation of each species must be given. The standard pressure is 1 atm and the standard temperature 25°C. The standard heat of formation will be denoted by ΔH_f°.

For example, the equation

$$H_2(g) + \tfrac{1}{2}O_2(g) = H_2O(\ell), \qquad \Delta H_f^\circ = -68{,}317 \text{ cal/gm-mole}$$

tells us that one mole of gaseous hydrogen upon reaction with half a mole of gaseous oxygen will give one mole of liquid water whose enthalpy will be 68.317 kcal less than that of the hydrogen and oxygen. It does not matter whether or not the reaction would take place at standard conditions because Hess' law assures us that provided we start and finish at standard temperature and pressure what we do in between does not affect the enthalpy change.

Since the enthalpy of water is less than that of the elements from which it was formed, the difference in energy is manifested by the liberation of 68.317 kcal for each mole of water formed. It is sometimes convenient to regard the heat as a product of the reaction and write

$$H_2(g) + \tfrac{1}{2}O_2(g) = H_2O(\ell) + 68{,}317 \text{ cal}$$

or in our convention

$$-H_2(g) - \tfrac{1}{2}O_2(g) + H_2O(\ell) + 68{,}317 \text{ cal} = 0. \qquad (3.2.1)$$

Such an equation is balanced thermally as well as in each elementary species (here, H and O) and can be manipulated freely as an equation. Thus if we had written the equation as $2H_2(g) + O_2(g) = 2\,H_2O(\ell)$, the enthalpy difference between the product (now two moles) and the elements would be 136.634 kcal. We get this by multiplying the whole equation by 2, namely:

$$-2H_2(g) - O_2(g) + 2H_2O(\ell) + 136{,}634 \text{ cal} = 0.$$

The standard heat of formation is still 68.317 kcal since two moles of liquid water are formed in this last reaction.

Again the standard heat of formation of water vapor is 57.798 kcal, and we would write this as

$$-H_2(g) - \tfrac{1}{2}O_2(g) + H_2O(g) + 57{,}798 \text{ cal} = 0. \qquad (3.2.2)$$

But by subtracting Eq. (3.2.2) from Eq. (3.2.1) we have

$$H_2O(\ell) - H_2O(g) + 10{,}519 \text{ cal} = 0.$$

This may be interpreted as the enthalpy of liquid water being 10.519 kcal/mol less than that of water vapor, and as giving the latent heat of water at 1 atm and 25°C.

3.3 Heat of Reaction

The heat of formation of a substance is just the heat of the reaction in which one mole of it is formed from elementary species. More generally the standard heat of any reaction is the difference between the total enthalpy of its products and that of its reagents at 1 atm and 25°C. For example,

$$PbO(s) + S(s) + \tfrac{3}{2}O_2(g) = PbSO_4(s), \qquad \Delta H° = -167{,}430 \text{ cal/gm-mole}$$

means that the enthalpy of one mole of lead sulfate is less than the sum of the enthalpies of the reactants by 167.43 kcal. This is the amount of heat that would be evolved if the reaction could be carried out at 1 atm and 25°C. Again we need not worry if this is impossible. Starting with the reactants at standard conditions, we could perform the reaction at any temperature and pressure and would get the same enthalpy change upon returning to standard conditions.

Again, it is informative to write the equation with the heat as a product

$$-\text{PbO(s)} - \text{S(s)} - \tfrac{3}{2}\text{O}_2(\text{g}) + \text{PbSO}_4(\text{s}) + 167{,}430 \text{ cal} = 0.$$

This equation is balanced thermally as well as in each element and may be multiplied by any constant and added to any other thermally balanced equation. If we denote the enthalpy of one mole of A_j by h_j, then, by definition, the heat of the reaction $\sum \alpha_j A_j = 0$ is

$$\Delta H = \sum_{j=1}^{S} \alpha_j h_j. \tag{3.3.1}$$

For, by our convention, the products of the reaction have positive stoichiometric coefficients, and the sum over the positive terms $\alpha_j h_j$ is the total enthalpy of the products. Similarly, the sum over the negative terms is the total enthalpy of the reactants and so the sign convention gives the difference. If ΔH is positive, the enthalpy of the products is greater than that of the reactants and heat will be absorbed as the reaction goes on; such a reaction is called *endothermic*. On the other hand, an *exothermic* reaction, for which ΔH is negative, evolves heat. The thermally balanced form of the reaction $\sum \alpha_j A_j = 0$ is thus

$$\sum_{j=1}^{S} \alpha_j A_j - \Delta H = 0, \tag{3.3.2}$$

and it may be freely manipulated and combined with others. If the reaction is assumed to take place under standard conditions we add a superscript, writing ΔH°.

The total enthalpy of a mixture of N_j moles of A_j, where $j = 1, 2, \ldots, S$, is

$$H = \sum_{j=1}^{S} N_j h_j, \tag{3.3.3}$$

for h_j is the partial molar enthalpy. But $N_j = N_{j0} + \alpha_j X$ so that

$$\begin{aligned}\frac{\partial H}{\partial X} &= \sum_{j=1}^{S} \left(\frac{\partial N_j}{\partial X} h_j + N_j \frac{\partial h_j}{\partial X}\right) \\ &= \sum_{j=1}^{S} \alpha_j h_j = \Delta H,\end{aligned} \tag{3.3.4}$$

because, as we shall show shortly, the second term within the parentheses vanishes. Thus the heat of reaction is the partial derivative of the total enthalpy with respect to the molar extent of reaction.

The vanishing of the second term is a feature that will recur in different forms and it is worth dealing with it once and for all. We shall refer to the following equation and its equivalents as Eq. (3.3A):

Sec. 3.3 Heat of Reaction

$$\sum_{j=1}^{S} N_j \frac{\partial h_j}{\partial N_k} = 0. \tag{3.3A}$$

To prove this, we observe that the enthalpy H is an extensive quantity and thus a homogeneous function of degree 1 in the variable N_j. The partial molal enthalpy h_j is intensive and a homogeneous function of degree zero. In fact, $h_j = \partial H/\partial N_j$ and $H = \sum N_j h_j$. The partial molal enthalpy h_j will depend on N_j as well as on pressure and temperature, but

$$h_k = \frac{\partial}{\partial N_k} \sum_{j=1}^{S} N_j h_j(P, T, N_1, \ldots, N_S)$$

$$= h_k + \sum_{1}^{S} N_j \frac{\partial h_j}{\partial N_k}.$$

The first term comes from the differentiation of the only term that contains N_j as a multiplicative factor, and the second from the dependence of each h_j on N_k. Thus,

$$\sum_{j=1}^{S} N_j \frac{\partial h_j}{\partial N_k} = 0,$$

which establishes Eq. (3.3A). This shows that h_k is a homogeneous function of degree zero, which satisfies the condition

$$\sum N_j \frac{\partial h_k}{\partial N_j} = \sum N_j \frac{\partial^2 H}{\partial N_j \partial N_k} \sum N_j \frac{\partial h_j}{\partial N_k} = 0$$

by Eq. (3.3A). The term in question in Eq. (3.3.4) is

$$\sum_j N_j \frac{\partial h_j}{\partial X} = \sum_j N_j \sum_k \frac{\partial h_j}{\partial N_k} \frac{\partial N_k}{\partial X} = \sum_k \alpha_k = \sum_j N_j \frac{\partial h_j}{\partial N_k} = 0.$$

We can also see that

$$\Delta H = \sum_{j=1}^{S} \alpha_j (\Delta H_f)_j. \tag{3.3.5}$$

We may write for the formation of each species the reaction

$$-\sum_{k=1}^{T} \epsilon_{jk} E_k + A_j - (\Delta H_f)_j = 0, \tag{3.3.6}$$

where ϵ_{jk} is the stoichiometric coefficient of the element E_k in the reaction forming one mole of A_j. Multiplying Eq. (3.3.6) by α_j and summing from 1 to S gives

$$-\sum_{k=1}^{T}\sum_{j=1}^{S} \alpha_j \epsilon_{jk} E_k + \sum_{j=1}^{S} \alpha_j A_j - \sum_{j=1}^{S} \alpha_j (\Delta H_f)_j = 0.$$

But the first term vanishes because the reaction is balanced in each elementary species and so $\sum_{j=1}^{S} \alpha_j \epsilon_{jk} = 0$ for all k (see Exercise 2.2.1). Comparing this equation with Eq. (3.3.2) shows that $\Delta H = \sum \alpha_j (\Delta H_f)_j$.

If U is the internal energy, V the volume, and P the pressure, the enthalpy $H = U + PV$. Hence at constant pressure

$$dH = dU + P\,dV,$$

and by the first law of thermodynamics this is $-dQ$, where Q is the heat released. Thus ΔH is the heat that would be absorbed by the reaction at constant pressure. To put it in another way, we could say that if the reaction goes to an infinitesimal extent dX at constant pressure the heat evolved would be $(-\Delta H)\,dX$. To calculate how heat is evolved in a constant volume reaction, we might suppose that the extent of reaction changes by dX at constant pressure and that the volume is restored isothermally. If v_j is the volume of one mole of A_j, the volume change will be $d(\sum N_j v_j) = (\sum \alpha_j v_j)\,dX = (\Delta V)\,dX$. The internal energy change associated with an isothermal volume change of dV is $(\partial U/\partial V)_T\,dV$ so that the second operation requires a heat removal of

$$\left(\frac{\partial U}{\partial V}\right)_T dV + P\,dV = \left[\left(\frac{\partial U}{\partial V}\right)_T + P\right](\Delta V)\,dX.$$

Hence, the net heat generated when the extent of reaction changes by dX at constant volume is

$$-\Delta U\,dX = -\left\{\Delta H - \left[\left(\frac{\partial U}{\partial V}\right)_T + P\right]\Delta V\right\}dX. \tag{3.3.7}$$

The free energy and the entropy changes associated with chemical reactions are similarly defined and manipulated. Thus, ΔG is the difference between the Gibbs free energy of the products and that of the reactants under standard conditions. The standard entropy change ΔS is related to ΔG and ΔH by

$$\Delta G = \Delta H - T\,\Delta S. \tag{3.3.8}$$

Because pressure and temperature are the natural independent variables for G, the Gibbs free energy will play the leading role in the discussion of equilibrium, to which we shall turn after considering the variation of the heat of reaction under other than standard conditions.

Exercise 3.3.1. Why is 167,430 cal not the heat of formation of $PbSO_4$ according to the reaction given at the beginning of this section?

Exercise 3.3.2. Calculate the standard heat of reaction for

$$CH_3OH - 2H_2 - CO = 0$$

from the heats of formation of CH_3OH, -48.08 kcal/gm-mole; and CO, -26.416 kcal/gm-mole.

Exercise 3.3.3. Calculate the heat of formation of $SO_3(g)$ from the following heats of reaction:

$SO_3(g) + 6H_2O(\ell) = H_2SO_4 \cdot 5H_2O(\ell)$, $\Delta H° = -45{,}013$;

$PbO(s) + S(s) + \tfrac{3}{2}O_2(g) = PbSO_4(s)$, $\Delta H° = -167{,}430$;

$$PbO(s) + H_2SO_4 \cdot 5H_2O(\ell)$$
$$= PbSO_4(s) + 6H_2O(\ell), \quad \Delta H° = -27{,}967.$$

Exercise 3.3.4. The *dissociation energy* of a molecule is the heat of the reaction by which it is split up into atoms (not molecules) of its elements. (For diatomic molecules it represents the strength of the bonds holding the atoms together.) Given that the dissociation energies of $H_2(g)$ and $Cl_2(g)$ are, respectively, 104.2 and 58.0 kcal/mole and that the heat of formation of $HCl(g)$ is -22.1 kcal/mole, find the dissociation energy of HCl.

Exercise 3.3.5. Show that the difference between the heats of reaction at constant volume and constant pressure is $\bar{a}RT$ for a reacting mixture of ideal gases.

3.4 Variation of the Heat of Reaction

From the definition of heat of reaction and the known relations for the thermodynamic functions it is easy to calculate ΔH at any pressure and temperature from its value at standard conditions. We have

$$\left(\frac{\partial H}{\partial T}\right)_P = C_P \quad \text{and} \quad \left(\frac{\partial h_j}{\partial T}\right)_P = c_{Pj}, \tag{3.4.1}$$

which are, respectively, the heat capacity of the mixture and the molar heat capacity of A_j at constant pressure. The standard relations are:

$$\left(\frac{\partial H}{\partial P}\right)_T = T\left(\frac{\partial S}{\partial P}\right)_T + V = V - T\left(\frac{\partial V}{\partial T}\right)_P = V(1-a), \tag{3.4.2}$$

where

$$a = \frac{T}{V}\left(\frac{\partial V}{\partial T}\right)_P = \left(\frac{\partial \ln V}{\partial \ln T}\right)_P \tag{3.4.3}$$

is a coefficient of thermal expansion (dimensionless). For an equation of state expressed by the compressibility factor

$$PV = ZNRT, \tag{3.4.4}$$

$$a = \left(\frac{\partial \ln Z}{\partial \ln T}\right)_P + 1,$$

and

$$\left(\frac{\partial H}{\partial P}\right)_T = -V\left(\frac{\partial \ln Z}{\partial \ln T}\right)_P. \tag{3.4.5}$$

Similarly, for the partial molar quantities we have

$$\left(\frac{\partial h_j}{\partial P}\right)_T = v_j(1 - a_j) = -v_j\left(\frac{\partial \ln Z_j}{\partial \ln T}\right)_P \tag{3.4.6}$$

If ΔH° denotes the standard value of the heat of reaction at 1 atm and 25°C and if ΔH is the value at some P and T,

$$\Delta H - \Delta H^\circ = \int_1^P \left(\frac{\partial \Delta H}{\partial P}\right)_T dP + \int_{298}^T \left(\frac{\partial \Delta H}{\partial T}\right)_P dT, \quad (3.4.7)$$

where the integrand of the first integral is evaluated at $T = 298°\text{K}$ and that of the second at P atm. From Eqs. (3.3.1), (3.4.1), and (3.4.6), the two integrands are

$$\sum \alpha_j v_j (1 - a_j) = -\sum \alpha_j v_j \left(\frac{\partial \ln Z_j}{\partial \ln T}\right)_P \quad (3.4.8)$$

and

$$\sum \alpha_j c_{Pj}. \quad (3.4.9)$$

For an ideal gas mixture, $a_j = 1$ and the pressure variation vanishes. Heat capacities of gases are often given as polynomials,

$$c_{Pj} = a_j + b_j T + c_j T^2,$$

so that for an ideal gas

$$\Delta H = \Delta H^\circ + a(T - 298) + \tfrac{1}{2}b(T^2 - 298^2) + \tfrac{1}{3}c(T^3 - 298^3),$$

where

$$a = \sum \alpha_j a_j, \quad b = \sum \alpha_j b_j, \quad c = \sum \alpha_j c_j.$$

3.5 Rate of Generation of Heat by Reaction

It is convenient to take the exothermic reaction as the "positive" case and talk about the rate of heat evolution or generation. At constant pressure $dQ = -dU - P\,dV = -d(U + PV) = -dH = (-\Delta H)\,dX$ so that

$$\frac{dQ}{dt} = -(\Delta H_m)\left(\frac{dX}{dt}\right) = (-\Delta H)R^*. \quad (3.5.1)$$

If V is the total volume, the rate of heat generation per unit volume is

$$q = \frac{1}{V}\frac{dQ}{dt} = (-\Delta H)r. \quad (3.5.2)$$

Similarly, for simultaneous reactions

$$q = \sum_{i=1}^R (-\Delta H)_i r_i. \quad (3.5.3)$$

If the pressure varies but the volume is held constant

$$q = (-\Delta U)r, \quad (3.5.4)$$

where $(-\Delta U)$ is given by Eq. (3.3.7).

3.6 Chemical Equilibrium

Consider a system of S chemical species with the reaction $\sum \alpha_j A_j = 0$ taking place and let it be maintained at constant temperature and pressure. The system will spontaneously change in the direction of increasing total entropy, reaching equilibrium when the entropy cannot increase any more. In an infinitesimal change, let the system release an amount of heat dQ and suffer an entropy change dS. The total entropy change of the system and thermostat is $dS + dQ/T$ and this must be positive except at equilibrium where it is zero. With dQ positive for heat released, dW the work done by the system, and dU the change in internal energy, the first law of thermodynamics is $dU + dQ + dW = 0$. Therefore, at constant P and T,

$$dG = d(U - TS + PV),$$
$$= dU - T\,dS + P\,dV,$$
$$= -(dQ + T\,dS) + P\,dV - dW.$$

If no work other than that against the external pressure is done by the system, then $dW = P\,dV$ and since $dQ + T\,dS \geq 0$,

$$dG \leq 0.$$

It is strictly negative for any spontaneous change and at equilibrium

$$dG = 0. \tag{3.6.1}$$

Now, if there are N_j moles of A_j, a change in the Gibbs free energy is given by

$$dG = -S\,dT + V\,dP + \sum \mu_j\,dN_j, \tag{3.6.2}$$

where μ_j is the chemical potential of A_j. At constant temperature and pressure ($dT = dP = 0$) and with $dN_j = \alpha_j\,dX$, the condition for equilibrium is

$$\sum_{j=1}^{S} \alpha_j \mu_j = 0. \tag{3.6.3}$$

Consider first an ideal gas mixture for which

$$\mu_j = \mu_j^\circ + RT \ln P_j, \tag{3.6.4}$$

where μ_j° is the chemical potential of A_j at unit partial pressure, P_j is the partial pressure of A_j, and μ_j° is only a function of temperature. Then substituting from Eq. (3.6.4) into Eq. (3.6.3) gives

$$RT \sum_{j=1}^{S} \alpha_j \ln P_j = -\sum_{j=1}^{S} \alpha_j \mu_j^\circ, \tag{3.6.5}$$

which defines the relation that must be satisfied at equilibrium.

At unit pressure, the total Gibbs free energy of the system is

$$G^\circ = \sum_{j=1}^{S} N_j \mu_j^\circ. \tag{3.6.6}$$

But

$$\frac{\partial G°}{\partial X} = \sum \alpha_j \mu_j° = \Delta G° \tag{3.6.7}$$

is the standard free energy change for the reaction. (N.B. We have used the fact that $\mu_j°$ is an intensive variable and the formula (3.3A) in doing this differentiation.) Therefore Eq. (3.6.5) can be written

$$\sum_{j=1}^{S} \alpha_j \ln P_j = -\frac{\Delta G°}{RT}. \tag{3.6.8}$$

This is usually rearranged by taking exponentials of both sides to give

$$\exp\left(\sum \alpha_j \ln P_j\right) = \exp\left(\sum \ln P_j^{\alpha_j}\right)$$
$$= P_1^{\alpha_1} P_2^{\alpha_2} \cdots P_S^{\alpha_S} = \exp - \frac{\Delta G°}{RT}. \tag{3.6.9}$$

Just as $\sum \alpha_j \ln P_j$ is a convenient notation for the sum $\alpha_1 \ln P_1 + \alpha_2 \ln P_2 + \cdots + \alpha_S \ln P_S$, so $\prod P_j^{\alpha_j}$ is a convenient notation for the product $P_1^{\alpha_1} P_2^{\alpha_2} \cdots P_S^{\alpha_S}$. Then Eq. (3.6.9) can be written

$$\prod_{j=1}^{S} P_j^{\alpha_j} = K, \tag{3.6.10}$$

where

$$K = \exp - \frac{\Delta G°}{RT}. \tag{3.6.11}$$

By Eq. (3.3.8) we see that this can be written

$$K = \left(\exp \frac{\Delta S°}{R}\right) \exp - \frac{\Delta H°}{RT}. \tag{3.6.12}$$

For nonideal mixtures, the partial pressure must be replaced by the partial fugacity and

$$\mu_j = \mu_j° + RT \ln f_j. \tag{3.6.13}$$

At equilibrium,

$$\prod_{j=1}^{S} f_j^{\alpha_j} = K(T) = \exp - \frac{\Delta G°}{RT}. \tag{3.6.14}$$

The relation between pressure and fugacity is often expressed by a fugacity coefficient γ,

$$f = \gamma P \tag{3.6.15}$$

$$\ln \gamma = \int_0^P \left(\frac{v}{RT} - \frac{1}{P}\right) dP. \tag{3.6.16}$$

The calculation of fugacity coefficients and the use of generalized charts are discussed in the standard thermodynamics texts. If the fugacity coefficients are known then in the nonideal case we find that

Sec. 3.6 Chemical Equilibrium

$$\prod P_j^{\alpha_j} = K(T) \prod \gamma_j^{-\alpha_j}. \tag{3.6.17}$$

It can be shown that

$$\frac{d}{dT}\left(\frac{\mu_j^\circ}{T}\right) = -\frac{h_j}{T^2},$$

so that differentiating $\ln K = -\sum \alpha_j \mu_j^\circ / RT$ gives

$$\frac{d}{dT} \ln K = \frac{\sum \alpha_j h_j}{RT^2} = \frac{\Delta H^\circ}{RT^2}. \tag{3.6.18}$$

Integrating this gives

$$K(T) = K^* \exp - \frac{\Delta H^\circ}{RT}, \tag{3.6.19}$$

and comparison with Eq. (3.6.12) shows that K^* depends on the entropy of reaction.

For ideal solutions

$$\mu_j = \mu_j^* + RT \ln x_j, \tag{3.6.20}$$

where μ_j^* is a function of temperature and pressure but not of composition. Thus

$$\prod x_j^{\alpha_j} = \exp - \frac{\sum \alpha_j \mu_j^*}{RT} \tag{3.6.21}$$

at equilibrium.

The equilibrium expression can be formed from any concentration measure and a suffix on K is a convenient way of making the distinction from K with no suffix which is calculated from fugacities. Thus Eq. (3.6.10) might be written

$$\prod P_j^{\alpha_j} = K_P = \frac{K}{K_\gamma}$$

by Eq. (3.6.17). (See also the following exercise.)

Exercise 3.6.1. Show that for gases

$$\prod x_j^{\alpha_j} = K(T) P^{-\alpha} \prod \gamma_j^{-\alpha_j} = K_x$$

and for ideal gases

$$\prod c_j^{\alpha_j} = K(T)(RT)^{-\alpha} = K_c.$$

Exercise 3.6.2. Determine the changes of free energy, enthalpy, and entropy at 298°K for the reaction $2\,NO_2 = N_2O_4$ given the data:

T:	273.0	291.3	322.9	346.6	372.8	
K_P:	65.0	13.8	1.25	0.296	0.075	(C.U.)

3.7 The Calculation of Homogeneous Equilibrium Compositions

The form of the equilibrium condition is distinctly nonlinear because it asserts that the product of the concentrations raised to the power of the stoichiometric coefficients equals some constant depending on temperature and pressure. In the last chapter we saw that the various measures of concentration varied linearly with the extent or as the quotient of two linear expressions, and we might expect that the interaction of linear and nonlinear conditions would tend to make the calculations difficult. Let us look first at the simple reaction $A_1 - A_2 - A_3 = 0$ whose stoichiometry we studied in Sec. 2.4. We shall use the molar concentrations, c_j, though any other measure of concentration could be used.

The equilibrium relationship is

$$\prod c_j^{\alpha_j} = K_c, \qquad (3.7.1)$$

where we use the suffix c to denote that the equilibrium constant has been calculated for the molar concentrations at the prescribed temperature and pressure. In the case of the reaction $A_1 - A_2 - A_3 = 0$ this is

$$\frac{c_1}{c_2 c_3} = K_c. \qquad (3.7.2)$$

In Fig. 3.2, which is similar to Fig. 2.1 but with the total number of moles

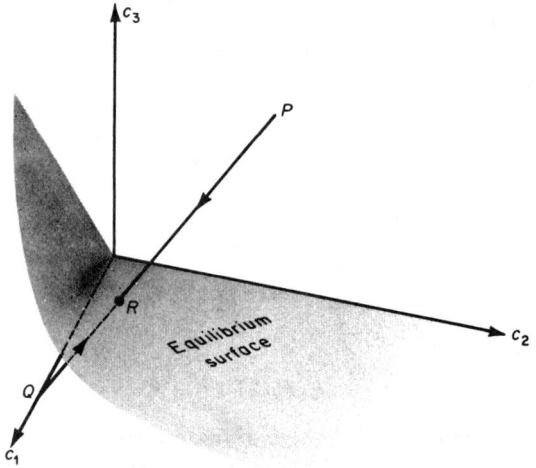

Fig. 3.2 Equilibrium surface for a single reaction $A_1 - A_2 - A_3 = 0$.

Sec. 3.7 The Calculation of Homogeneous Equilibrium Compositions   41

N_j replaced by the concentration c_j, the equilibrium relation, Eq. (3.7.2), is the hyperbolic surface. The reaction path is again a straight line

$$c_1 = c_{10} + \xi, \quad c_2 = c_{20} - \xi, \quad c_3 = c_{30} - \xi. \tag{3.7.3}$$

For example, if $c_{10} = 0$ and $c_{20} = c_{30} = 1$, point P represents the starting composition. The reaction will proceed so that the point representing the composition moves down the line PQ. However, the composition point cannot get to Q, for it first meets the equilibrium surface at R. If the initial composition had been $c_{10} = 1$ and $c_{20} = c_{30} = 0$, the point representing the composition would have started at Q and moved along the same reaction path to R. Clearly the position of R depends on the position of the equilibrium surface and hence on the value of K_c. If we substitute from Eq. (3.7.3) into Eq. (3.7.2) we have a quadratic equation for ξ, the extent of reaction at equilibrium:

$$c_{10} + \xi = K_c(c_{20} - \xi)(c_{30} - \xi)$$

or

$$\xi^2 - (c_{20} + c_{30} + K_c^{-1})\xi + (c_{20}c_{30} - c_{10}K_c^{-1}) = 0.$$

The solution is

$$\xi = \tfrac{1}{2}(c_{20} + c_{30} + K_c^{-1}) \pm \tfrac{1}{2}\{(c_{20} + c_{30} + K_c^{-1})^2 - 4c_{20}c_{30} + 4c_{10}K_c^{-1}\}^{1/2}, \tag{3.7.4}$$

and we need to ask which sign should be taken. If the initial composition were $c_{10} = 1$ and $c_{20} = c_{30} = 0$, then ξ would have to be negative for c_2 and c_3 to be positive; we should therefore take the negative sign in the solution of the quadratic equation. This is confirmed if $c_{10} = 0$ and $c_{20} = c_{30} = 1$, for then ξ must be positive and less than 1; but if the positive sign were taken in Eq. (3.7.4), ξ would be greater than 1.

In this elementary case, we see that there is a unique solution to the problem of finding the equilibrium extent of reaction. We can substitute from the relation

$$c_j = c_{j0} + \alpha_j \xi \tag{3.7.5}$$

for concentrations in the general reaction $\sum \alpha_j A_j = 0$ into the equilibrium relation

$$\prod c_j^{\alpha_j} = K_c. \tag{3.7.6}$$

This gives an equation for the equilibrium extent, namely

$$\prod_{j=1}^{S} (c_{j0} + \alpha_j \xi)^{\alpha_j} = K_c. \tag{3.7.7}$$

Only in special cases will this turn out to be as simple as a linear or quadratic equation so that it will be generally necessary to find the solution by some numerical procedure. In this case we need to be sure that there is one and only one solution; we shall show that this is so and see how the equilibrium extent is affected by change of conditions.

For any given reaction and reference composition, the left-hand side of Eq. (3.7.7) is a function of ξ only, i.e.,

$$F(\xi) = \prod c_j^{\alpha_j} = \prod (c_{j0} + \alpha_j \xi)^{\alpha_j}. \qquad (3.7.8)$$

Now ξ is bounded above by $\xi_{\max}$, the extent at which the least abundant reactant is exhausted, and below by $\xi_{\min}$, the extent (perhaps negative) at which one of the products would have zero concentration. Since the products have positive stoichiometric coefficients one of the concentrations in the numerator of Eq. (3.7.8) vanishes when $\xi = \xi_{\min}$ and so $F(\xi_{\min}) = 0$. On the other hand, the reactants have negative stoichiometric coefficients, so that at $\xi = \xi_{\max}$ one of the concentrations in the denominator is zero and $F(\xi_{\max})$ is infinite. If we could show that $F(\xi)$ increases monotonically in $\xi_{\min} \leq \xi \leq \xi_{\max}$, it is clear from Fig. 3.3 that there could be only one value of ξ for which $F(\xi) = K_c$ and so there would be a unique solution for Eq. (3.7.7). But

$$\ln F(\xi) = \ln \prod (c_{j0} + \alpha_j \xi)^{\alpha_j} = \sum \alpha_j \ln (c_{j0} + \alpha_j \xi).$$

Hence

$$\frac{d}{d\xi} \ln F(\xi) = \frac{F'(\xi)}{F(\xi)} = \sum \frac{\alpha_j^2}{c_{j0} + \alpha_j \xi} = \sum \frac{\alpha_j^2}{c_j}, \qquad (3.7.9)$$

and $F'(\xi)$ is always positive, for F and all the c_j are positive and the α_j are squared. It follows that the solution to Eq. (3.7.7) is unique as illustrated in Fig. 3.3.

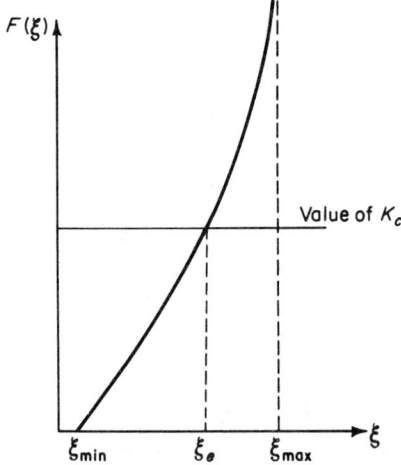

Fig. 3.3 Illustrating the solution $\xi = \xi_e$ of Eq. (3.7.7).

Sec. 3.7 The Calculation of Homogeneous Equilibrium Compositions   43

From this figure we may also see how variations in the reference composition and in the equilibrium constant affect the equilibrium extent. If we differentiate $F(\xi)$ partially with respect to a particular component of the initial composition, say c_{k0}, we have (by logarithmic differentiation as before)

$$\frac{\partial F(\xi)}{\partial c_{k0}} = F(\xi)\frac{\alpha_k}{c_k}, \tag{3.7.10}$$

and this has the sign of α_k. Thus if k denotes a reactant species ($\alpha_k < 0$) an increase in c_{k0} will move the whole curve downwards. For a fixed K_c, which is independent of the reference composition, the intersection $\xi = \xi_e$ will be moved to the right. Hence an increase in the initial or reference concentration of one of the reactants increases the equilibrium extent, whereas an increase in the concentration of one of the products ($\alpha_k > 0$) will decrease the equilibrium extent.

Again it is clear from the figure that if K_c is increased the value of ξ at the intersection will be increased. Now for an endothermic reaction increasing the temperature increases the value of K_c, because

$$\frac{d}{dT}\ln K_c = \frac{d}{dT}\ln (RT)^{-\bar{\alpha}}K(T) = \frac{-\bar{\alpha}}{T} + \frac{\Delta H}{RT^2};$$

the second of these terms usually dominates the first. Hence for an endothermic reaction increasing the temperature increases the equilibrium extent and, conversely, for an exothermic reaction an increase in temperature produces a decrease in equilibrium extent.

To see how pressure affects the equilibrium we may think in terms of ideal gases because the effect is not marked in liquids at normal pressures. Setting $\gamma_j = 1$, where $j = 1, 2, \ldots, S$ in the first formula of Exercise 3.6.1, and using Eq. (2.6.3) we have the equation

$$G(\xi') = \prod x_j^{\alpha_j} = \prod \left(\frac{x_{j0} + \alpha_j \xi'}{1 + \bar{\alpha}\xi'}\right)^{\alpha_j} = K(T)P^{-\bar{\alpha}}. \tag{3.7.11}$$

It can be shown, though with slightly more difficulty (see Exercise 3.7.3), that $G(\xi')$ is a monotonically increasing function of ξ' and therefore has the same shape as $F(\xi)$. Now if $\bar{\alpha} > 0$ there is a net increase in the total number of moles by reaction and an increase of pressure will decrease the right-hand side. Thus an increase in pressure decreases the equilibrium extent when $\bar{\alpha} > 0$ and, conversely, when there is a net decrease in the total number of moles ($\bar{\alpha} < 0$) an increase of pressure increases the equilibrium extent. Except where pressure has a pronounced effect on the fugacity coefficients γ_j, the same conclusions will hold for nonideal systems. All these changes are in accord with LeChatelier's principle.

It is also useful to picture the equilibrium relation as a curve in the plane of extent and temperature. Figure 3.4 has been drawn to give a portrait of the equilibrium of the second order, reversible, exothermic reaction whose kinetic properties we shall study below. The reaction is $-A_1 - A_2 + A_3 + A_4$

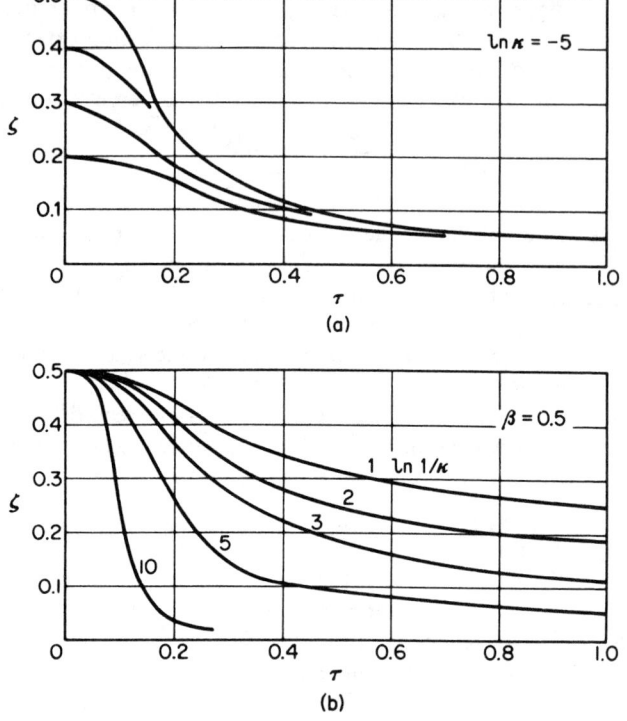

Fig. 3.4 Equilibrium portrait of the exothermic second order reaction: (a) variation with β (note that $\zeta = \beta$ at $\tau = 0$); (b) variation with κ.

$= 0$ and the reference composition is assumed to be free of products, $c_{30} = c_{40} = 0$. We do not assume that $c_{10} = c_{20}$, i.e., that the reactants are in stoichiometric proportions, but set

$$\beta = \frac{c_{10}}{c_{10} + c_{20}} \qquad \zeta = \frac{\xi}{c_{10} + c_{20}}. \qquad (3.7.12)$$

In terms of these dimensionless quantities

$$c_1 = (c_{10} + c_{20})(\beta - \zeta), \qquad c_2 = (c_{10} + c_{20})(1 - \beta - \zeta),$$
$$c_3 = c_4 = (c_{10} + c_{20})\zeta.$$

If we also take a dimensionless temperature

$$\tau = \frac{RT}{-\Delta H} \qquad (3.7.13)$$

the equilibrium constant is

$$K(T) = \exp\left(\frac{\Delta S}{R} - \frac{\Delta H}{RT}\right) = \kappa \exp\frac{1}{\tau}. \qquad (3.7.14)$$

Sec. 3.7 The Calculation of Homogeneous Equilibrium Compositions   45

Thus the equilibrium extent in its dimensionless form $\zeta = \zeta_e(\tau)$ is the solution of

$$\frac{\zeta^2}{(\beta - \zeta)(1 - \beta - \zeta)} = \kappa \exp \frac{1}{\tau}. \tag{3.7.15}$$

Figure 3.4(a) shows the solutions of this equation for $\ln \kappa = -5$ and for various β; since $\zeta_e(0) = \beta$, the value of β for each curve is its intercept at the left. No significant change of form results from a change of β. However, keeping $\beta = 0.5$ and changing κ, as in Fig. 3.4(b), gives a much more significant variation of shape.

It is worth noting that, though it may often be difficult to solve equations such as Eqs. (3.7.7), (3.7.11), or (3.7.15) for ξ or ξ' in terms of T [i.e., for $\xi_e(T)$ and $\xi'_e(T)$], it is not hard to solve for the temperature at which a given extent is the equilibrium extent. Thus for Eq. (3.7.15)

$$\tau_e(\zeta) = \left[\ln \frac{\zeta^2}{\kappa(\beta - \zeta)(1 - \beta - \zeta)}\right]^{-1}. \tag{3.7.16}$$

In general, if we take ΔS and ΔH to be relatively independent of temperature

$$T_e = \frac{-\Delta H}{-\Delta S + R \sum \alpha_j \ln f_j} \tag{3.7.17}$$

gives the equilibrium temperature for any composition.

If we start from a certain temperature and composition and allow the reaction to proceed adiabatically, it is less easy to calculate the equilibrium composition because the temperature varies with the extent and so both sides of the equation are changing. If the pressure is constant, the partial molar enthalpy is a function only of temperature and composition and so is the total enthalpy per unit volume

$$H = \sum c_j h_j(T, c_1, \cdots, c_S).$$

Now, if the reaction is adiabatic

$$dH = \sum c_j \frac{\partial h_j}{\partial T} dT + \sum \frac{\partial c_j}{\partial \xi} h_j d\xi + \sum \sum c_j \frac{\partial h_j}{\partial c_k} \frac{\partial c_k}{\partial \xi} d\xi = 0,$$

and the double sum vanishes by Eq. (3.3A). But $\partial h_j/\partial T$ is c_{Pj}, the molar heat capacity, and $\partial c_j/\partial \xi = \alpha_j$, and so the first two sums give $\sum c_j c_{Pj} = C_P$, the heat capacity per unit volume, and $\sum \alpha_j h_j = \Delta H$, the heat of reaction. Hence

$$\frac{dT}{d\xi} = \frac{-\Delta H}{C_P}, \tag{3.7.18}$$

showing that the temperature increases with extent for an exothermic reaction and decreases for an endothermic one as we should expect.

It is clear from Fig. 3.5 that the equilibrium is unique, for Eq. (3.7.18) can be integrated to give the adiabatic path of the reaction. In the exothermic

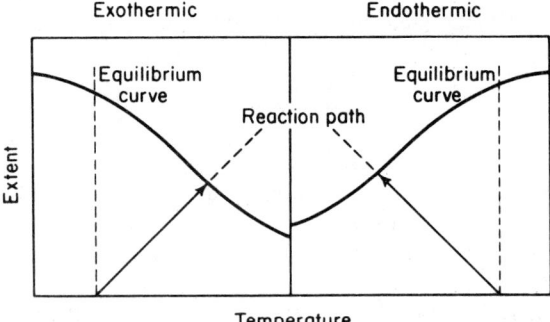

Fig. 3.5 Adiabatic reaction paths for exothermic and endothermic reversible reactions.

case, this is a path from the initial temperature and extent upward and to the right; it will intersect the equilibrium curve in a unique point giving the equilibrium composition and temperature. In the endothermic case, the reaction path goes upward and to the left, but since the equilibrium curve goes upward to the right, the intersection is again unique. Such a diagram might be used to get a rough value of the equilibrium extent which could then be refined by a numerical technique. This is even more practical because the ratio $(-\Delta H/C_P)$ is often tolerably constant over a wide range of temperature. For gases, it is often better to work with ξ'', the extent per unit mass, for $(-\Delta H)/(C_P/\rho)$ is much more nearly constant and

$$\frac{dT}{d\xi''} = \rho \frac{dT}{d\xi} = \rho \frac{-\Delta H}{C_P}. \tag{3.7.19}$$

Exercise 3.7.1. Find the equilibrium concentration of A_1 in the reaction $A_1 - A_2 - A_3 = 0$, by expressing c_2 and c_3 in terms of c_1. Show that the result is identical with the value derivable from Eq. (3.7.4.).

Exercise 3.7.2. If $g(\xi) = \prod (c_{j0} + \alpha_j \xi)^{\beta_j}$ show that its derivative is $g'(\xi) = g(\xi) \sum (\alpha_j \beta_j / c_j)$, the sum and the product being over $j = 1, 2, \ldots, S$. (This is a trivial generalization of the result in the text but one that we shall need later.)

Exercise 3.7.3. Show that the function $G(\xi')$ defined by Eq. (3.7.11) is monotone increasing. *Hint*: You may need Cauchy's inequality:

$$(\sum a_j^2)(\sum b_j^2) \geq (\sum a_j b_j)^2.$$

Exercise 3.7.4. If $S = 2$ we may take α_1 positive and α_2 negative. In the first order case $\alpha_1 = 1$, $\alpha_2 = -1$, show that

$$\xi_e(T) = \frac{c_{20} K_c(T) - c_{10}}{K_c(T) + 1}.$$

Sec. 3.8 Equilibrium of Simultaneous and Heterogeneous Reactions

Exercise 3.7.5. At 800°K the value of K for the reaction $NH_3 - \frac{1}{2}N_2 - \frac{3}{2}H_2 = 0$ is approximately 0.0032. The value of K_γ at this temperature depends on pressure as follows:

P(atm):	250	350	450
K_γ:	0.8	0.75	0.7

If the reference composition by mole percent is 12% inert gases, 22% N_2, and 66% H_2, find the equilibrium mole percent of ammonia at the three pressures.

Exercise 3.7.6. Benzene may be obtained by the dehydrogenation of hexane over a catalyst at a pressure of 1 atm: $C_6H_{14} = C_6H_6 + 4H_2$. Neglecting other reactions, estimate the minimum temperature for 95% conversion of hexane to benzene.

Standard free energies of formation (kcal/mole) are:

	100°C	400°C	800°C	
Benzene	33.5	45.5	62.7	
Hexane	9.7	53.1	114.0	(C.U.)

Exercise 3.7.7. Calculate the maximum conversion of ethylene to alcohol by catalytic vapor-phase hydration at 300°C and 30 atm using an initial steam-ethylene mole ratio of 10. Assume the vapors behave as ideal gases. The following thermodynamic data are available:

$$C_2H_4 + H_2O \rightleftharpoons C_2H_5OH,$$

$$K_p = 6.8 \times 10^{-2} \text{ atm}^{-1} \text{ at } 145°C,$$

$$\Delta H^\circ_{298} = -10,940 \text{ cal/gm-mole}.$$

Molal heat capacities (cal gm-mol^{-1}°K^{-1}):

$C_2H_4(g)$ $C_p = 2.83 + 28.60(10^{-3})T - 8.73(10^{-6})T^2$

$H_2O(g)$ $C_p = 7.26 + 2.30(10^{-3})T + 0.28(10^{-6})T^2$

$C_2H_5OH(g)$ $C_p = 6.99 + 39.74(10^{-3})T - 11.93(10^{-6})T^2$.   (C.U.)

3.8 Equilibrium of Simultaneous and Heterogeneous Reactions

The example of simultaneous reactions used in Sec. 2.4 may be used to determine what is involved in the equilibrium of the two reactions. Again, working in concentrations rather than total moles, we have for the reactions

$$-A_1 + A_2 = 0, \quad -A_2 + A_3 = 0$$

the relations
$$c_1 = c_{10} - \xi_1, \quad c_2 = c_{20} + \xi_1 - \xi_2, \quad c_3 = c_{30} + \xi_2. \tag{3.8.1}$$
The point representing the composition is again confined to the plane
$$c_1 + c_2 + c_3 = c_{10} + c_{20} + c_{30} = c_0 \text{ (say)} \tag{3.8.2}$$
shown in Fig. 3.6 as ABC. The two equilibrium conditions are
$$\frac{c_2}{c_1} = K_{c1} \quad \text{and} \quad \frac{c_3}{c_2} = K_{c2}. \tag{3.8.3}$$

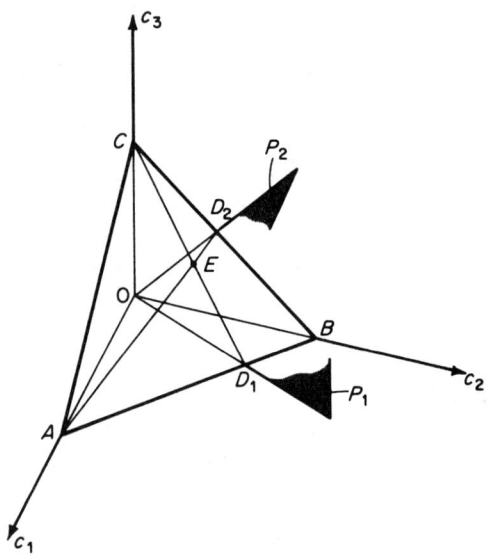

Fig. 3.6 Equilibrium and reaction planes for the two reactions $A_1 \rightleftharpoons A_2 \rightleftharpoons A_3$.

These are the planes P_1 and P_2 intersecting the plane ABC in the lines CD_1 and AD_2, respectively. Thus point E at the intersection of the straight lines satisfies both the stoichiometric Eq. (3.8.2) and the two equilibrium conditions, Eq. (3.8.3), and so represents the equilibrium composition.

If we substitute from Eq. (3.8.1) into Eq. (3.8.3) we obtain the simultaneous equations:
$$(1 + K_{c1})\xi_1 - \xi_2 = K_{c1}c_{10} - c_{20}$$
$$-K_{c2}\xi_1 + (1 + K_{c2})\xi_2 = K_{c2}c_{20} - c_{30},$$
whence
$$\xi_{1e} = c_{10} - \frac{c_0}{1 + K_{c1} + K_{c1}K_{c2}}$$
$$\xi_{2e} = \frac{K_{c1}K_{c2}c_0}{1 + K_{c1} + K_{c1}K_{c2}} - c_{30}.$$

The form of these solutions suggests that the algebra might have been simpler had we used concentrations c_1 and c_3 as unknowns. Multiplying the two equilibrium relations together we have

$$K_{c1}K_{c2}c_1 - c_3 = 0,$$

and substituting $c_2 = c_0 - c_1 - c_3$ into the first equation gives

$$(1 + K_{c1})c_1 + c_3 = c_0,$$

whence

$$c_{1e} = \frac{c_0}{1 + K_{c1} + K_{c1}K_{c2}}$$

$$c_{3e} = \frac{K_{c1}K_{c2}c_0}{1 + K_{c1} + K_{c1}K_{c2}}.$$

This brings out a point that we wish to stress, namely that in particular problems it may be better to use certain of the concentrations rather than to adhere rigidly to the use of extents.

We have tacitly assumed that with simultaneous reactions the equilibrium conditions for each reaction must be satisfied simultaneously and independently. To see why this is so let us go back to the extensive Gibbs free energy change

$$dG = S\,dT + V\,dP + \sum_{j=1}^{S} \mu_j\,dN_j. \tag{3.8.4}$$

For the simultaneous reactions

$$\sum_{j=1}^{S} \alpha_{ij}A_j = 0, \qquad i = 1, 2, \ldots, R$$

we have

$$dN_j = \sum_{i=1}^{R} \alpha_{ij}\,dX_i.$$

For equilibrium at constant temperature and pressure

$$dG = \sum_{j=1}^{S} \mu_j \sum_{i=1}^{R} \alpha_{ij}\,dX_i = 0,$$

and this can be rearranged as

$$\sum_{i=1}^{R} dX_i \sum_{j=1}^{S} \alpha_{ij}\mu_j = 0. \tag{3.8.5}$$

The reactions are independent so that there is no possibility of the sum $\sum \alpha_{ij} dX_i$ vanishing for some particular set of dX_i. If Eq. (3.8.5) is to be true therefore

$$\sum_{j=1}^{S} \alpha_{ij}\mu_j = 0, \qquad i = 1, 2, \ldots, R, \tag{3.8.6}$$

and this leads straight to the simultaneous equilibrium conditions

$$\prod_{j=1}^{S} f_j^{\alpha_{ij}} = K_i(T) = \exp\frac{-\sum_{j=1}^{S}\alpha_{ij}\mu_j^\circ}{RT} \qquad (3.8.7)$$

The solution of such equations is now rather more complicated but it can be proved that there is a unique solution giving the composition at equilibrium.

If a reacting species is involved in the reaction as a pure liquid or as a solid its fugacity can be taken to be unity and the corresponding term does not appear in the product. For example, in the reduction of carbon dioxide over solid carbon $2\,CO - C - CO_2 = 0$, the equilibrium relationship would be

$$\frac{f_{CO}^2}{f_{CO_2}} = K.$$

If a solid decomposes into another solid and a gas,

$$A_1(g) + \alpha_2 A_2(s) + \alpha_3 A_3(s) = 0,$$

the reaction will proceed until the pressure of the gas equals the equilibrium pressure. This pressure is known as the *pressure of decomposition* or *dissociation pressure*.

Exercise 3.8.1. Explore the equilibrium of the reactions $A_1 \rightleftharpoons A_2$, $2A_2 \rightleftharpoons A_3$ along the same lines as are used in the opening part of this section.

Exercise 3.8.2. Producer gas of composition 20% CO, 15% H_2, 5% CO_2, and 60% N_2 is burned with air (79% N_2 and 21% O_2) in the molar ratio of 1 : 5. The principal reactions are $H_2O - H_2 - \tfrac{1}{2}O_2 = 0$ $(K_1 = 150)$ and $CO_2 - CO - \tfrac{1}{2}O_2 = 0$ $(K_2 = 32)$. Find the equilibrium composition. Using this as the first approximation, take into account the reaction $NO - \tfrac{1}{2}O_2 - \tfrac{1}{2}N_2 = 0$ $(K_3 = 0.05)$.

Exercise 3.8.3. Assuming ideal behavior, calculate the percentage of CO_2 that is reduced by the reaction $2CO - CO_2 - C = 0$ at 100°C and a pressure of 1 atm. The free energies of formation of CO, CO_2, and C at this temperature are $-76{,}062$, $-138{,}078$, and -2771 cal/gm-mole, respectively. Show also that if x is the equilibrium mole fraction of CO, then $(1-x)/x^2$ is proportional to the pressure.

Exercise 3.8.4. Calcium carbonate dissociates into calcium oxide and carbon dioxide,

$$-CaCO_3(s) + CaO(s) + CO_2(g) = 0.$$

The carbon dioxide can further dissociate according to

$$-CO_2(g) + CO(g) + \tfrac{1}{2}O_2(g) = 0.$$

If the dissociation pressure of $CaCO_3$ is 3.87 atm at 1000°C and 0.22 atm at 800°C and the equilibrium constant for the dissociation of CO_2 is $10^{-5} (atm)^{1/2}$ at 1200°C, what is the partial pressure of oxygen at the equilibrium of the two reactions at 1200°C?

REFERENCES

A critical appraisal of the many and varied volumes on thermodynamics would be out of place here. The following can be recommended without implying judgment of others not mentioned.

K. G. Denbigh, *The Principles of Chemical Equilibrium*. New York: Cambridge University Press, 1955.

O. A. Hougen, K. M. Watson, and R. A. Ragatz, *Chemical Process Principles*, Part II: *Thermodynamics*, 2nd ed. New York: John Wiley & Sons, Inc., 1959.

N. A. Hall and W. E. Ibele, *Engineering Thermodynamics*. Englewood Cliffs, N. J.: Prentice-Hall, Inc., 1960.

Extensive tables of entropies, free energies, and heats of formation are to be found in Chap. 25 of the Hougen, Watson and Ragatz book.

The student in need of a quick brush-up of thermochemistry and equilibrium calculations can profitably use the programmed units of:

G. M. Barrow, M. E. Kenney, J. D. Lassila, R. D. Litte, and W. E. Thompson, *Understanding Chemistry*. New York: W. A. Benjamin, Inc., 1967.

NOTATION

(In the dimensions, cal has been used to denote energy units and °K for temperature units.)

A_j jth chemical species
a thermal expansion coefficients $(\partial \ln V/\partial \ln T)_P$
C_P total heat capacity, $\sum c_j c_{Pj}$ (cal/°K vol.)
c_{Pj} molar heat capacity (cal/°K mole)
c_j molar concentration A_j (moles/vol.)
E_k kth chemical element
$F(\xi)$ function defined by Eq. (3.7.8)
f_j fugacity of A_j (pressure)
G Gibbs free energy (cal)
$G(\xi')$ function defined by Eq. (3.7.11)
ΔG Free energy change of reaction (cal/mole)

H	Enthalpy (cal)
ΔH	Heat of reaction (cal/mole)
ΔH_f	Heat of formation (cal/mole)
h_j	partial molar enthalpy (cal/mole)
J	$-\Delta H/C_P$ (°K vol./mole)
K	equilibrium constant, $\exp -\Delta G/RT$
K_c	$\prod c_j^{\alpha_j}$ at equilibrium. Suffixes P, n, and γ can also be used.
K_{ci}	equilibrium constant for ith reaction
K^*	pre-exponential factor of K, $\exp \Delta S/R$
N_j	number of moles of A_j present
P	total pressure (pressure)
P_j	partial pressure of A_j (pressure)
Q	heat generated or transferred (cal)
q	rate of heat generation per unit volume (cal/time vol.)
R	gas constant (cal/°K mole)
R	number of independent reactions
r	rate of reaction (moles/vol. time)
S	number of species present
ΔS	change in entropy by reaction (cal/°K mole)
T	temperature (°K)
$T_e(\xi)$	equilibrium temperature at extent (°K)
U	internal energy (cal)
V	volume (vol.)
ΔV	volume change by reaction $\sum \alpha_j v_j$
v_j	partial molar volume of A_j (vol./mole)
W	work done by system (cal)
X	extent of reaction (moles)
Z	compressibility factor
α_j, α_{ij}	stoichiometric coefficients
$\bar{\alpha}, \bar{\alpha}_i$	$\sum_j \alpha_j, \sum_j \alpha_{ij}$
β	ratio $c_{10}/(c_{10} + c_{20})$ in Eq. (3.7.12)
γ	activity coefficient, fugacity coefficient
ϵ_{jk}	number of atoms of E_k in molecule of A_j
ζ	dimensionless extent $\xi/(c_{10} + c_{20})$ in Eq. (3.7.12)
κ	dimensionless pre-exponential factor in Eq. (3.7.14)
ξ	extent of reaction (moles/vol.)
$\xi_e(T)$	equilibrium extent (moles/vol.)
μ_j	chemical potential of A_j (cal/mole)
τ	dimensionless temperature, $RT/-\Delta H$, in Eq. (3.7.13)

Endnote (pp. 32–33) The heat of mixing is not correctly accounted for in the several relations given. For perfect mixtures this makes no matter, but if H is defined by Eq. 3.3.3 then $\partial H/\partial X = \Delta H + \Delta H_m$, ΔH_m = heat of mixing. The h_j used in the definition (3.31) is the enthalpy of a mole of pure A_j. A different symbol, say $\bar{h}_j$, should be used for the partial molar enthalpy, then $\Delta H_m = \Sigma \alpha_j (\bar{h}_j - h_j)$. With single phase mixtures ΔH_m is usually small.

Reaction Rates 4

4.1 What Is Needed and What Can Be Obtained

The nature of a need is defined by an objective. A study such as the present one has a twofold objective. On the theoretical side, our objective is to understand and predict the significant features of a reactor's behavior. Practically, the demand is for methods of design and control of actual reactors. As has been previously emphasized, the underlying theme of this book is the theoretical aspect; but the analytical insight obtained is to serve the practical objective by giving the fundamental understanding and by providing methods of design and suggestions for control.

The reaction rate expression $r(\xi, T, P)$ provides all that is needed theoretically. We could go quite far by assuming only certain properties of the function r without even recording an explicit formula for it. In practice, of course, there are common types of expression for r which have a sound basis in the physical chemistry of reaction, and they are used to build up a feeling for the way in which the reaction rate will behave. In this chapter, we are concerned with the development of this "feeling" for behavior of rates.

On the practical side, the reaction rate expression provides all that is needed for the design and control of a given type of reactor. However, the expression $r(\xi, T, P)$ with a reliable evaluation of all its constants is seldom available. More often, there is only an approximate evaluation of constants based on a modicum of research and development work. Sometimes there is less than this, and the construction of a reactor becomes an exercise in the art of experienced guessing and scale-up. The health of the chemical industry is a tribute to the skillful practice of this art, but the paucity of background information emphasizes the need for a general understanding of reaction and reactor behavior. The student can at least begin to get a feel for this from reaction rate expressions conceived in the conjunction of chemical theory and experiment and developed in the spirit of applied mathematics. But he should be warned that this will give him only the beginning of wisdom; it can never presume to displace the knowledge that comes of experience.

In summary, all that is needed to develop a coherent description of reactor analysis is provided by a formula for $r(\xi, T, P)$ that is in accord with chemical realities. What can be obtained in an industrial situation may, for reasons of time and money, fall far short of this precision; but whatever expression can be obtained must play the same role in practical design as does r in the theoretical analysis.

4.2 Homogeneous Reaction Rates

Before developing the general expression that is often used for homogeneous reactions let us look at a simple and by now familiar reaction. Our standard way of writing the reaction $A_2 + A_3 \rightleftharpoons A_1$ is

$$A_1 - A_2 - A_3 = 0, \tag{4.2.1}$$

where the species A_2 and A_3 are the *reactants* and A_1 is the *product*. We know all about the stoichiometry of this reaction (Sec. 2.4) and about its equilibrium (Sec. 3.7) and wish presently to describe its kinetics. Consider first the situation when the reactants are present in concentrations c_2 and c_3 but no product has yet been formed. The rate of reaction will depend on the concentrations of the reactants and on temperature. If it is a reaction in the liquid phase it may also depend on the solvent, the pH, or some other parametric quantity; if in the gas phase, on pressure or some other thermodynamic variable. For the moment however we shall only consider the influence of concentration and temperature. Moreover, since the concentrations are not arbitrary but are given in terms of the extent by

$$c_1 = \xi, \qquad c_2 = c_{20} - \xi, \qquad c_3 = c_{30} - \xi, \tag{4.2.2}$$

we have only two independent variables, ξ and T. The start of the reaction is given here by $\xi = 0$, and it is governed throughout by the equation

$$r = \frac{d\xi}{dt}. \tag{4.2.3}$$

In saying that the reaction rate is a function of the initial concentrations of A_2 and A_3, we are asserting that at any given temperature there is a function $f(c_{20}, c_{30})$ such that when $\xi = 0$, $r = f(c_{20}, c_{30})$. Just what this function is will have to be determined experimentally and for the moment we will not question how this is best done. In general, however, we expect this initial reaction rate to increase with increasing c_{20} and c_{30}. Let us suppose that experiments show that the initial reaction rate r_0 is proportional to the product of the two concentrations:

$$r_0 = kc_{20}c_{30}. \tag{4.2.4}$$

Such a reaction is said to be of the second order overall (since the sum of the powers of the concentrations is two) and of the first order with respect to A_2 or A_3 individually (since c_{20} and c_{30} are each raised to the first power).

When there is no product present, the reaction can only go in one direction, but as soon as some of the product is formed the reaction rate slows down. It does so for two reasons. In the first place, if the reaction is at all reversible the net reaction rate is the difference between the forward reaction rate (the rate at which the product is being formed) and the back reaction rate (the rate at which the product is dissociating). We may write

$$r = r_f - r_b \tag{4.2.5}$$

and note that as soon as r_b is greater than zero it will tend to slow down the reaction. Equilibrium will be attained when $r_f = r_b$. In the second place, r_f will be a function of the concentration of the reactants and by analogy with Eq. (4.2.4) (where $r_0 = r_f$, since there can be no back reaction if no product is formed) we will suppose

$$r_f = kc_2c_3 = k(c_{20} - \xi)(c_{30} - \xi). \tag{4.2.6}$$

Now as the reaction proceeds and ξ increases, these concentrations decrease and so does r_f. It is important to emphasize that there is no a priori reason for writing $r_f = kc_2c_3$; it is a perfectly valid and a not unusual form of dependence, but for any particular reaction it must be tested experimentally.

We would expect the rate of the back reaction to increase with an increasing concentration of the product, which presumably would make its dissociation easier. Let us assume that it has been found that

$$r_b = k'c_1. \tag{4.2.7}$$

Constants k and k' are determined experimentally at a given temperature and are themselves functions of temperature as we shall see shortly. Combining Eqs. (4.2.5), (4.2.6), and (4.2.7) gives

$$r = kc_2c_3 - k'c_1, \tag{4.2.8}$$

which is described as second order in the forward direction (first order with respect to each reactant individually) and first order in the reverse direction. If we denote the fact that k and k' are functions of temperature T by writing them $k(T)$ and $k'(T)$, then we have

$$r(\xi, T) = k(T)(c_{20} - \xi)(c_{30} - \xi) - k'(T)(c_{10} + \xi). \tag{4.2.9}$$

[We have generalized the stoichiometric Eqs. (4.2.2) very slightly by not insisting that $c_{10} = 0$.]

This, then, is the kind of homogeneous reaction rate expression that is useful and we have to become thoroughly familiar with its properties. It is of course only defined in the region for which all the concentrations are positive [i.e., for $\xi \geq -c_{10}$ and $\xi \leq \min(c_{20}, c_{30})$], and we notice that the rate is strictly positive at the lower limit and strictly negative at the upper. This shows that the solution of the equation

$$\frac{d\xi}{dt} = r(\xi, T) \qquad (4.2.10)$$

can never lead into a region of negative concentration, a condition that is certainly necessary if this equation is to represent a real chemical reaction. If either k or k' is zero, the reaction is said to be irreversible for it will only go in one direction. In this case, the rate may be zero when one of the concentrations is zero, but it will still not lead into regions of negative concentration.

It is also clear from Eq. (4.2.9) that the reaction rate decreases as the extent increases, for

$$\frac{\partial r}{\partial \xi} = -k(T)[(c_{30} - \xi) + (c_{20} - \xi)] - k'(T) < 0.$$

The rate will become zero when

$$k(T)c_2 c_3 - k'(T)c_1 = 0,$$

that is, when

$$\frac{c_1}{c_2 c_3} = \frac{k(T)}{k'(T)}. \qquad (4.2.11)$$

But this has exactly the same form as the equilibrium condition, Eq. (3.7.2),

$$\frac{c_1}{c_2 c_3} = K_c(T),$$

and shows that the equilibrium constant is the ratio of the forward and back reaction rate constants.

We wish now to show how all these properties of the rate of this particular reaction are found in a quite general reaction rate expression. Consider the reaction

$$\sum_{j=1}^{S} \alpha_j A_j = 0 \qquad (4.2.12)$$

for which

$$c_j = c_{j0} + \alpha_j \xi. \qquad (4.2.13)$$

We expect the forward reaction rate to be proportional to the product of certain of the concentrations raised to certain powers. Now we can write such an expression in the form

$$r_f = k c_1^{\beta_1} c_2^{\beta_2} \cdots c_S^{\beta_S} = k \prod_{j=1}^{S} c_j^{\beta_j}, \qquad (4.2.14)$$

Sec. 4.2 Homogeneous Reaction Rates

for if a particular c_j does not really appear we have only to set the corresponding $\beta_j = 0$ and that factor becomes unity. If β_j is positive, then increasing the concentration of A_j increases the forward reaction rate. We expect this to be true for the reactant species, to which we have conventionally assigned a negative α_j. On the other hand, if the concentration of a product species ($\alpha_j > 0$) does affect the forward reaction rate, we would expect that increasing concentration would decrease the rate; this implies that the corresponding β_j is negative. Thus in either case

$$\alpha_j \beta_j \leq 0. \tag{4.2.15}$$

There are some exceptions to this general expectation in which the formation of the product actually speeds up the reaction. These are known as autocatalytic effects and we shall mention them later; for the most part however, the inequality (4.2.15) is a useful rule.

In a similar way we can write

$$r_b = k' c_1^{\gamma_1} c_2^{\gamma_2} \cdots c_S^{\gamma_S} = k' \prod_{j=1}^{S} c_j^{\gamma_j} \tag{4.2.16}$$

and, since the roles of product and reactant are reversed with respect to the back reaction, we expect

$$\alpha_j \gamma_j \geq 0 \tag{4.2.17}$$

in the absence of autocatalytic effects. Combining Eqs. (4.2.5), (4.2.14), and (4.2.16) we have the general expression

$$r = k \prod_{j=1}^{S} c_j^{\beta_j} - k' \prod_{j=1}^{S} c_j^{\gamma_j}. \tag{4.2.18}$$

In many cases, we can regard the reference composition as fixed in a particular situation and then the reaction rate becomes a function of extent and temperature,

$$r(\xi, T) = k(T) \prod_{j=1}^{S} (c_{j0} + \alpha_j \xi)^{\beta_j} - k'(T) \prod_{j=1}^{S} (c_{j0} + \alpha_j \xi)^{\gamma_j}. \tag{4.2.19}$$

In the simple example we studied first we had $\alpha_1 = 1$, $\alpha_2 = \alpha_3 = -1$, $\beta_1 = 0$, $\beta_2 = \beta_3 = 1$, and $\gamma_1 = 1$, $\gamma_2 = \gamma_3 = 0$. This form of kinetic expression is widely used to correlate data even when the reaction is not homogeneous. The *order* of the forward reaction with respect to A_j is β_j and that of the back reaction is γ_j. The overall order of the forward reaction is $\bar{\beta} = \sum \beta_j$ and that of the back reaction $\bar{\gamma} = \sum \gamma_j$; the larger of $\bar{\beta}$ and $\bar{\gamma}$ values might be called the order of the whole reaction, but this is clearly imprecise. If $k' = 0$, the reaction is said to be *irreversible*. If $r_f \gg r_b$ so that the rate is approximately that of a forward reaction we say that the reaction is *far from equilibrium*.

From Eq. (4.2.18) we see that $r = 0$ when

$$\prod_{j=1}^{S} c_j^{\gamma_j - \beta_j} = \frac{k}{k'}. \tag{4.2.20}$$

But we already know an equilibrium condition

$$\prod_{j=1}^{S} c_j^{\alpha_j} = K_c. \quad (4.2.21)$$

Since the left-hand sides are functions only of composition and the right-hand sides only of temperature, it is very tempting to conclude that

$$\gamma_j - \beta_j = \alpha_j \quad (4.2.22)$$

and

$$\frac{k}{k'} = K_c. \quad (4.2.23)$$

Clearly, this allows the two equations to be consistent and if we had put $\gamma_j - \beta_j = n\alpha_j$ and $k/k' = K_c^n$ we could have argued that, since the equation $\sum \alpha_j A_j = 0$ is unchanged when multiplied by n, we lose no generality by taking $n = 1$. In fact this is quite true though it is by no means a trivial job to prove it and the proof given in my *Introduction to the Analysis of Chemical Reactors* is erroneous. Here we shall be content simply to assert that it can always be arranged so that Eq. (4.2.22) [and hence Eq. (4.2.23)] is true.

Differentiating the logarithm of Eq. (4.2.23) with respect to temperature gives

$$\frac{d}{dT} \ln k - \frac{d}{dT} \ln k' = \frac{\Delta H}{RT^2}.$$

This leads to the suggestion that the rate constants k and k' have the same kind of temperature dependence as their ratio K_c. Thus if

$$k = A \exp -\frac{E}{RT}, \qquad k' = A' \exp -\frac{E'}{RT}, \quad (4.2.24)$$

we would have

$$\frac{A}{A'} = K^*, \qquad \Delta H = E - E' \quad (4.2.25)$$

since

$$K_c = K^* \exp -\frac{\Delta H}{RT}.$$

This form was proposed by Arrhenius in the last century and though modern kinetic theory has shown that there may be another factor of some power of T, there is scarcely a reaction rate that is not adequately correlated by this form. The quantity E is called the *activation energy* and A the *pre-exponential* or *frequency factor*.

The Arrhenius rate "constant" varies from 0, when $T = 0$, to A as $T \longrightarrow \infty$. Since

$$\frac{dk}{dT} = \frac{E}{RT^2} k, \qquad \frac{d^2 k}{dT^2} = \left(\frac{E}{RT^2}\right)^2 \left(1 - \frac{2RT}{E}\right) k, \quad (4.2.26)$$

the steepest slope of the curve occurs at $T = E/2R = \frac{1}{4}E$. Since E is commonly of the order of 10^4 cal/mol, this point of inflection is at an extremely high temperature and normally reaction temperatures fall at the very beginning of the curve shown in Fig. 4.1 where k is increasing rapidly. In the figure the ordinate is k/A and, since A is often very large, the value of k is reasonable, although $k/A = \exp - (E/RT)$ may be extremely small. The fact that K is increasing rapidly under normal conditions lies behind the old rule of thumb that the reaction rate doubles itself for every $10°K$ temperature rise. In a strict sense, the rule is complete nonsense though it embodies a certain feel for the rapidity of increase of reaction rates. Figure 4.2 has been drawn

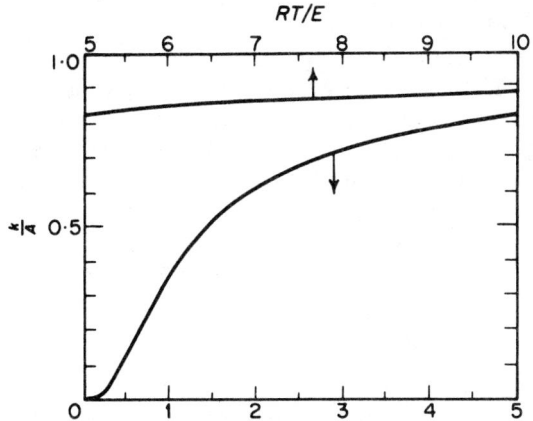

Fig. 4.1 Variation of the Arrhenius rate constant $k(T) = A \exp - (E/RT)$.

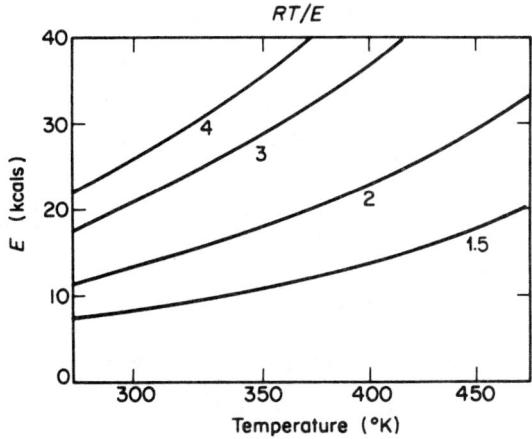

Fig. 4.2 Values of $k(T + 10)/k(T)$ for various E and T.

to give another way of feeling the shape of the Arrhenius expression. The contours are of the ratio $k(T + 10)/k(T)$ in the plane of E and T. Thus if $E = 20$ kcal, an increase of temperature from $400°$K to $410°$K would double the reaction rate of an irreversible reaction or of one far from equilibrium. The kind of function of temperature that would double every so and so many degrees is of course a positive exponential, $B \exp CT$. Now this function can be fitted to the Arrhenius expression in the neighborhood of $T = T_0$ (see Exercise 4.2.1) and may be a tolerable approximation for a few degrees on either side, but it has quite a different shape from $A \exp -E/RT$. For example, it has a value $B > 0$, not zero, when $T = 0$ and it tends to infinity, instead of A, as T becomes very large. This approximation should therefore be used with great caution.

Let us emphasize the following points in closing this description of a very generally useful reaction rate expression. First, the orders of reaction β_j and γ_j have no necessary connection with the stoichiometric coefficients but must be determined experimentally. Second, they are, however, constrained by $\gamma_j - \beta_j = \alpha_j$ and in the absence of autocatalytic effects by $\alpha_j \beta_j \leq 0, \alpha_j \gamma_j \geq 0$. For an elementary step of a reaction, however, the order and stoichiometry are much more closely related by $\beta_j = -\alpha_j$ for all reactant species, $\gamma_j = \alpha_j$ for all product species, and the other β_j and γ_j are zero. This is given by the somewhat clumsy formulae

$$\beta_j = \tfrac{1}{2}(|\alpha_j| - \alpha_j), \qquad \gamma_j = \tfrac{1}{2}(|\alpha_j| + \alpha_j). \tag{4.2.27}$$

Third, the Arrhenius form of the rate constant is sufficiently reliable that any significant departure from its temperature dependence in fitting data would lead one to question the rate law that was being fitted. Fourth, though not rigorously justifiable, this expression has been used to fit data from heterogeneous reactions. However, a rate law of the form

$$r = f(\xi, T)(k \prod c_j^{\beta_j} - k' \prod c_j^{\gamma_j})$$

is often more appropriate as we shall see below. For the moment we need only to point out that if the function f is always positive then the equilibrium conditions are the same as before.

Exercise 4.2.1. Show that in the neighborhood of the temperature T_0, k can be approximated by $B \exp CT$ where

$$B = A \exp -\frac{2E}{RT_0} \quad \text{and} \quad C = \frac{E}{RT_0^2}.$$

Estimate the range over which this approximation would be good to within 5%, if $T_0 = 600°$K and $E = 30$ kcal.

Exercise 4.2.2. Show that $r(\xi, T)$ can be written in the form

$$r = A^{1+p} A'^{-p} K_c^{-p} \prod c_j^{\beta_j}(1 - K_c^{-1} \prod c_j^{\alpha_j}).$$

Sec. 4.2 Homogeneous Reaction Rates

Exercise 4.2.3. Write out the following reactions in standard notation, giving values for as many constants as possible:

(i) Irreversible second order decomposition of acetaldehyde to methane and carbon monoxide;
(ii) Reversible saponification of isopropyl acetate, first order with respect to each reactant in the forward direction;
(iii) $NO_2 + O_3 \rightleftharpoons NO_3 + O_2$, first order with respect to NO_2 and O_3;
(iv) Reversible formation of phosgene $COCl_2$ from CO and Cl_2, forward reaction first order with respect to CO and of order $\frac{3}{2}$ with respect to Cl_2.

Exercise 4.2.4. The following data on the initial rate of the hydrolysis of ethyl acetate by sodium hydroxide were obtained. The reaction is $A_1 + A_2 - A_3 - A_4 = 0$ where $A_3 =$ EtAc and $A_4 =$ NaOH. Concentrations are measured in moles per liter ($c_{10} = c_{20} = 0$) and the rate is in moles per liter per second. The temperature is 25°C.

c_{30}	c_{40}	r_0
0.1	0.1	1.050×10^{-3}
0.05	0.1	0.524×10^{-3}
0.05	0.05	0.261×10^{-3}
0.1	0.05	0.526×10^{-3}

What conclusions would you draw about the order of the forward reaction? Estimate k and write down a reaction rate expression which is consistent with equilibrium.

Exercise 4.2.5. Walles and Platt [*Ind. Eng. Chem.*, **59**, 41 (June 1967)] have suggested that the autocatalytic reaction $A \longrightarrow B + C$ really proceeds by three reactions:

(i) A slow first order background or starting reaction,
$A \longrightarrow B + C$ (rate constant k_1);
(ii) A second order reaction forming a complex E,
$A + B \longrightarrow E$ (rate constant k_2);
(iii) A first order decomposition of the complex E,
$E \longrightarrow 2B + C$ (rate constant k_3).

How many of these reactions are independent? Write down the rates of each reaction and the rate of formation of each species. Hence, or otherwise, show that $(a - b + 2c)$ and $(b - c + e)$ are both constant, where the lower case letters denote the concentrations of the corresponding species.

4.3 Variation of Reaction Rate with Extent and Temperature

We want to develop a proper feel for the shape of the reaction rate expression

$$r(\xi, T) = Ae^{-(E/RT)} \prod_{j=1}^{S}(c_{j0} + \alpha_j\xi)^{\beta_j} - A'e^{-(E'/RT)} \prod_{j=1}^{S}(c_{j0} + \alpha_j\xi)^{\gamma_j}. \quad (4.3.1)$$

In the simple example considered in the last paragraph, we saw that r decreased as ξ increased. Let us show that in the absence of any autocatalytic effects, this is always the case. It is useful to recall (see Exercise 3.7.2) that

$$\frac{\partial}{\partial \xi} \ln \prod c_j^{\beta_j} = \frac{\partial}{\partial \xi} \sum \beta_j \ln(c_{j0} + \alpha_j \xi) = \sum \frac{\alpha_j \beta_j}{c_j}.$$

Hence

$$\frac{\partial}{\partial \xi} \prod c_j^{\beta_j} = \prod c_j^{\beta_j} \sum \frac{\alpha_k \beta_k}{c_k}$$

and

$$\frac{\partial}{\partial \xi} \prod c_j^{\gamma_j} = \prod c_j^{\gamma_j} \sum \frac{\alpha_k \gamma_k}{c_k}.$$

Let us abbreviate by writing $\prod$ for $\prod c_j^{\beta_j}$, $\prod'$ for $\prod c_j^{\gamma_j}$, and $\sum$ and $\sum'$ for the sums $\sum \alpha_j \beta_j / c_j$ and $\sum \alpha_j \gamma_j / c_j$. Thus

$$r = k\prod - k\prod'$$

and

$$\frac{\partial r}{\partial \xi} = k \prod \sum - k' \prod' \sum'. \quad (4.3.2)$$

Now all the terms k, k', $\prod$, $\prod'$, and c_j are positive and if indeed $\alpha_j \beta_j \leq 0$ and $\alpha_j \gamma_j \geq 0$ then $\sum \leq 0$ and $\sum' \geq 0$. In fact, these must be strict inequalities since otherwise all the terms would vanish. If follows that

$$\frac{\partial r}{\partial \xi} < 0. \quad (4.3.3)$$

This is a very reasonable consequence of the way in which we have set things up. Since we have set $\alpha_j > 0$ for the products, the direction of reaction which is of interest is that of increasing ξ and the reaction slows down as it proceeds toward completion or equilibrium.

For the most part, the reference composition will be kept constant but we may see its influence too by partially differentiating with respect to c_{j0}. We have then

$$\frac{\partial r}{\partial c_{j0}} = (k\prod)\frac{\beta_j}{c_j} - (k'\prod')\frac{\gamma_j}{c_j}. \quad (4.3.4)$$

Sec. 4.3 Variation of Reaction Rate with Extent and Temperature

If we multiply this by α_j and use the inequalities (4.2.15) and (4.2.17) for the absence of autocatalysis, we see that

$$\alpha_j \frac{\partial r}{\partial c_{j0}} < 0. \tag{4.3.5}$$

It follows that increasing the reference concentration of a reactant will increase the rate, while increasing the concentration of a product will decrease it.

Before considering the effect of temperature in the general rate expression let us take the very simple example of $A \rightleftharpoons B$. Since the sum of the concentrations a and b of A and B is constant, $a + b = a_0 + b_0$. For simplicity, let $a_0 + b_0 = 1$ so that we can write

$$a = 1 - \xi, \quad b = \xi. \tag{4.3.6}$$

If the reaction is of the first order in both directions we have

$$r(\xi, T) = Ae^{-(E/RT)} (1 - \xi) - A'e^{-(E'/RT)} \xi. \tag{4.3.7}$$

This is a particularly simple form because of its linearity in ξ; we could solve for ξ in terms of r and T and hence plot contours of constant r in the plane of ξ and T.

If the reaction is irreversible, $A' = 0$ and we have only the first term, so that a contour of constant r is given by

$$\xi = 1 - \frac{r\, e^{E/RT}}{A}. \tag{4.3.8}$$

The contours are shown in Fig. 4.3. For any value of $r(< A)$ there is a temperature $T = E/R \ln(A/r)$ at which $\xi = 0$ and the contour rises from this point to the horizontal asymptote $\xi = 1 - r/A$ as $T \longrightarrow \infty$. The "equilibrium" of the irreversible reaction is reached when the reactant is exhausted, namely on the line $\xi = 1$.

If the reaction is endothermic, we know that the equilibrium curve, on which $r = 0$, is such that $\xi_e(T)$ in-

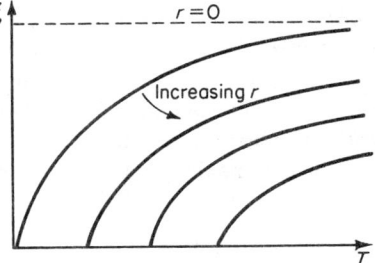

Fig. 4.3 Contours of constant reaction rate for an irreversible reaction.

creases as T increases. Now differentiating Eq. (4.3.7) partially with respect to T gives

$$\frac{\partial r}{\partial T} = \frac{1}{RT^2} [EA\, e^{-E/RT} (1 - \xi) - E'A'\, e^{-E'/RT} \xi]$$
$$= \frac{1}{RT^2} (Er_f - E'r_b). \tag{4.3.9}$$

where r_f and r_b denote the forward and back reaction rates as in the previous section. If the reaction is endothermic, $E - E' = \Delta H > 0$ and so $E > E'$. But when the reaction rate is positive we must have $r_f > r_b$ and so $Er_f > E'r_b$, a fortiori. Thus $\partial r/\partial T > 0$ and it therefore follows that the contours of constant r must always run upwards and to the right in the plane of ξ and T. This is confirmed by solving Eq. (4.3.7) for the contour,

$$\xi = \frac{A e^{-E/RT} - r}{A e^{-E/RT} + A' e^{-E'/RT}}, \qquad (4.3.10)$$

which shows that the contour for $r < A$ starts at the same temperature, $E/R \ln (A/r)$, as in the irreversible case and goes to the asymptote

$$\xi = \frac{(1 - r/A)}{(1 + A'/A)}$$

as $T \longrightarrow \infty$. The contours are shown in Fig. 4.4. In Fig. 4.5, the surface $r(\xi, T)$ is shown with contours of constant r. It is easy to visualize for its intersection with the back vertical plane, $\xi = 0$, has the form of the forward rate constant, while its intersection with the base plane is the equilibrium curve, $\xi_e(T)$. Since it is linear in ξ, the sections by planes parallel to $T = 0$ are straight lines, and the surface is generated by straight lines joining the two intersection curves.

For an exothermic reaction, the situation is more interesting for here $E' > E$. Now in the region of positive r, $r_f > r_b$ but as we approach equilibrium r_f decreases and r_b increases until they are equal on the equilibrium line, $\xi = \xi_e(T)$. It follows that there is some locus just short of equilibrium where $r_f = (E'/E)r_b$, since E'/E is greater than 1. But, by Eq. (4.3.9), $\partial r/\partial T$ vanishes on this locus, i.e., for a given extent the reaction rate is a maximum. This is reasonable, for at low temperatures the reaction rate is very small because of the smallness of the rate constants k and k', while at the equilibrium temperature r again approaches zero. We should therefore expect it to have a maximum at some temperature short of equilibrium, which we

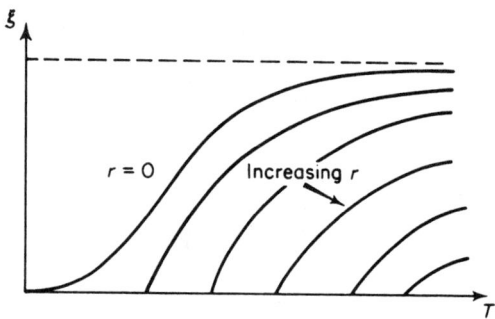

Fig. 4.4 Contours of constant reaction rate for an endothermic reversible reaction.

Sec. 4.3 Variation of Reaction Rate with Extent and Temperature   65

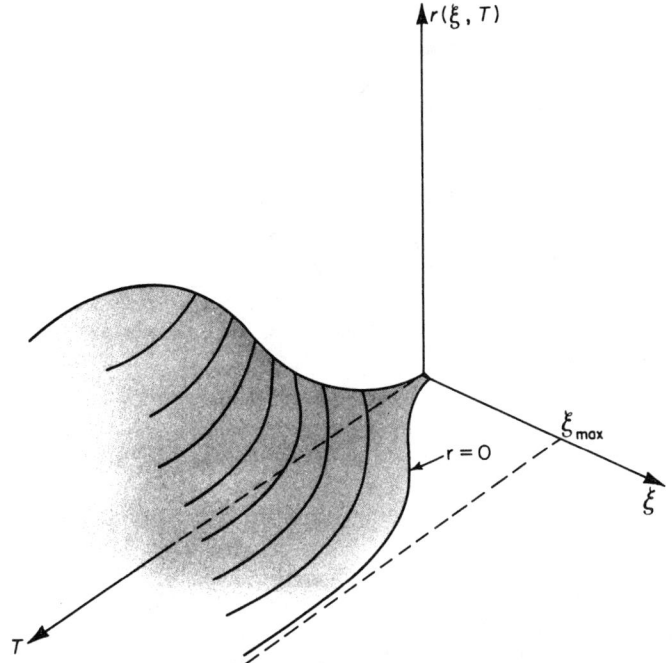

Fig. 4.5 The surface $r(\xi, T)$ for an endothermic reversible reaction with contours of constant r. (Note that there is some distortion in the vertical scale, but the character of the shape is unchanged.)

will denote by $T_m(\xi)$. By setting $\partial r/\partial T = 0$ in Eq. (4.3.9) we have

$$e^{-(E-E')/RT} = \frac{E'A'}{EA} \frac{\xi}{1-\xi}$$

and hence

$$T_m(\xi) = \frac{-\Delta H}{R \ln [E'K^*\xi/E(1-\xi)]} \qquad (4.3.11)$$

since $\Delta H = E - E'$ and $K^* = A'/A$. Comparing this with the equilibrium temperature

$$T_e(\xi) = \frac{-\Delta H}{R \ln [K^*\xi/(1-\xi)]}, \qquad (4.3.12)$$

we see that

$$T_m(\xi) < T_e(\xi), \qquad (4.3.13)$$

since $\ln (E'/E) > 0$. Figure 4.6 shows contours of constant r in the ξ, T plane; Γ_e is the equilibrium curve and the broken line is the locus of maximum reaction rate, Γ_m. It is the existence of this optimal temperature at which the reaction rate is greatest for any given composition that makes the exo-

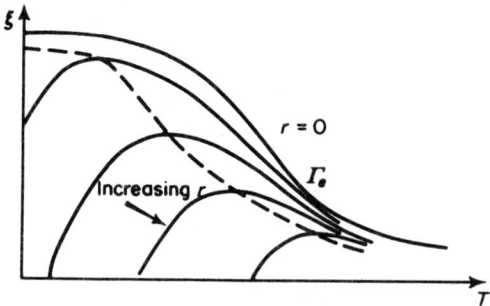

Fig. 4.6 Contours of constant reaction rate for an exothermic reversible reaction.

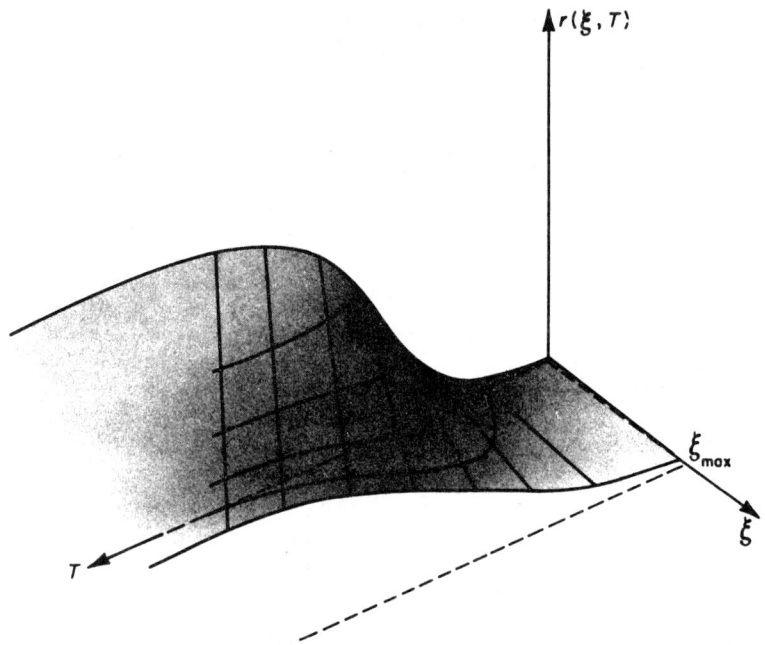

Fig. 4.7 The surface $r(\xi, T)$ for an exothermic reversible reaction with contours of constant r. (Note that there is some distortion in the vertical scale, but the character of the shape is unchanged.)

thermic reaction rate expression so much more interesting than the endothermic one. Figure 4.7 illustrates the surface shape in this case. Again it is generated by straight lines in planes of constant T and has the same intersection with the back plane $\xi = 0$. But this time the equilibrium curve goes in the other direction and $\xi_e(T)$ decreases as T increases. This puts a twist into the surface and gives rise to its more interesting behavior.

Sec. 4.3 Variation of Reaction Rate with Extent and Temperature

Precisely the same kind of reasoning can be applied to the general rate expression

$$r(\xi, T) = k(T) \prod c_j^{\beta_j} - k'(T) \prod c_j^{\gamma_j},$$

for we again have

$$\frac{\partial r}{\partial T} = \frac{E}{RT^2} k \prod - \frac{E'}{RT^2} k' \prod'$$

$$= \frac{1}{RT^2} (E r_f - E' r_b). \quad (4.3.14)$$

Thus if the reaction is irreversible ($r_b = 0$) or endothermic ($E > E'$), $\partial r/\partial T > 0$ and r increases continually with temperature. However if the reaction is exothermic ($E' > E$) the derivative will vanish at

$$T_m(\xi) = -\frac{\Delta H}{R} \left[\ln \frac{E'A'}{EA} \prod (c_{j0} + \alpha_j \xi)^{\alpha_j} \right]^{-1},$$

$$= -\frac{\Delta H}{R} \left[\ln \frac{E'}{E} - \ln K^* + \sum \alpha_j \ln c_j \right]^{-1}, \quad (4.3.15)$$

$$< -\frac{\Delta H}{R} \left[-\ln K^* + \sum \alpha_j \ln c_j \right]^{-1} = T_e(\xi).$$

Both $T_m(\xi)$ and $T_e(\xi)$ decrease as ξ increases. Using the abbreviation

$$p = \frac{E}{-\Delta H} = \frac{E}{E' - E} \quad (4.3.16)$$

so that $E'/-\Delta H = p + 1$, we have

$$e^{-E/RT} = (e^{-\Delta H/RT})^{-p}, \quad e^{-E'/RT} = (e^{-\Delta H/RT})^{-(1+p)}.$$

Now Eq. (4.3.14) shows that at $T = T_m(\xi)$

$$e^{-(\Delta H/RT)} = \frac{A'}{A} \frac{E'}{E} \prod c_j^{\alpha_j}, \quad (4.3.17)$$

so that the maximum reaction rate for a given ξ is

$$r_m(\xi) = r(\xi, T_m(\xi)) = A e^{-E/RT} \left(\prod c_j^{\beta_j} \right) \left(1 - \frac{\prod c_j^{\alpha_j}}{K^c} \right)$$

$$= A \left(\frac{A'E'}{AE} \prod c_j^{\alpha_j} \right)^{-p} \left(\prod c_j^{\beta_j} \right) \left(1 - \frac{E}{E'} \right),$$

$$= \frac{A^{1+p}}{A'^p} \frac{p^p}{(1+p)^{1+p}} \prod c_j^{\beta_j - p\alpha_j}, \quad (4.3.18)$$

$$= \frac{[A \prod c_j^{\beta_j}/(1+p)]^{1+p}}{(A' \prod c_j^{\gamma_j}/p)^p}.$$

By taking logarithms and differentiating we see that

$$\frac{dr_m(\xi)}{d\xi} = r_m(\xi) \sum \frac{\alpha_j \beta_j - p\alpha_j^2}{c_j} < 0 \quad (4.3.19)$$

so that $r_m(\xi)$ decreases as ξ increases.

It follows that all the diagrams we have drawn for the simple first order reaction will be much the same in character for the general reaction. The sections by planes of constant T will no longer be straight lines but they will certainly be curves sloping downward in the same way since $\partial r/\partial \xi < 0$.

The autocatalytic reaction has been mentioned above and though it is somewhat rare it will serve to complete our description of the shape of reaction rates. Suppose (following Walles and Platt, see Exercise 4.2.5) we have an irreversible reaction $A \longrightarrow B + C$, whose rate increases not only with increasing concentration of A (denoted by a) but also with increasing concentration of B (denoted by b). Thus
$$r = kab$$
and since by stoichiometry $a = a_0 - \xi$, $b = b_0 + \xi$ we have
$$r(\xi, T) = k(T)(a_0 - \xi)(b_0 + \xi).$$
(In our standard notation, we would write the reaction $-A_1 + A_2 + A_3 = 0$ and $r = kc_1c_2$. Then
$$\alpha_1 = -1, \qquad \alpha_2 = \alpha_3 = 1, \qquad \beta_1 = \beta_2 = 1, \qquad \beta_3 = 0$$
but the condition $\alpha_j \beta_j < 0$ is not satisfied for $j = 2$.) We notice that
$$\frac{\partial r}{\partial \xi} = k(a_0 - b_0 - 2\xi)$$
and that if $a_0 > b_0$ this will be positive at the beginning of the reaction and negative later on. Figure 4.8 shows the contours of constant reaction rate for such a reaction.

It is clear that if b_0 were zero, the reaction would not even start and the suggestion has been made that there is a slow background reaction $A \longrightarrow B + C$ which provides the first small amount of B. If this is rapidly taken up by A to form a complex E according to the reaction $A + B \longrightarrow E$, and this complex dissociates to give $2B + C$, then we would have a mecha-

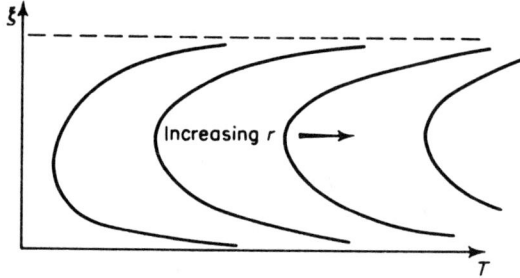

Fig. 4.8 Contours of constant reaction rate for an autocatalytic reaction.

nism whereby the reaction could start and would proceed autocatalytically. (See Exercises 4.2.5 and 4.3.4 and the reference given therein.)

Exercise 4.3.1. In the case $K_c = K^* \exp - \Delta H/RT$ show that

$$\frac{1}{T_m(\xi)} - \frac{1}{T_e(\xi)} = \frac{R}{-\Delta H} \ln \frac{E'}{E}.$$

Exercise 4.3.2. Show that

$$-\frac{dT_m}{d\xi} = \frac{RT_m^2}{-\Delta H} \sum \frac{\alpha_j^2}{c_j}.$$

Exercise 4.3.3. By calculating $\partial^2 r/\partial T^2$ and examining its sign when $\partial r/\partial T = 0$, show that the rate at $T = T_m(\xi)$ is indeed a maximum for a constant ξ when the reaction is exothermic.

Exercise 4.3.4. If the rate constants k_2 and k_3 in Exercise 4.2.5 are very much larger than k_1 and if the concentration of E is small and varies slowly, show that the reaction rate would be approximately $r = kab$ and express k in terms of k_1, k_2, and k_3.

4.4 A Portrait of the Second Order Reversible Reaction

It is of interest to follow up the study of the equilibrium of the second order reaction (Sec. 3.7) to obtain a parametric portrait of the reaction rate. As always in such studies, it pays to make the variables dimensionless. The reaction is $-A_1 - A_2 + A_3 + A_4 = 0$, and we shall suppose it is first order with respect to each of the reactants A_1 and A_2 in the forward direction. Thus,

$$r = A e^{-E/RT} c_1 c_2 - A' e^{-E'/RT} c_3 c_4. \quad (4.4.1)$$

As before, we have dimensionless extent, concentration, and temperature

$$\zeta = \frac{\xi}{c_{10} + c_{20}}, \quad \beta = \frac{c_{10}}{c_{10} + c_{20}}, \quad \tau = \frac{RT}{-\Delta H},$$

and by using the abbreviation of Eq. (4.3.16), we have

$$r = A(c_{10} + c_{20})^2 e^{-p/\tau}(\beta - \zeta)(1 - \beta - \zeta) - A'(c_{10} + c_{20})^2 e^{-(1+p)/\tau} \zeta^2.$$

It follows that $r/A'(c_{10} + c_{20})^2$ will be a dimensionless reaction rate, and with $\kappa = A/A'$ we have

$$\rho(\zeta, \tau) = \frac{r(\xi, T)}{A'(c_{10} + c_{20})^2} = \kappa e^{-p/\tau}(\beta - \zeta)(1 - \beta - \zeta) - e^{-(1+p)/\tau} \zeta^2. \quad (4.4.2)$$

The method of calculating this function is outlined in the exercise below.

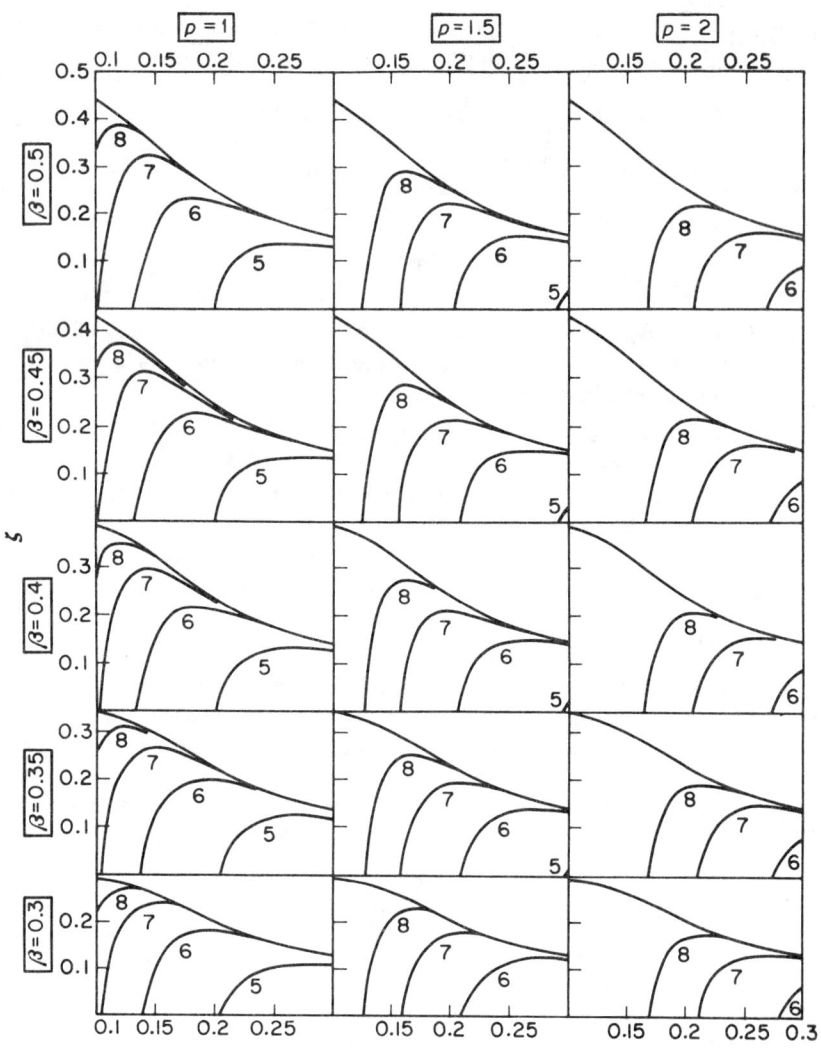

Fig. 4.9 Rate portrait of the second order reaction. Variation with β and p; $\kappa = e^{-5}$. The value of ρ (ζ, τ) on curve labeled n is 10^{-n}.

Figure 4.9 shows the variation of rate with p and the initial proportion of the reactants β; $\kappa = e^{-5}$. The curves are those of constant $\rho = 10^{-5}$, 10^{-6}, 10^{-7}, and 10^{-8}, labeled with the appropriate power of 10^{-1}. Reading down the columns, the effect of the initial proportions is evidently slight.

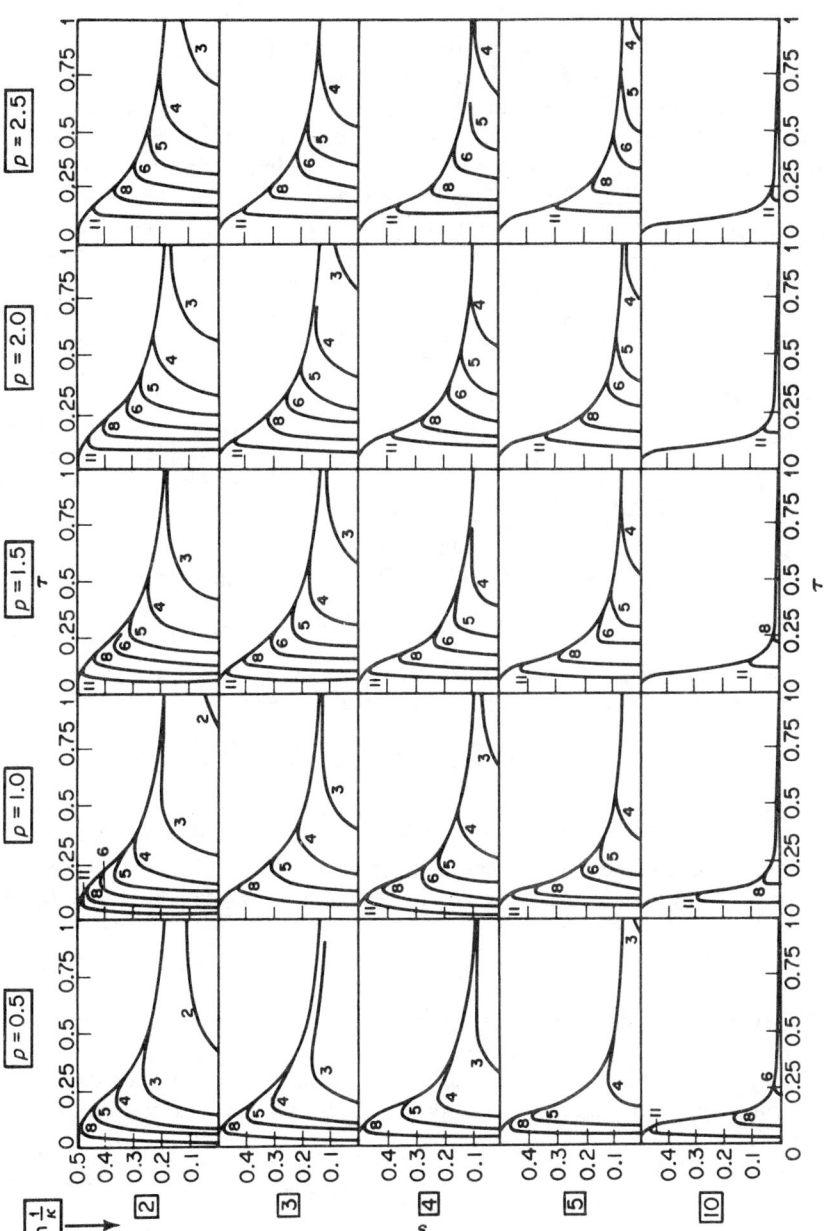

Fig. 4.10 Rate portrait of the second order reaction. Variation with κ and p; $\beta = 0.5$. The value of $\rho(\zeta, \tau)$ on the curve labeled n is 10^{-n}.

When $\beta = \frac{1}{2}$ (i.e., the reactants are in stoichiometric proportions), the reaction rate for any ζ, τ is as large as possible, but a change from the ratio 1 : 1 to 7 : 3 does not decrease the rate much. In Fig. 4.10, the initial proportions are fixed at $\beta = \frac{1}{2}$, and a wider variation of p and κ is given; the labels are again powers of 10^{-1}. For a fixed κ, an increase in p decreases the reaction rate without changing the equilibrium. For a fixed p, decreasing κ decreases the reaction rate by the bearing down of the equilibrium restriction.

The optimal temperature is given by

$$\tau_m(\zeta) = \left[\ln \frac{1+p}{\kappa p} \frac{\zeta^2}{(\beta - \zeta)(1 - \beta - \zeta)}\right]^{-1}, \qquad (4.4.3)$$

and the corresponding reaction rate is

$$\rho_m(\zeta) = \frac{p^p}{(1+p)^{1+p}} \frac{\{\kappa(\beta - \zeta)(1 - \beta - \zeta)\}^{1+p}}{\zeta^{2p}}. \qquad (4.4.4)$$

Exercise 4.4.1. What is the range of values of ζ for which the reaction rate given by Eq. (4.2.2) is negative for all temperatures? What is the range of values of ζ for which $\rho(\zeta, \tau)$ is positive for all temperatures?

Exercise 4.4.2. If $\zeta = 0$, at what temperature is $\partial \rho / \partial \tau$ greatest?

Exercise 4.4.3. If B and C stand, respectively, for the quantities $\kappa e^{-p/\tau}$ and $e^{-(1+p)/\tau}$, show that a contour of constant ρ is given by

$$\zeta = \frac{B - \{B^2 - 4(B - C)[B\beta(1 - \beta) - \rho]\}^{\frac{1}{2}}}{2(B - C)}.$$

What is the value of ζ when $\tau = 1/\ln(1/\kappa)$?

Exercise 4.4.4. Use the same technique of rendering the variables dimensionless to study the first order reversible reaction $A_1 - A_2 = 0$, $\beta_1 = \gamma_2 = 0$, and $\beta_2 = \gamma_1 = 1$ with $c_{10} = 0$.

4.5 Reaction Rates near Equilibrium

Near equilibrium, reaction rates become approximately linear in their deviation from equilibrium conditions. Suppose ξ_e, T_e is a point on the equilibrium curve of

$$r(\xi, T) = A e^{-E/RT} \prod (c_{j0} + \alpha_j \xi)^{\beta_j} - A' e^{-E'/RT} \prod (c_{j0} + \alpha_j \xi)^{\gamma_j} \qquad (4.5.1)$$

and

$$c_{je} = c_{j0} + \alpha_j \xi_e \qquad (4.5.2)$$

are the corresponding equilibrium concentrations. At equilibrium, the two

Sec. 4.5 Reaction Rates near Equilibrium

terms expressing the forward and backward reaction rates are equal and opposite. Denote their common value by P_e and let

$$x = \xi_e - \xi \quad y = T_e - T. \tag{4.5.3}$$

Then, by using Eqs. (4.3.2) and (4.3.3)

$$\begin{aligned}\left(\frac{\partial r}{\partial \xi}\right)_e &= P_e \sum \frac{\alpha_j \beta_j}{c_{je}} - P_e \sum \frac{\alpha_j \gamma_j}{c_{je}}, \\ &= -P_e \sum \frac{\alpha_j^2}{c_{je}},\end{aligned} \tag{4.5.4}$$

and by using Eq. (4.3.14) we get

$$\left(\frac{\partial r}{\partial T}\right)_e = \frac{E - E'}{RT_e^2} P_e = \frac{\Delta H}{RT_e^2} P_e. \tag{4.5.5}$$

Hence by a Taylor expansion about ξ_e, T_e

$$r(\xi, T) = x P_e \sum \frac{\alpha_j^2}{c_{je}} + y \frac{(-\Delta H)}{RT_e^2} P_e + O(x^2, y^2), \tag{4.5.6}$$

where $O(x^2, y^2)$ denotes terms of order x^2, xy, y^2 or higher powers.

If the approach to the equilibrium point is ultimately tangential to the line

$$y = mx,$$

then near equilibrium

$$\frac{dx}{dt} = -\frac{d\xi}{dt} = -r = -\kappa_e x, \tag{4.5.7}$$

where

$$\kappa_e = P_e \left(\sum \frac{\alpha_j^2}{c_{je}} + m \frac{-\Delta H}{RT_e^2} \right). \tag{4.5.8}$$

In particular, for an isothermal approach $m = 0$ and

$$\kappa_{is} = P_e \sum \frac{\alpha_j^2}{c_{je}}, \tag{4.5.9}$$

while for an adiabatic approach $m = -\Delta H/C_p = J$, and

$$\kappa_{ad} = P_e \left(\sum \frac{\alpha_j^2}{c_{je}} + \frac{-\Delta H}{C_p T_e} \frac{-\Delta H}{RT_e} \right). \tag{4.5.10}$$

Notice that κ is always positive, both for exothermic and endothermic reactions. The solution of Eq. (4.5.7), when $\xi = \xi_0$ at $t = t_0$ and $\xi_0 - \xi_e$ is sufficiently small, is

$$x = x_0 e^{-\kappa_e(t-t_0)}$$

or

$$\xi = \xi_e - (\xi_e - \xi_0) e^{-\kappa_e(t-s_0)}, \tag{4.5.11}$$

so that the approach to equilibrium is always an exponential decay and never a damped oscillation. This is a general characteristic of the approach to equilibrium in reaction systems. The *principle of microscopic reversibility* (also called the *principle of detailed balancing*) asserts that in a complex reacting system at equilibrium each individual reaction must be at equilibrium. This excludes the possibility of continuous cycles in which for example $A \longrightarrow B \longrightarrow C \longrightarrow A$, the rates being such as to keep the concentrations constant. It can be shown as a consequence of this that all the extents must approach their equilibrium values by an exponential decay and not by a damped oscillation. This does not prevent some or all of the concentrations from going through maxima or minima on their way to their final equilibrium values. It does mean, however, that the number of such extrema is limited (there can be at most $R - 1$ for R independent actions), whereas a damped oscillation has an infinite number of maxima and minima.

4.6 Reaction Mechanisms

Though we have emphasized that the study of reaction mechanisms for their own sake is not our primary concern, it is worth looking at two simple cases. In the first, we shall see how the kind of reaction rate expression which we have studied can arise; in the second, how a more general expression is needed.

The formation of phosgene (A_1) from carbon monoxide (A_2) and chlorine (A_3) is a classic example. If the reaction were an elementary step resulting from the collision of one molecule of A_2 with one of A_3 to give a molecule of A_1 we should expect the forward reaction rate to be kc_2c_3. In fact, the reaction rate far from equilibrium is found to be

$$r = kc_2 c_3^{3/2} \tag{4.6.1}$$

so that there is need for some explanation. It has therefore been suggested that besides the three species A_1, A_2, and A_3 there are also present chlorine atoms (denoted by A_4) and an intermediate COCl (denoted by A_5). The reaction is further supposed to take place in three steps:

(i) $Cl_2 \rightleftharpoons 2Cl$ or $2A_4 - A_3 = 0$;
(ii) $Cl + CO \rightleftharpoons COCl$ or $A_5 - A_4 - A_2 = 0$;
(iii) $COCl + Cl_2 \rightleftharpoons COCl_2 + Cl$ or $A_1 + A_4 - A_3 - A_5 = 0$.

If reactions (i) and (ii) were so fast as compared to (iii) that they could always be held to be at equilibrium, then we would have

$$\frac{c_4^2}{c_3} = K_1 \quad \text{and} \quad \frac{c_5}{c_2 c_4} = K_2.$$

But these equations can be solved to give c_4 and c_5 in terms of c_2 and c_3,

$$c_4 = K_1^{1/2} c_3^{1/2}, \quad c_5 = K_2 K_1^{1/2} c_2 c_3^{1/2}. \tag{4.6.2}$$

Now we may assume that the rate of the third step is $k_3 c_3 c_5$ in the forward direction and substituting for c_5 from Eq. (4.6.2) we get

$$r_f = k_3 K_2 K_1^{1/2} c_2 c_3^{3/2} = k c_2 c_3^{3/2}. \tag{4.6.3}$$

This agrees with the observations embodied in Eq. (4.6.1) and we see that $k = k_3 K_2 K_1^{1/2}$. If the back reaction of (iii) is also of the second order and has a rate $k_3' c_1 c_4$, Eq. (4.6.2) gives

$$r_b = k_3' K_1^{1/2} c_1 c_3^{1/2} = k' c_1 c_3^{1/2}. \tag{4.6.4}$$

Thus the complete reaction rate expression would be

$$r = r_f - r_b = k c_2 c_3^{3/2} - k' c_1 c_3^{1/2}, \tag{4.6.5}$$

and in terms of our standard nomenclature $\beta_1 = 0$, $\beta_2 = 1$, $\beta_3 = \frac{3}{2}$, $\gamma_1 = 1$, $\gamma_2 = 0$, and $\gamma_3 = \frac{1}{2}$. If we denote the equilibrium constant for the third reaction by $K_3 = k_3/k_3'$, then the equilibrium constant for the whole reaction is

$$K = \frac{k}{k'} = \frac{k_3 K_2 K_1^{1/2}}{k_3' K_1^{1/2}} = K_2 K_3.$$

This corresponds to the fact that the overall reaction is the sum of steps (ii) and (iii) and the first reaction is only needed to provide the chlorine atoms to start the reaction. Notice also that the back reaction is autocatalytic in the sense that the chlorine it produces increases the back reaction rate; accordingly $\alpha_3 \gamma_3 = -\frac{1}{2} < 0$.

The first example was one in which the ultimate rate expression was of the same form as that for each step. We turn now to an example where a different kind of expression is obtained. Reactions of the type $A \longrightarrow B$ are sometimes catalyzed by an enzyme E. This appears to form a complex C with reactant A by the reaction

$$A + E \rightleftharpoons C,$$

and it is this complex which breaks down irreversibly into the product B and the original enzyme,

$$C \longrightarrow B + E.$$

Using lower case letters for the concentrations of the corresponding species we can write the rates of the two reactions as $k_1 ae - k_1' c$ and $k_2 c$, respectively. Then we have

$$\frac{da}{dt} = -k_1 ae + k_1' c$$

$$\frac{db}{dt} = k_2 c,$$

$$\frac{dc}{dt} = k_1 ae - (k_1' + k_2) c,$$

$$\frac{de}{dt} = -k_1 ae + (k_1' + k_2) c,$$

and we may suppose that at the start of the reaction $a = a_0$, $b = c = 0$, and $e = e_0$. We also notice that $d(e + c)/dt = 0$, and so $(e + c)$ is constant and therefore equal to its initial value, e_0. Substituting $e = e_0 - c$ in the first and third equations gives

$$\frac{da}{dt} = -k_1 e_0 a + (k_1 a + k_1')c \tag{4.6.6}$$

$$\frac{dc}{dt} = k_1 e_0 a - (k_1 a + k_1' + k_2)c. \tag{4.6.7}$$

This is a pair of nonlinear equations and no simple solution can be written down, but they can be reduced to a single equation by making a so-called pseudo–steady state hypothesis. This is the assumption that, when e_0 is much smaller than a_0, the concentration c is never very large and varies very slowly. We can then set $dc/dt = 0$. This hypothesis appears to be very "pseudo" indeed at first sight, but it can in fact be justified by what is known as the *singular perturbation theory of differential equations*. Setting $dc/dt = 0$ in Eq. (4.6.7) we can solve for c in terms of a:

$$c = \frac{k_1 e_0 a}{k_1 a + k_1' + k_2}. \tag{4.6.8}$$

Substituting this in Eq. (4.6.6) and in the equation for b gives

$$\frac{da}{dt} = -\frac{db}{dt} = -\frac{ka}{K_M + a}, \tag{4.6.9}$$

$$K_M = \frac{k_1' + k_2}{k_1}, \quad k = k_2 e_0. \tag{4.6.10}$$

Thus the stoichiometry of the overall reaction $A \longrightarrow B$ is observed, but the rate expression is a rational function of the concentration a rather than a simple polynomial. The constant K_M is known among biochemists as the *Michaelis constant*.

In both these cases, reactions $A_1 - A_2 - A_3 = 0$ and $B - A = 0$ can be regarded as *entire* because we have been able to find expressions for their rates, Eqs. (4.6.5) and (4.6.9), which involve only the observable concentrations (c_1, c_2, c_3, b, and a). To do so, however, some further reactions have been introduced and certain hypotheses made—in the first case that one of the reactions was slow compared to the others, in the second that the concentration of enzyme was small and constant. This is the general procedure of the chemical kineticist, who then devises experiments to prove the reactions he has introduced and the hypotheses he has made.

Exercise 4.6.1. If r_0 is the value of the reaction rate Eq. (4.6.9) when $a = a_0$, sketch r_0 as a function of a_0 indicating the slope for small a_0 and the asymptote. Show how a plot of $1/r_0$ against $1/a_0$ can be used to determine the two constants.

Exercise 4.6.2. The following data represent the hydrolysis of acetylcholine in the presence of an esterase. Determine K_M and k.

a_0(mole/l)   0.002   0.005   0.01   0.02
r_0(mole/l sec) 0.889   1.333   1.60   1.78

Exercise 4.6.3. The two steps $A + B \underset{2}{\overset{1}{\rightleftharpoons}} C$ and $C + D \overset{3}{\longrightarrow} E + A$ are proposed as a mechanism for the entire reaction $B + D \longrightarrow E$; k_1, k_2, and k_3 are the constants appropriate to the reactions as numbered. Reactions 1 and 3 are second order but reaction 2 is first order. If reaction 3 is very much slower than reactions 1 and 2, so that the first reaction is always at equilibrium, show that the rate of formation of E is

$$\left(\frac{k_1 k_3 a_0}{k_2}\right) \frac{bd}{1 + (k_1 b/k_2)},$$

where a_0 is the sum of the concentrations of A and C. (The lower case letters represent concentrations of species denoted by the corresponding capital letter.)

Exercise 4.6.4. If the hypothesis is made that c is constant, show that the rate is

$$\left(\frac{k_1 k_3 a_0}{k_2}\right) \frac{bd}{1 + (k_1 b/k_2) + (k_3 d/k_2)}.$$

Exercise 4.6.5. Show that if the second step is regarded as reversible with a back reaction rate constant k_4 and if the hypothesis of constant c is made, then

$$r = \frac{(k_1 k_2 a_0/k_2)bd - (k_4 a_0)e}{1 + (k_1/k_2)b + (k_3/k_2)d + (k_4/k_2)e}.$$

4.7 Heterogeneous Reaction Rate Expressions

Though the reaction rate expression Eq. (4.3.1) is sometimes used to correlate heterogeneous reaction rates it really has only a theoretical basis for homogeneous reactions. Even with these, as we have seen in the previous section, the mechanism may lead to a rational function of the concentrations rather than a purely polynomial one. In Chap. 6 we shall meet similar expressions. They often fall into the type

$$r(\xi, T) = \frac{k \prod c_j^{\beta_j} - k' \prod c_j^{\gamma_j}}{(1 + \sum k_j^* c_j)^p}. \qquad (4.7.1)$$

For example, if we identify B, D, and E in Exercise 4.6.5 with A_1, A_2, and A_3, respectively, we have $\beta_1 = \beta_2 = \gamma_3 = 1$, $\beta_3 = \gamma_1 = \gamma_2 = 0$, $p = 1$, $k = k_1 k_3 a_0/k_2$, $k' = k_4 a_0$, $k_1^* = k_1/k_2$, $k_2^* = k_3/k_2$, and $k_3^* = k_4/k_2$. More generally

$$r(\xi, T) = f(\xi, T)(k \prod c_j^{\beta_j} - k' \prod c_j^{\gamma_j}), \qquad (4.7.2)$$

where $f(\xi, T)$ is a positive function for the whole range of ξ and T. In many cases the factor f does not so distort the form of the surface given by $(k\Pi - k'\Pi')$ as to change its character completely and the general picture we have built up for the different reactions remains valid.

4.8 Reaction Rates in Other Concentration Measures

We have been using the concentration unit c of moles per unit volume in the preceding discussion. It is well to be reminded that this is not the only unit, nor need it be the most appropriate one. Thus, for the mass concentration ρ we have

$$\rho_j = m_j c_j, \qquad (4.8.1)$$

where m_j is the molecular weight of A_j. With $r = k \prod c_j^{\beta_j} - k' \prod c_j^{\gamma_j}$ we have

$$r = (k \prod m_j^{-\beta_j}) \prod \rho_j^{\beta_j} - (k' \prod m_j^{-\gamma_j}) \prod \rho_j^{\gamma_j} \qquad (4.8.2)$$

and

$$\rho_j = \rho_{j0} + \alpha_j m_j \xi. \qquad (4.8.3)$$

For the mass fractions $g_j = \rho_j/\rho$ and the reaction rate r'', in terms of which

$$\frac{dg_j}{dt} = \alpha_j m_j r'', \qquad (4.8.4)$$

we have

$$r'' = \frac{r}{\rho} = (k \prod m_j^{-\beta_j})\rho^{\bar{\beta}-1} \prod g_j^{\beta_j} - (k' \prod m_j^{-\gamma_j})\rho^{\bar{\gamma}-1} \prod g_j^{\gamma_j}, \qquad (4.8.5)$$

where

$$g_j = g_{j0} + \alpha_j m_j \xi'' \qquad (4.8.6)$$

and

$$\bar{\beta} = \sum \beta_j, \qquad \bar{\gamma} = \sum \gamma_j. \qquad (4.8.7)$$

In terms of partial pressure of an ideal gas,

$$c_j = \frac{p_j}{RT} \qquad (4.8.8)$$

and

$$r = k(RT)^{-\bar{\beta}} \prod p_j^{\beta_j} - k'(RT)^{-\bar{\gamma}} \prod p_j^{\gamma_j} \qquad (4.8.9)$$

Alternatively, we have the mole fraction (see Sec. 2.6)

$$x_j = c_j \frac{RT}{P}, \qquad (4.8.10)$$

and

$$r = k\left(\frac{P}{RT}\right)^{\bar{\beta}} \prod x_j^{\beta_j} - k'\left(\frac{P}{RT}\right)^{\bar{\gamma}} \prod x_j^{\gamma_j},$$
$$= \left(\frac{P}{RT}\right)^{\bar{\beta}} \left[k(1 + \bar{\alpha}\xi')^{-\bar{\beta}} \prod (x_{j0} + \alpha_j\xi')^{\beta_j} \right. \quad (4.8.11)$$
$$\left. - k'\left(\frac{P}{RT}\right)^{\bar{\alpha}}(1 + \bar{\alpha}\xi')^{-\bar{\gamma}} \prod (x_{j0} + \alpha_j\xi')^{\gamma_j}\right].$$

In these expressions, the terms in the square brackets show how the Arrhenius constants are modified with further factors of temperature and pressure. It is evident from Eq. (4.8.11) that for $\bar{\beta} > 0$ and $\bar{\alpha} < 0$ the reaction rate will increase with pressure, for both the factor $P^{\bar{\beta}}$ in front increases and the equilibrium is shifted favorably by the decrease of P^{α}.

4.9 Classification and Orders of Magnitude of Reaction Rates

Let us summarize some of the terms introduced in this chapter and give some idea of the order of magnitude of some of the constants.

The only definitive way to classify a reaction is to give its complete rate expression but it is sometimes useful to use looser expressions. We have met most of these terms before but will collect them together here.

(i) *Reversibility.* If the back reaction rate is zero or negligible, the reaction is called *irreversible*. True irreversibility with $k' = 0$ is seldom found, but it may be an acceptable approximation. All other reactions are *reversible*.

(ii) *Order.* Chemists sometimes accuse the engineers of introducing the term "order of reaction" to their own confusion, and certainly it is a term to be used with care. We have defined β_j to be the order of the forward reaction with respect to A_j, and this is quite precise for the reaction rate expression Eq. (2.4.2). Similarly, γ_j is the order of the back reaction with respect to A_j. The sums $\sum \beta_j = \bar{\beta}$ and $\sum \gamma_j = \bar{\gamma}$ are often called the orders of the forward and back reactions, respectively, though perhaps the term overall order of the forward (or back) reaction would be a useful distinction. It is important to remember that the order of a reaction is an empirical quantity which must be determined by experiment.

(iii) *Molecularity.* An equation such as 2 NO + O_2 − 2 NO_2 = 0 for the dissociation of NO_2 might be imagined to be a description of a process in which two NO_2 molecules, on coming together, react to form three molecules, two of NO and one of O_2. In this case, the reaction would be called bimolecular with respect to NO_2. This, however, does not follow as an immediate consequence of writing down the stoichiometric equation. It could have been written $NO + \frac{1}{2}O_2 - NO_2 = 0$ and it is conceivable (at any rate to the layman) that a NO_2 molecule spontaneously disintegrates into a NO molecule and an oxygen atom, and that the latter combines instantaneously to form the molecule O_2. In that case, the reaction would be unimolecular with respect to NO_2, and only experiment can decide between these alternatives. Since the stoichiometric equation is arbitrary up to a constant multiplier it can always be written to reflect the molecularity. In an *elementary step* of a reaction mechanism it is assumed that the reactants only appear in the expression for the forward reaction rate and $\beta_j = -\alpha_j$ for them. Likewise, the products only appear in the back reaction and $\gamma_j = \alpha_j$ for the products. The β_j for the products and γ_j for the reactants are zero. Such a reaction is often called "simple," but this adjective is vastly overworked in all branches of science and has lost many of its original overtones in general usage. We might say that the stoichiometry reflects the molecularity or that molecularity and order coincide. Some examples are given in the table at the end of this section.

(iv) *Simultaneity.* We have treated simultaneous reactions in quite a general way by writing them as $\sum \alpha_{ij} A_j = 0$, but it is sometimes convenient to recognize simple types. Thus reactions $A \longrightarrow B$ and $A \longrightarrow C$ would be described as *concurrent*, whereas $A \longrightarrow B \longrightarrow C$ would be *consecutive*. In *chain* reactions it is often possible to recognize *initiation, propagation,* and *termination* steps. For example, in a polymerization such as $nM_1 \longrightarrow M_n$, the *n*-mer, M_n, is formed from n molecules of monomer M_1. But this will perhaps be initiated by the activation of monomer M_1 to active monomer P_1, $M_1 \longrightarrow P_1$. The reaction is propagated by the formation of active polymers $P_{n-1} + M_1 \longrightarrow P_n$ and terminated by $P_{n-1} + M_1 \longrightarrow M_n$.

(v) *Orders of magnitude.* In general, it is not possible or useful to say what is meant by a fast or slow reaction, but we can give some idea of the orders of magnitude. The following are some second order reactions with the rates given in l/mole sec at 30°C.

Fast	HCl + NaOH	approximately 10^{11}
Medium	CO_2 + 2 NaOH	1.24×10^4
Slow	$HCOOCH_3$ + NaOH	4.7×10^1

It is possible to give some broad generalizations about the orders of magnitude of the frequency factor A in gas phase reactions, but it must be emphasized that they are very broad.

Unimolecular	10^{13}–10^{14}	sec^{-1}
Bimolecular	10^{14}–10^{15}	$cm^3 mole^{-1} sec^{-1}$
Termolecular	10^{15}–10^{16}	$cm^6 mole^{-2} sec^{-1}$

Sec. 4.9 Classification and Orders of Magnitude of Reaction Rates

These generalities are subject to many qualifications—the bimolecular value of A, for example, applies to the interaction of two atoms. For the interaction of more complex structures it may be much smaller. This is sometimes expressed by a steric factor multiplied by the frequency factor. Thus if an atom and polyatomic molecule interact, the value of A will be reduced by a factor of betwen 10^{-1}–10^{-3}; if two polyatomic molecules are involved, the steric factor may be as small as 10^{-4}–10^{-6}. For reactions in solution, the effects of solvents are important. Even less can be said about activation energies without going into a much more detailed analysis. Semiempirical methods, combining some thermodynamics with other features, are used to correlate data.

The following table gives some examples of reactions of different molecularities with the values of A and E in $cm^{3(m-1)}mole^{1-m}sec^{-1}$ and kcal mole^{-1}, respectively, for the forward reaction.

REACTION	A	E
1. Unimolecular ($m = 1$):		
(i) Decompositions:		
cyclo-$C_4H_8 \rightarrow 2\, C_2H_4$	4.2×10^{15}	62.5
$CH_3CH_2Cl \rightarrow C_2H_4 + HCl$	3.9×10^{14}	60.8
(ii) Isomerizations:		
cyclo-$C_4H_6 \rightarrow$ 1,3-C_4H_6	2.5×10^{13}	32.9
$CH_3NC \rightarrow CH_3CN$	4.0×10^{13}	38.4
2. Bimolecular ($m = 2$):		
(i) Exchange reactions:		
$NO + O_3 \rightarrow NO_2 + O_2$	7.9×10^{11}	2.5
$NO + N_2O \rightarrow NO_2 + N_2$	2.5×10^{14}	50.0
(ii) Association reactions:		
$2\, C_2F_4 \rightarrow$ cyclo-C_4F_8	1.0×10^{11}	25.4
$C_2H_4 +$ 1,3-$C_4H_6 \rightarrow$ cyclo$= C_6H_{10}$	3.2×10^{10}	27.5
3. Termolecular reactions ($m = 3$):		
$2\, NO + O_2 \rightarrow 2\, NO_2$	1.04×10^9	-1.1
$I + I + M \rightarrow I_2 + M$ (M = He)	43.2×10^{15}	0

REFERENCES

4.2. The chemical basis for the reaction rate expression is found in many texts, e.g.:

M. Boudart, *The Kinetics of Chemical Processes*. Englewood Cliffs, N. J.: Prentice-Hall, Inc., 1968.

A. A. Frost and R. G. Pearson, *Kinetics and Mechanism*, 2nd ed. New York: John Wiley & Sons, Inc., 1961.

4.6. The above books are very relevant to this section. For the proper justification of pseudo-steady state hypotheses see:

J. R. Bowen, A. Acrivos, and A. K. Oppenheim, "Singular perturbation refinement to the quasi-steady state approximation in chemical kinetics," *Chem. Eng. Sci.*, **18**, 177 (1963).

F. G. Heineken, H. M. Tsuchiya, and R. Aris, "On the mathematical status of the pseudo-steady state hypothesis of biochemical kinetics," *Math. Biosciences*, **1**, 95 (1967).

4.9. A comprehensive work on chemical kinetics is:

S. W. Benson, *The Foundations of Chemical Kinetics*. New York: McGraw-Hill Book Co., 1960.

Many extrathermodynamic relationships are given in:

J. E. Leffler and E. Grunwald, *Rates and Equilibria of Organic Reactions*. New York: John Wiley & Sons, Inc., 1963.

NOTATION

A, A'	frequency factors in Arrhenius constants k, k'
A_j	jth chemical species
c_j, c_{j0}	concentration and reference concentration of A_j
E, E'	activation energies in Arrhenius constants k, k'
$f(\xi, T)$	factor in general rate expression
ΔH	heat of reaction
K^*	A/A', pre-exponential factor in equilibrium constant
k, k'	Arrhenius rate constants of forward and back reactions
$k_j^*(T)$	constants in a form of $f = [1 + \sum k_j(T)c_j]^{-1}$
P_e	value of forward (or back) reaction rate at equilibrium
p	$E/-\Delta H$
R	gas constant
r	reaction rate, $r = r_f - r_b$

Notation

r_f, r_b	rates of forward and back reactions
$r_m(\xi)$	maximum reaction rate for given composition
T_e	equilibrium temperature
$T_m(\xi)$	temperature at which reaction rate is greatest
x	$\xi_e - \xi$
y	$T_e - T$

Lower case roman letters are used in some examples for the concentrations of the species denoted by the corresponding capitals.

α_j	stoichiometric coefficient of A_j
$\bar{\alpha}$	$\sum \alpha_j$
β	$c_{10}/(c_{10} + c_{20})$ (Sec. 4.4)
β_j	order of forward reaction with respect to A_j
$\bar{\beta}$	$\sum \beta_j$
γ_j	order of back reaction with respect to A_j
$\bar{\gamma}$	$\sum \gamma_j$
ζ	dimensionless extent $\xi/(c_{10} + c_{20})$ (Sec. 4.4)
κ_{ad}	decay constant for adiabatic approach to equilibrium
κ_e	decay constant for general approach to equilibrium
κ_{is}	decay constant for isothermal approach to equilibrium
ξ	extent
ξ_e	equilibrium extent
ρ	dimensionless reaction rate, $r/A'(c_{10} + c_{20})^2$ (Sec. 4.4)
$\rho_m(\zeta)$	maximum ρ for given composition (Sec. 4.4)
τ	dimensionless temperature, $RT/(-\Delta H)$ (Sec. 4.4)
$\tau_m(\zeta)$	values of τ for maximum ρ

The Progress of the Reaction in Time 5

5.1 Introduction

Our concern in the last chapter was to get a feel for the way in which the reaction rate varies with composition and temperature. Here we wish to see how the composition varies in time as the reaction proceeds isothermally in a batch reactor. When we come to discuss different types of reactor we shall have to deal with variations of temperature and hence of the rate constants, but here they will be assumed to be constant throughout the reaction. The reaction rate will depend only on the composition, but this of course will vary during the reaction and we shall have to solve differential equations. Sometimes we shall work with the extent of reaction, sometimes with concentrations of reactants or products. No apology is made for this variety of approach since it is important that the student be versatile with the use of different variables and develop an eye for those that will give the simplest form of a solution. The use of the extent is a routine matter, useful for avoiding mistakes in complex situations, but in simpler cases it is often possible to write down the differential equations for concentrations by inspection. From Sec. 5.2 onward, the rate constants will be denoted by lower case k's with a variety of suffixes, the concentrations by c_j or the lower case letter corresponding to the species.

Sec. 5.1 Introduction

Since the stock in trade of sophomore year differential equations has often accumulated a few cobwebs by the senior year it is worth pausing to dust them off. The following is a superficial review of a few common types of differential equation and the notation is specific to this section.

First order equations: The general differential equation of the first order is

$$\frac{dy}{dx} = f(x, y). \tag{5.1.1}$$

Its general solution is a one-parameter family of curves $y = y(x, c)$, but in physical problems a particular solution is needed and this is picked out by the initial condition

$$y = y_0 \quad \text{when} \quad x = x_0. \tag{5.1.2}$$

There are two special forms of f that we shall frequently encounter:

(i) *The separable equation:*

$$f(x, y) = g(x) h(y) \tag{5.1.3}$$

(ii) *The linear equation:*

$$f(x, y) = -p(x) y + q(x) \tag{5.1.4}$$

where g, p, and q are functions only of x and h is a function only of y.

I. We may write Eq. (5.1.1) with f given by Eq. (5.1.3)

$$\frac{1}{h(y)} \frac{dy}{dx} = g(x)$$

and integrate both sides with respect to x from x_0 to a variable point x. Then

$$\int_{x_0}^{x} \frac{1}{h(y')} \frac{dy'}{dx'} dx' = \int_{x_0}^{x} g(x') \, dx'$$

and the left-hand side is just the integral $\int dy'/h(y')$ with the integration variable changed to x. Changing it back to y and recalling that $y = y_0$ when $x = x_0$ we have for $y(x)$ the equation

$$\int_{y_0}^{y(x)} \frac{dy'}{h(y')} = \int_{x_0}^{x} g(x') \, dx'. \tag{5.1.5}$$

Whether this equation can be solved explicitly for y as a function of x will depend on the complexity of the functions g and h.

II. Here we write Eqs. (5.1.1) and (5.1.4) as

$$\frac{dy}{dx} + p(x)y = q(x) \tag{5.1.6}$$

and denote by $P(x)$ the indefinite integral

$$P(x) = \int_{x_0}^{x} p(x'') \, dx''. \tag{5.1.7}$$

Multiplying both sides by exp $P(x)$, we have

$$e^{P(x)} \frac{dy}{dx} + p(x) e^{P(x)} y = \frac{d}{dx} [e^{P(x)} y(x)] = q(x) e^{P(x)}.$$

Again, both sides may be integrated with respect to x and

$$[e^{P(x)} y(x)]_{x_0}^{x} = e^{P(x)} y(x) - e^{P(x_0)} y_0 = \int_{x_0}^{x} q(x') e^{P(x')} dx'.$$

Since $P(x_0) = 0$, we have explicitly

$$\begin{aligned} y(x) &= y_0 e^{-P(x)} + e^{-P(x)} \int_{x_0}^{x} q(x') e^{P(x')} dx' \\ &= y_0 e^{-P(x)} + \int_{x_0}^{x} q(x') e^{-[P(x)-P(x')]} dx', \end{aligned} \quad (5.1.8)$$

where

$$P(x) - P(x') = \int_{x'}^{x} p(x'') dx''.$$

A third equation that we shall need as a reference is the linear equation with constant coefficients.

III.

$$a \frac{d^2 y}{dx^2} + 2b \frac{dy}{dx} + cy = 0. \quad (5.1.9)$$

This needs two conditions if the solution is to be specified uniquely. Sometimes these are both at one end of an interval, e.g.,

$$y(x_0) = y_0, \quad \left(\frac{dy}{dx}\right)_{x=x_0} = y'_0, \quad (5.1.10)$$

and at other times at both ends, e.g.,

$$y(x_0) = y_0, \quad \left(\frac{dy}{dx}\right)_{x=x_1} = y'_1 \quad (5.1.11)$$

or

$$\left(-\frac{a}{2b} \frac{dy}{dx} + y\right)_{x=x_0} = y_0, \quad \left(\frac{dy}{dx}\right)_{x=x_1} = 0. \quad (5.1.12)$$

In all cases the solution consists of the linear combination of two exponentials, e^{mx}. Now if e^{mx} satisfies Eq. (5.1.9) we must have

$$am^2 + 2bm + c = 0,$$

and m has one of the two values

$$\left.\begin{matrix} m_1 \\ m_2 \end{matrix}\right\} = -\frac{b}{a} \pm \left[\left(\frac{b}{a}\right)^2 - \left(\frac{c}{a}\right)\right]^{1/2}. \quad (5.1.13)$$

The general solution is

$$y(x) = A_1 e^{m_1 x} + A_2 e^{m_2 x}, \quad (5.1.14)$$

where the constants A_1 and A_2 have to be determined by satisfying the two boundary conditions. For example, the conditions of Eq. (5.1.10),

$$y(x_0) = A_1 e^{m_1 x_0} + A_2 e^{m_2 x_0} = y_0$$

$$\left(\frac{dy}{dx}\right)_{x=x_0} = A_1 m_1 e^{m_1 x_0} + A_2 m_2 e^{m_2 x_0} = y_0',$$

give two simultaneous equations for A_1 and A_2 and lead to the solution

$$y(x) = \frac{(y_0' - m_2 y_0)}{m_1 - m_2} e^{m_1(x-x_0)} + \frac{m_1 y_0 - y_0'}{m_1 - m_2} e^{m_2(x-x_0)}. \quad (5.1.15)$$

If $b^2 - ac > 0$, both m_1 and m_2 are real. If, further, (b/a) and (c/a) are both positive, then m_1 and m_2 are both negative and $y \longrightarrow 0$ as $x \longrightarrow \infty$. If either (b/a) or (c/a) or both are negative, at least one of the roots will be positive and $y \longrightarrow +\infty$ or $-\infty$ as $x \longrightarrow \infty$. If $b^2 < ac$, the roots are complex conjugate and it is often convenient to write

$$\lambda = \frac{b}{a} \quad \text{and} \quad \mu = \frac{(ac - b^2)^{1/2}}{a}.$$

Then the solution takes the form

$$y(x) = A e^{-\lambda x} \cos(\mu x + \alpha),$$

where A and α have to be determined. If $\lambda > 0$ then $y \longrightarrow 0$ as $x \longrightarrow \infty$.

A related pair of equations is

IV.

$$\left. \begin{aligned} \frac{dy}{dx} &= \alpha y + \beta z, & y &= y_0 \\ \frac{dz}{dx} &= \gamma y + \delta z, & z &= z_0 \end{aligned} \right\} \quad \text{when } x = x_0. \quad (5.1.16)$$

If we differentiate the first equation with respect to x, we have a third equation and can eliminate z and dz/dx between them. Then

$$\frac{d^2 y}{dx^2} - (\alpha + \delta)\frac{dy}{dx} + (\alpha\delta - \beta\gamma)y = 0, \quad (5.1.17)$$

so that

$$m_1, m_2 = \tfrac{1}{2}(\alpha + \delta) \pm \tfrac{1}{2}[(\alpha - \delta)^2 + 4\beta\gamma]^{1/2} \quad (5.1.18)$$

and A_1 and A_2 have to be chosen so that $y = y_0$ and $y' = \alpha y_0 + \beta z_0$ at $x = x_0$. The same equation would be obtained for z if we eliminated y and the constants would have to be chosen to satisfy $z = z_0$ and $z' = \gamma y_0 + \delta z_0$ at $x = x_0$.

It is instructive also to try a solution $y = A e^{mx}$ and $z = B e^{mx}$. Then on substituting we get the equations

$$(-m + \alpha)A + \beta B = 0$$
$$\gamma A + (-m + \delta)B = 0$$

for A and B. The only way in which A and B can differ from zero is for the determinant of these equations to vanish. But this determinant is just the quadratic $m^2 - (\alpha + \delta)m + (\alpha\delta - \beta\gamma) = 0$ whose roots are given by Eq. (5.1.18).

Exercise 5.1.1. Treat the equation
$$\frac{d^2y}{dx^2} = \lambda^2 y,$$
subject to
$$\frac{dy}{dx} = 0 \quad \text{at} \quad x = 0$$
$$y = 1 \quad \text{at} \quad x = \ell,$$
as a special case of Eqs. (5.1.9) and (5.1.11) and hence solve it. (We shall meet this equation in Chap. 6.)

Exercise 5.1.2. Treat the equation
$$\frac{1}{P}\frac{dy^2}{dx^2} - \frac{dy}{dx} + Qy = 0,$$
subject to
$$-\frac{1}{P}\frac{dy}{dx} + y = y_0 \quad \text{at} \quad x = 0$$
$$\frac{dy}{dx} = 1 \quad \text{at} \quad x = 1,$$
as a special case of Eqs. (5.1.9) and (5.1.12) and hence solve it. (We shall meet this equation in Chap. 9.)

Exercise 5.1.3. The solution of the equation
$$\frac{dy}{dx} = q - py, \quad y(0) = y_0,$$
with constant p and q, is worth having at your finger tips. Solve it and grasp it—not just by learning it by heart, but by really seeing what each term means.

5.2 The First Order Reaction

(i) *Irreversible first order reaction.* For the typical first order irreversible dissociation we may write $A \longrightarrow$ products, for, since it is irreversible, its rate is unaffected by the number or concentration of its products. If c is the concentration of A, the statement that the reaction is first order is just a way of saying that the rate of decrease of c is proportional to c itself, or

Sec. 5.2 The First Order Reaction

$$\frac{dc}{dt} = -kc. \tag{5.2.1}$$

If c_0 is the value of c at $t = 0$ we have, by method I of Sec. 5.1,

$$\int_{c_0}^{c(t)} \frac{dc'}{c'} = \ln \frac{c(t)}{c_0} = -kt \tag{5.2.2}$$

or

$$c(t) = c_0 e^{-kt}. \tag{5.2.3}$$

The concentration of A thus decreases exponentially to zero and the concentrations of the product or products increase proportionally to their stoichiometric coefficients and the amount of A that has been used up, namely,

$$c_0 - c = c_0(1 - e^{-kt}).$$

The time course of c and $c_0 - c = \xi$ are seen in Fig. 5.1.

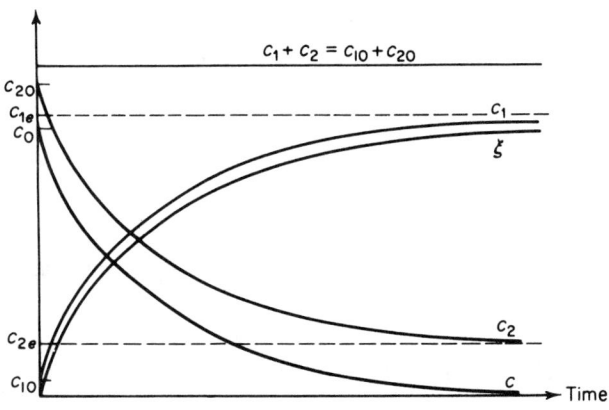

Fig. 5.1 Variation of concentrations and extent with time during the course of a first order reversible reaction.

(ii) *Reversible first order reaction.* Consider the reaction $A_1 - A_2 = 0$ in which A_2 reacts to form A_1; it will be first order if

$$r = kc_2 - k'c_1. \tag{5.2.4}$$

This reaction provides the simplest nontrivial way of comparing three ways of treating the problem. First, we use the notion of extent: here

$$c_1 = c_{10} + \xi, \quad c_2 = c_{20} - \xi \tag{5.2.5}$$

so that

$$r = (kc_{20} - k'c_{10}) - (k + k')\xi$$
$$= r_0 - (k + k')\xi;$$

and
$$\frac{d\xi}{dt} = r = r_0 - (k + k')\xi, \tag{5.2.6}$$

where $\xi = 0$ when $t = 0$. This equation can be written as

$$\int_0^{\xi(t)} \frac{-(k + k')\,d\xi'}{r_0 - (k + k')\xi'} = -(k + k')t,$$

and it gives

$$\ln\left(1 - \frac{k + k'}{r_0}\xi\right) = -(k + k')t$$

or

$$\xi(t) = \frac{r_0}{k + k'}[1 - e^{-(k+k')t}]. \tag{5.2.7}$$

We obtain $c_1(t)$ and $c_2(t)$ by substituting in Eq. (5.2.5). We have chosen to do this by separating variables and integrating, although it could have been written down immediately by using the solution of Exercise 5.1.3.

Second, we use concentrations and observe from the stoichiometry of the reaction that $c_1 + c_2$ is an invariant. Thus,

$$c_1 = c_{10} + c_{20} - c_2 \tag{5.2.8}$$

and

$$r = -\frac{dc_2}{dt} = kc_2 - k'(c_{10} + c_{20} - c_2),$$
$$= (k + k')c_2 - k'(c_{10} + c_{20}),$$

with $c_2 = c_{20}$ when $t = 0$. Again, the equation is separable and yields

$$c_2(t) = (c_{10} + c_{20})\frac{k'}{k + k'} + \frac{r_0}{k + k'}e^{-(k+k')t}. \tag{5.2.9}$$

From Eq. (5.2.8)

$$c_1(t) = (c_{10} + c_{20})\frac{k}{k + k'} - \frac{r_0}{k + k'}e^{-(k+k')t}, \tag{5.2.10}$$

and it is easily confirmed that these expressions are consistent with Eqs. (5.2.5) and (5.2.7).

Third, since the reaction is reversible, let us find the equilibrium composition. The equilibrium concentrations c_{1e} and c_{2e} must satisfy Eq. (5.2.8) and make the rate zero, i.e.,

$$0 = kc_{2e} - k'c_{1e} \tag{5.2.11}$$

and

$$c_{1e} + c_{2e} = c_{10} + c_{20}.$$

Thus,

$$c_{1e} = \frac{k}{k + k'}(c_{10} + c_{20}), \qquad c_{2e} = \frac{k'}{k + k'}(c_{10} + c_{20}). \tag{5.2.12}$$

Sec. 5.2   The First Order Reaction   91

Subtract Eq. (5.2.11) from Eq. (5.2.4) and denote differences from the equilibrium composition, $c_2 - c_{2e} = c_{1e} - c_1$, by c. Then

$$r = k(c_2 - c_{2e}) - k'(c_1 - c_{1e}) = (k + k')c$$

and

$$\frac{dc}{dt} = -r = -(k + k')c.$$

Thus

$$c = c_0 e^{-(k+k')t}, \qquad (5.2.13)$$

where

$$c_0 = c_{20} - c_{2e} = c_{1e} - c_{10} = \frac{kc_{20} - k'c_{10}}{k + k'} = \frac{r_0}{k + k'}. \qquad (5.2.14)$$

It is evident that Eqs. (5.2.12), (5.2.13), and (5.2.14) agree with Eqs. (5.2.9) and (5.2.10).

Figure 5.1 shows how the various quantities vary with time. In all cases the combination $(k + k')t$ appears, so that the sum $(k + k')$ is a measure of the speed with which the system approaches equilibrium, whereas it is the ratio k/k' that governs the equilibrium state. To establish that the reaction is first order, it must be shown that one of the preceding expressions fits the experimental data with sufficient accuracy.

Exercise 5.2.1 The slope of the curve $c_1(t)$ at the point where it intersects the curve $c_2(t)$ (Fig. 5.1) is one-fifth of its slope at $t = 0$. The time of intersection is 50 sec. The tangent of the curve $c_1(t)$ at this point of intersection carried back to the time axis intersects it at $t = -50$ sec. Find k and k'.

Exercise 5.2.2. Data on the concentration of A as a function of time are taken for the reaction $A \rightleftarrows B$ for two initial compositions a_0 and b_0 but for the same total concentration $a_0 + b_0$. Show that whether reversible or not, the difference between $a(t)$ for the two runs is an exponential function of time for first order reactions. Hence, or otherwise, establish that the following two sets of data correspond to a first order reaction and determine k and k'.

Case I. $a_0 + b_0 = 1$

Time, t	0	0.05	0.1	0.15	0.2	0.25
Run 1, $a(t)$	1.0	0.646	0.432	0.301	0.221	0.174
Run 2, $a(t)$	0.8	0.525	0.358	0.256	0.194	0.157

Case II. $a_0 + b_0 = 1$

Time, t	0	0.05	0.1	0.15	0.2	0.25
Run 1, $a(t)$	1.0	0.606	0.368	0.223	0.135	0.082
Run 2, $a(t)$	0.8	0.486	0.294	0.178	0.108	0.066

Exercise 5.2.3. A reversible first order reaction $A \underset{k_2}{\overset{k_1}{\rightleftharpoons}} B$ takes place at constant temperature. At time $t = 0$, the reaction mixture is pure A and the fractions of A left at times T, $2T$, and $3T$ are measured. If these are 0.8, 0.7, and 0.65, respectively, show that the equilibrium concentration of B will be 40% of the initial concentration of A.

5.3 The General Irreversible Reaction

If the dissociation $A \longrightarrow$ products is of the nth order, the concentration of A satisfies

$$\frac{dc}{dt} = -kc^n. \tag{5.3.1}$$

Again, using the separability of the equation (type I), we have

$$\left(\frac{1}{c(t)}\right)^{n-1} - \left(\frac{1}{c_0}\right)^{n-1} = k(n-1)t$$

or

$$c(t) = c_0[1 + kc_0^{n-1}(n-1)t]^{-1/(n-1)}, \qquad n \neq 1. \tag{5.3.2}$$

The formula breaks down for the first order case, though the well-known limit as $1/(n-1) \longrightarrow \infty$ gives the exponential. The dimensions of k are $(\text{mole/vol})^{1-n} \, (\text{time})^{-1}$, so that $kc_0^{n-1}t$ is a dimensionless time θ. The time course of the reaction can be represented as in Fig. 5.2 by plotting c/c_0 against θ,

$$\frac{c}{c_0} = [1 + (n-1)\theta]^{-1/(n-1)}. \tag{5.3.3}$$

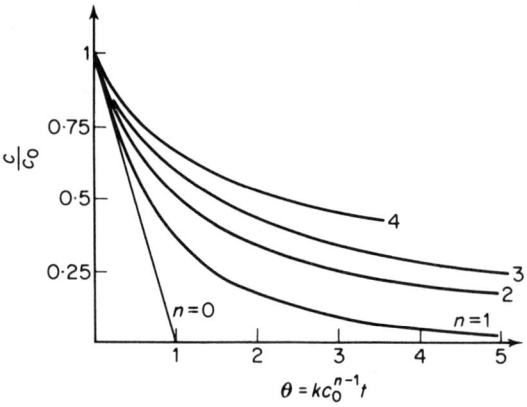

Fig. 5.2 Decrease in concentration of a substance dissociating by an nth order irreversible reaction.

Sec. 5.3 The General Irreversible Reaction

As an example of the use of concentration measures other than that of moles per unit volume consider the gas reaction $A \longrightarrow B + C$, say the decomposition of acetaldehyde vapor into methane and carbon monoxide. If the reaction proceeds at constant volume and temperature, it may be followed by the increase in pressure. Suppose N_0 moles of A are introduced into a volume V at temperature T so that the initial pressure is $P_0 = N_0 RT/V$. When the gross extent of reaction is X there will be $(N_0 + X)$ total moles present and the pressure will be $P = (N_0 + X)RT/V$. Hence $X = (P - P_0)V/RT$. If the reaction is second order and so proportional to the square of the concentration $(N_0 - X)/V$, we have

$$r = \frac{1}{V}\frac{dX}{dt} = k\frac{(N_0 - X)^2}{V^2}. \tag{5.3.4}$$

This equation can be integrated in the usual way to give

$$\frac{kt}{V} = \int_0^x \frac{dX}{(N_0 - X)^2} = \frac{1}{N_0 - X} - \frac{1}{N_0}$$
$$= \frac{1}{N_0}\frac{P - P_0}{2P_0 - P}.$$

This last equation could be rearranged to show how the pressure depends on the time, namely,

$$P = P_0 \frac{1 + (2kP_0 t/RT)}{1 + (kP_0 t/RT)}, \tag{5.3.5}$$

or it could be used to determine k, namely,

$$k = \frac{RT(P - P_0)}{P_0(2P_0 - P)t}. \tag{5.3.6}$$

Exercise 5.3.1. Let t_α denote the time to reach a concentration $c = c_0/\alpha$ in an irreversible nth order reaction. If $\beta = \alpha^2$, show that

$$n = 1 + \frac{\ln(t_\beta/t_\alpha) - 1}{\ln \alpha}.$$

Exercise 5.3.2. Use the following data to find n and k.

t(sec)	0	12	84
c(mole/l)	0.2	0.1	0.025

Exercise 5.3.3. The irreversible reaction $A \longrightarrow B$ obeys the Michaelis-Menten rate law [see Eq. (4.6.9)]

$$\frac{da}{dt} = -\frac{ka}{K_M + a}$$

and $a = a_0$ when $t = 0$. Show that
$$K_M \ln \frac{a_0}{a} + a_0 - a = kt$$
and deduce that after a long time a good approximation is given by
$$a = a_0 \exp - \frac{kt - a_0}{K_M}.$$

Exercise 5.3.4. An experimenter thinks that he is measuring the initial rate of reaction r_0 in the previous exercise (see Exercise 4.6.1), but is actually measuring the rate r' after a very short time τ. Show that if he plots $1/r'$ against $1/a_0$ and uses the slope and intercept to estimate K_M and k, then he will have errors of $2k\tau$ and $k^2\tau/K_M$ respectively in these estimates, if the values of a_0 that he uses are all much smaller than K_M. What can you say about his accuracy if all his values of a_0 are much larger than K_M?

Exercise 5.3.5. Acetaldehyde vapor is put in a flask at a temperature of 518°C and a pressure of 363 mm Hg and begins to decompose into CH_4 and CO. The increase of pressure ΔP (mm Hg) at subsequent times t (sec) is:

t	42	105	242	480	840	1440
ΔP	34	74	134	194	244	284.

Test to see if this is a second order reaction and calculate the rate constant k in l/mole sec. (AIChE)

5.4 The General Homogeneous Reaction

For the general rate expression with constant temperature
$$\frac{d\xi}{dt} = r(\xi) = k \prod (c_{j0} + \alpha_j \xi)^{\beta_j} - k' \prod (c_{j0} + \alpha_j \xi)^{\gamma_j} \quad (5.4.1)$$
we have immediately by the separation of variables
$$t = \int_0^\xi \frac{d\xi'}{r(\xi')}. \quad (5.4.2)$$

(In this equation, ξ' is the dummy variable of integration not the dimensionless extent of reaction.) This integral may have to be integrated numerically but there are cases where an analytic solution is possible. If all the β_j and γ_j are integers, $r(\xi)$ is a polynomial and can in principle be represented as a product $k^*(\xi - \xi_1) \cdots (\xi - \xi_n)$ where k^* is a constant involving k, k', and the α_j. At least one of the ξ_j will be real and positive and the least such value is the equilibrium extent. Then the integrand can be split into partial fractions and the integral becomes a sum of logarithms. It is not

Sec. 5.4 The General Homogeneous Reaction

always possible to invert this function and to get an explicit formula for $\xi(t)$.

As an example, let us consider the second order reaction we have studied in Secs. 4.4 and 3.7. Here $-A_1 - A_2 + A_3 + A_4 = 0$ and if $c_{30} = c_{40} = 0$,

$$r = kc_1c_2 - k'c_3c_4 = k(c_{10} - \xi)(c_{20} - \xi) - k'\xi^2. \quad (5.4.3)$$

Making the variables dimensionless as before and adding a dimensionless time by setting

$$\zeta = \frac{\xi}{c_{10} + c_{20}}, \quad \beta = \frac{c_{10}}{c_{10} + c_{20}}, \quad \theta = k'(c_{10} + c_{20})t, \quad K = \frac{k}{k'},$$

we have

$$\begin{aligned}\frac{d\zeta}{d\theta} &= K(\beta - \zeta)(1 - \beta - \zeta) - \zeta^2, \\ &= K\beta(1 - \beta) - K\zeta + (K - 1)\zeta^2, \\ &= (K - 1)(\zeta_1 - \zeta)(\zeta_2 - \zeta),\end{aligned} \quad (5.4.4)$$

where

$$\left.\begin{matrix}\zeta_1 \\ \zeta_2\end{matrix}\right\} = \frac{K \pm \{K^2[1 - 4\beta(1 - \beta)] + 4K\beta(1 - \beta)\}^{1/2}}{2(K - 1)}. \quad (5.4.5)$$

The roots ζ_1 and ζ_2 are real and $\zeta_1 > \zeta_2$, if $K > 1$, but $\zeta_1 < 0 < \zeta_2$, if $K < 1$. In the exceptional case $K = 1$ there is only one root, $\zeta = \beta(1 - \beta)$. Thus in all cases ζ_2, given by the negative sign in Eq. (5.4.5), is the least positive root and $(\zeta_1 - \zeta_2)$ and $(K - 1)$ have the same sign. Now Eq. (5.4.4) can be written as

$$(K - 1)(\zeta_1 - \zeta_2)\int_0^\theta d\theta' = \int_0^\zeta \frac{(\zeta_1 - \zeta_2)\,d\zeta'}{(\zeta_1 - \zeta')(\zeta_2 - \zeta')}$$

$$= \int_0^\zeta \frac{d\zeta'}{\zeta_2 - \zeta'} - \int_0^\zeta \frac{d\zeta'}{\zeta_1 - \zeta'},$$

i.e.,

$$(K - 1)(\zeta_1 - \zeta_2)\theta = \ln\frac{(\zeta_1 - \zeta)\zeta_2}{(\zeta_2 - \zeta)\zeta_1} \quad (5.4.6)$$

or

$$\zeta(\theta) = \zeta_2 \frac{1 - e^{-\lambda\theta}}{1 - (\zeta_2/\zeta_1)e^{-\lambda\theta}},$$

$$\lambda = (K - 1)(\zeta_1 - \zeta_2) = \{K^2[1 - 4\beta(1 - \beta)] + 4K\beta(1 - \beta)\}^{1/2}. \quad (5.4.7)$$

From this it is clear that as θ (i.e., time) increases, ζ increases from zero to its equilibrium value ζ_2 [see Eq. (3.7.15)]. Figure 5.3 shows θ as a function of ζ/ζ_2 for $\beta = \frac{1}{2}$ and various K. In reflecting on the relation between this graph in which p does not appear, and Fig. 4.10 where the rate is obviously

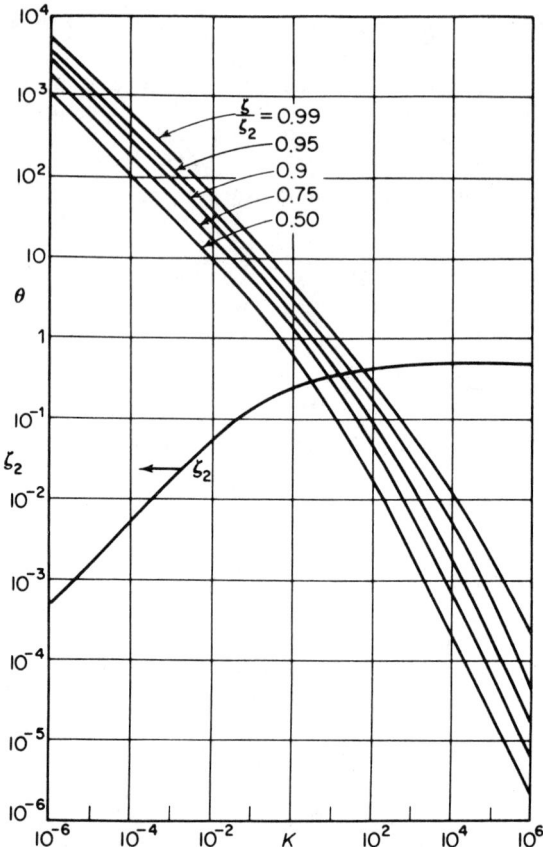

Fig. 5.3 Times to 50, 75, 90, 95, and 99% completion of the second order reversible reaction.

dependent on p, it must be remembered that the dimensionless time involves the temperature, for

$$\theta = k'(c_{10} + c_{20})t = [A'(c_{10} + c_{20})t] \exp -\frac{1+p}{\tau} \qquad \tau = \frac{RT}{-\Delta H}.$$

Exercise 5.4.1. Use Fig. 5.3. to estimate the time to 90% completion at 300°K, 400°K, and 500°K of the second order reaction with: $c_{10} = c_{20} = 0.5$ mole/l; $A = 10^9$ and $A' = 2 \times 10^{10}$ l/mole min; $E = 15$, and $E' = 20$ kcal/mole.

Exercise 5.4.2. Consider the second order reversible reaction $2A \rightleftharpoons B + C$ and formulate the differential equation for the concentration a of A. Solve this equation for the case $a_0 = 1$, $b_0 = c_0 = 0$.

Sec. 5.4 The General Homogeneous Reaction

Exercise 5.4.3. The decomposition of hydrogen iodide is a reversible reaction of second order in both directions

$$HI \underset{k_2}{\overset{k_1}{\rightleftharpoons}} \frac{1}{2} H_2 + \frac{1}{2} I_2.$$

A 200-l flask was filled with pure HI at 1.24 atm. and 683°K. The decomposition was followed photometrically by measuring the absorption of light by the iodine produced in the reaction. The optical density given is proportional to the iodine concentration. Immediately after the last reading was taken the flask was chilled and was found to contain 1.17 gm of iodine (at. wt. 127). Show that the data obtained agree with the reaction scheme and estimate the values of k_1 and k_2 in mole^{-1} min^{-1}.

Time t (min)	42	118	230	397	680	770	940
Optical density	0.81	2.13	3.66	5.04	6.00	6.18	6.21

(C.U.)

Exercise 5.4.4. Find for the reaction

$$H_2 + I_2 \underset{k_2}{\overset{k_1}{\rightleftharpoons}} 2 HI$$

the time required to produce 1 mole of HI in a 1-l vessel, when 2 moles of hydrogen are reacted with 1 mole of iodine at 550°C. At this temperature, the second order rate constants k_1 and k_2 are 1.25×10^{-4} and 0.25×10^{-4} l mole^{-1} sec^{-1}, respectively. (C.U.)

Exercise 5.4.5. The following data were obtained for the rate of conversion of trypsinogen into trypsin, the concentration of the latter being given in arbitrary units:

t (hr)	0	1	2	3	4	4.5	5	5.5	6	8	9.5	∞
Conc.	trace	1.5	3	7.5	16	20	27	40.5	47	67	68.5	75.

Determine whether or not the data are consistent with the assumption that the reaction is autocatalytic, so that the velocity can be expressed as

$$-\frac{da}{dt} = kab,$$

where a = concentration of trypsinogen,

b = concentration of trypsin. (C.U.)

Exercise 5.4.6. The reversible second order reaction $A_1 + A_2 - A_3 - A_4 = 0$ proceeds isothermally at a rate $r = 9c_3c_4 - c_1c_2$, c_j being the concentration of A_j. A mixture of equal concentrations of A_3 and A_4 starts to react at time $t = 0$, and at time t_μ the concentration of A_1 has reached a fraction μ of its equilibrium value. Show that

$$\frac{t_{3/4} - t_{1/4}}{t_{1/2}} = \frac{\log 15 - \log 7}{\log 15 - \log 10}.$$

5.5 Concurrent Reactions of Low Order

The simplest pair of concurrent reactions is that in which A dissociates either into B or into C. For example, an alcohol ($C_nH_{2n+1}OH$) can decompose into an aldehyde ($C_nH_{2n}O$) and hydrogen or into an olefin (C_nH_{2n}) and water. This simple model often serves in much more complicated situations for want of any better. Thus A might be a high boiling crude petroleum, B a lower boiling fraction, and C a tarry waste product. Even though an extremely complicated set of reactions is taking place, it may be sufficient to model it by the reaction system

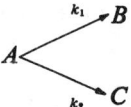

For first order irreversible reactions with rate constants as indicated,

$$\frac{da}{dt} = -(k_1 + k_2)a, \tag{5.5.1}$$

$$\frac{db}{dt} = k_1 a, \qquad \frac{dc}{dt} = k_2 a. \tag{5.5.2}$$

The first of these equations is immediately integrable and gives

$$a = a_0 \exp -(k_1 + k_2)t. \tag{5.5.3}$$

The other equations can then be solved by integrating both sides:

$$\begin{aligned} b &= b_0 + a_0 \frac{k_1}{k_1 + k_2}(1 - e^{-(k_1+k_2)t}) \\ c &= c_0 + a_0 \frac{k_2}{k_1 + k_2}(1 - e^{-(k_1+k_2)t}). \end{aligned} \tag{5.5.4}$$

The increases in the concentrations of B and C are proportional to k_1 and k_2, respectively, and the three concentrations behave as shown in Fig. 5.4.

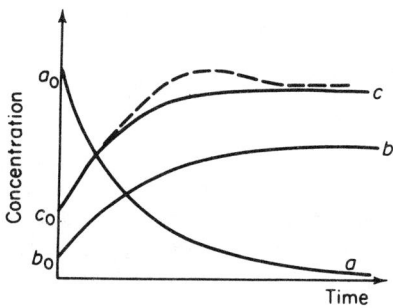

Fig. 5.4 Variation of concentrations with time for the concurrent reactions $A \longrightarrow B$, $A \longrightarrow C$.

Sec. 5.5 Concurrent Reactions of Low Order

If the reactions are reversible, the equations are slightly less easy to solve, though the behavior of the solutions is entirely similar. Using primes to denote the back reaction rate constant, we have the equations

$$\frac{da}{dt} = -(k_1 + k_2)a + k_1'b + k_2'c,$$

$$\frac{db}{dt} = k_1 a - k_1' b, \qquad (5.5.5)$$

$$\frac{dc}{dt} = k_2 a - k_2' c.$$

Clearly, $(a + b + c)$ is constant and by choosing the units suitably we can make $a + b + c = 1$. The equilibrium concentrations a_e, b_e, and c_e must satisfy the simultaneous equations obtained by setting to zero all the time derivatives, namely

$$0 = -(k_1 + k_2)a_e + k_1' b_e + k_2' c_e,$$

$$0 = k_1 a_e - k_1' b_e, \qquad (5.5.6)$$

$$0 = k_2 a_e - k_2' c_e.$$

The last two equations give

$$b_e = K_1 a_e, \qquad c_e = K_2 a_e,$$

where $K_i = k_i/k_i'$ and $i = 1, 2, \ldots$, and since $a_e + b_e + c_e = 1$ we have

$$a_e = \frac{1}{1 + K_1 + K_2}, \qquad b_e = \frac{K_1}{1 + K_1 + K_2}, \qquad c_e = \frac{K_2}{1 + K_1 + K_2}. \qquad (5.5.7)$$

It is convenient to set

$$x = a - a_e, \qquad y = b - b_e, \qquad z = c - c_e. \qquad (5.5.8)$$

Notice that $x + y + z = 0$, so that $x = -y - z$. Subtracting Eq. (5.5.6) from Eq. (5.5.8) gives exactly the same equations with x, y, and z replacing a, b, and c. Substituting for x in the last equations gives

$$\frac{dy}{dt} = -(k_1 + k_1')y - k_1 z$$

$$\frac{dz}{dt} = -k_2 y - (k_2 + k_2')z. \qquad (5.5.9)$$

This pair of equations is precisely of the type IV reviewed in Sec. 5.1, so that we know that the solutions are linear combinations of $\exp m_1 t$ and $\exp m_2 t$, where m_1 and m_2 are the roots of the equation

$$m^2 + m(k_1 + k_1' + k_2 + k_2') + (k_1 k_2' + k_2 k_1' + k_1' k_2') = 0.$$

It is not hard to show that the roots of this equation are both real and negative. The initial values of y and z are $-(b_e - b_0)$ and $-(c_e - c_0)$, respectively, hence the initial value of dy/dt is $(k_1 + k_1')(b_e - b_0) + k_1(c_e - c_0)$. Using Eq. (5.1.15) with t playing the role of x, $x_0 = 0$, and the given values of y_0 and y_0', we have

$$b = b_e + (b_e - b_0)\frac{(k_1 + k_1' + m_2)e^{m_1 t} - (k_1 + k_1' + m_1)e^{m_2 t}}{(m_1 - m_2)}$$
$$+ (c_e - c_0)\frac{k_1}{m_1 - m_2}(e^{m_1 t} - e^{m_2 t}). \quad (5.5.10)$$

Similarly,

$$c = c_e + (b_e - b_0)\frac{k_2}{m_1 - m_2}(e^{m_1 t} - e^{m_2 t})$$
$$+ (c_e - c_0)\frac{(k_2 + k_2' + m_2)e^{m_1 t} - (k_2 + k_2' + m_1)e^{m_2 t}}{m_1 - m_2}. \quad (5.5.11)$$

The course of the reaction will again resemble that shown in Fig. 5.4 for the irreversible case, the asymptotes being horizontal lines with concentrations a_e, b_e, and c_e. In this case, there does exist the possibility of overshooting in one of the concentrations, as shown by the broken lines in Fig. 5.4. Whether or not this happens will depend on the rate constants and the initial conditions, but the reality of the roots m_1 and m_2 ensures that there can be only one maximum or minimum in the curve for $b(t)$ or $c(t)$.

Other cases could be considered, but enough has been said to demonstrate two of the principal methods.

Exercise 5.5.1. Consider the concurrent irreversible reactions $A \xrightarrow{k_1} B$, $A \xrightarrow{k_2} C$ where the orders of the reaction are respectively p_1 and p_2, for the cases

$$\begin{array}{ccc} p_1 & 1 & 2 \\ p_2 & 2 & 2. \end{array}$$

5.6 Consecutive First Order Reactions

An important class of reactions has as its most elementary example the consecutive reactions $A \xrightarrow{k_1} B \xrightarrow{k_2} C$. Again, it is simplest to work in the concentrations a, b, and c, for which the equations are

$$\frac{da}{dt} = -k_1 a, \quad (5.6.1)$$

$$\frac{db}{dt} = k_1 a - k_2 b, \quad (5.6.2)$$

$$\frac{dc}{dt} = k_2 b, \quad (5.6.3)$$

with $a = a_0$, $b = b_0$, and $c = c_0$ when $t = 0$. The first equation is by now only too familiar and gives

$$a(t) = a_0 e^{-k_1 t}. \tag{5.6.4}$$

Substituting into Eq. (5.6.2), we get a linear equation (type II) of the form of Eq. (5.1.6),

$$\frac{db}{dt} + k_2 b = k_1 a_0 e^{-k_1 t}. \tag{5.6.5}$$

From the method of solution of Eq. (5.1.8) given in the first section of this chapter

$$b(t) = b_0 e^{-k_2 t} + \frac{a_0 k_1 (e^{-k_1 t} - e^{-k_2 t})}{k_2 - k_1}, \tag{5.6.6}$$

and by integrating Eq. (5.6.3) we get

$$c(t) = c_0 + b_0(1 - e^{-k_2 t}) \\ + \frac{a_0[(1/k_1)(1 - e^{-k_1 t}) - (1/k_2)(1 - e^{-k_2 t})]}{(1/k_1) - (1/k_2)}. \tag{5.6.7}$$

For $b_0 = c_0 = 0$ we may take the invariant $a + b + c = a_0 = 1$; the form of the curves is as shown in Fig. 5.5. In these figures, k is the ratio k_2/k_1 and $\tau = k_1 t$ is the dimensionless time, so that

$$a = e^{-\tau} \qquad b = \frac{e^{-\tau} - e^{-k\tau}}{k - 1}. \tag{5.6.8}$$

(The curve of a is the same for all and is shown only in three of the sections.) Now b clearly has a maximum, for it is zero to begin with and tends to zero again as the irreversible reaction approaches completion. By differentiating Eq. (5.6.8) we see that this occurs when

$$\tau = \tau_m = \frac{\ln k}{k - 1} \tag{5.6.9}$$

and that the maximum value of b is

$$b_m = k^{k/(1-k)}. \tag{5.6.10}$$

It is evident from Fig. 5.5 that as k increases, τ_m and b_m both decrease and the peaks become relatively sharper.

This example also introduces a method that is sometimes used to get approximate solutions. It is the sort of hypothesis that we have met above in Sec. 4.6; but here, with so elementary an exact solution, we have a chance to test its validity. If $k \gg 1$ so that the second reaction is much faster than the first, we expect that B will dissociate almost as rapidly as it is formed and never rise to more than a very small concentration. Suppose the concentration of B were small and constant; then, from Eq. (5.6.2) its value would be a/k. Of course, a varies, so that it is inconsistent to suppose that b is both strictly constant and also a small, constant fraction of a. However,

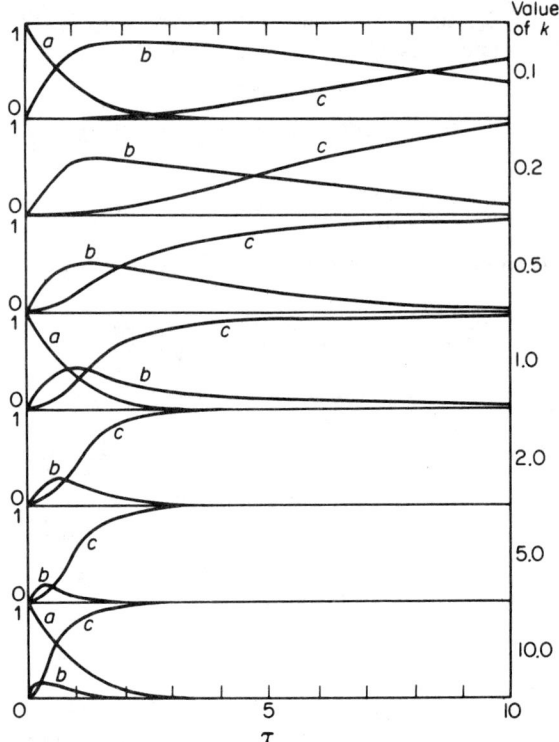

Fig. 5.5 Variation of concentrations with dimensionless time for the first order consecutive reactions $A \rightarrow B \rightarrow C$. (Figure based on calculations by Konowalow, Hirschfelder, and others: see references.)

from Fig. 5.5 we see that for a large k, the concentration of B quickly rises to its maximum and thereafter falls off in the same way as the concentration of A. The hypothesis will, we hope, be a reasonable approximation over a large part of the reaction history. If it were true, we would have (taking $a_0 = 1$)

$$\frac{dc}{dt} = k_2 b = k_1 a = k_1 e^{-k_1 t}$$

or, if $c_0 = 0$, $c = 1 - e^{-k_1 t}$. In dimensionless time τ we have the approximate c, say $\bar{c}(\tau)$, given by

$$\bar{c}(\tau) = 1 - e^{-\tau},$$

and this has to be compared with the exact value

$$c(\tau) = 1 - \frac{k e^{-\tau} - e^{-k\tau}}{k - 1}.$$

The difference $\bar{c} - c = (e^{-\tau} - e^{-k\tau})/(k-1)$ is just the exact value b given by Eq. (5.6.8). This shows clearly that the hypothesis will give good results for $k \gg 1$ but that it will be seriously in error for $k < 1$.

Another method of solving these equations is to seek a linear combination of the first two equations. Thus, if we put $g = a + \lambda b$ and take Eq. (5.6.1) plus λ times Eq. (5.6.2), we have

$$\frac{dg}{dt} = -k_1(1-\lambda)a - k_2\lambda b,$$
$$= -k_2 g,$$

if $\lambda = 1 - k$. Then

$$a = a_0 e^{-k_1 t} \quad \text{and} \quad g = g_0 e^{-k_2 t},$$

where $g_0 = a_0 + (1-k)b_0$; this again gives the solutions given by Eqs. (5.6.4.) and (5.6.6).

The integration of sets of linear equations can be multiplied to distraction by considering linear chains and branching chains and so on. However, the really significant features of linear systems come into play when the reactions are reversible; these will be discussed in a descriptive way in the next section.

Exercise 5.6.1. Obtain the limiting forms of Eqs. (5.6.6), (5.6.7), (5.6.8), (5.6.9), and (5.6.10) as $k \longrightarrow 1$.

Exercise 5.6.2. Write the equations for the series of first order irreversible reactions

$$A_1 \xrightarrow{k_1} A_2 \xrightarrow{k_2} A_3 \xrightarrow{k_3} \cdots A_N \xrightarrow{k_N} A_{N+1},$$

letting c_n denote the concentration of A_n. If $c_{10} = 1$ and $c_{n0} = 0$ where $n = 2, 3, \ldots$, show that

$$c_1(t) = e^{-k_1 t}, \qquad c_n(t) = k_{n-1} \int_0^t c_{n-1}(s) e^{-k_n(t-s)} ds.$$

Exercise 5.6.3. If in this sequence of reactions $k_1 = k_2 = \ldots = k_N = k$, $c_{10} = 1$, and $c_{n0} = 0$ where $n > 1$, show that

$$c_n(t) = \frac{(kt)^{n-1}}{(n-1)!} e^{-kt}.$$

Prove that this (unlikely) possibility could be established if it were found that the times of the maxima of successive c_n's were in strict arithmetic progression.

Exercise 5.6.4. Consider $A \longrightarrow B \longrightarrow C$ when the first reaction is of the second order, and express the concentrations in terms of the commonly occurring functions and the exponential integral $Ei(y) = \int_{-\infty}^{y} e^t \, dt/t$. (See Appendix.)

Exercise 5.6.5. Find the concentrations of X, Y, P, and Q as functions of time for Denbigh's reaction system

$$A \xrightarrow{k_1} X \xrightarrow{k_3} Y$$
$$\downarrow k_2 \quad\; \downarrow k_4$$
$$P \qquad Q$$

All reactions are first order, irreversible.

Exercise 5.6.6. When isopropylbenzene is alkylated by n-butylene in the presence of hydrogen fluoride the reaction scheme is $A \longrightarrow B \longrightarrow C$ with A = isopropylbenzene, B = isopropyl-sec-butylbenzene, C = isopropyldi-sec-butylbenzene. From the following data estimate k_1 and k_2.

t	0	0.5	1.0	1.5	2.0	2.23	2.5	3.0
a	1	0.60	0.37	0.22	0.14	0.107	0.08	0.05
b	0	0.38	0.58	0.68	0.71	0.715	0.71	0.69
c	0	0.02	0.05	0.10	0.15	0.178	0.21	0.26

The time units are arbitrary; the concentrations are in mole fractions.

5.7 Systems of First Order Reactions

It would be beyond our present scope to try to cover the detail of the theory of systems of first order reactions, but the basic ideas are so simple and elegant that it seems a pity that they should pass completely unnoticed. A simple example will show the main features. We shall take the reversible sequential reaction $A \rightleftharpoons B \rightleftharpoons C$ and add a further reaction $C \rightleftharpoons A$, thus turning it into the triangular system

$$\begin{array}{c} A \\ k_1' \swarrow \;\; k_1 \;k_3 \;\; \searrow k_3' \\ B \underset{k_2'}{\overset{k_2}{\rightleftharpoons}} C \end{array}$$

The species A, B, and C might be three isomers or the scheme might be a model of a more complicated situation. The third reaction is not of course stoichiometrically independent of the other two, but it may represent a distinct chemical reaction and in that sense be kinetically independent. Here is a case where we shall not want to get the smallest number of equations but will exploit the symmetrical aspect of the problem.

There will be an equilibrium state of the system attained when all the reactions are separately at equilibrium. Thus, if $K_i = k_i/k_i'$ where $i = 1, 2$, and 3, the three equilibrium constants are

$$\frac{b_e}{a_e} = K_1, \qquad \frac{c_e}{b_e} = K_2, \qquad \frac{a_e}{c_e} = K_3. \tag{5.7.1}$$

But multiplying the terms on the left together gives 1; therefore

$$K_1 K_2 K_3 = 1 \quad \text{or} \quad k_1 k_2 k_3 = k_1' k_2' k_3'. \tag{5.7.2}$$

This means that not all the rate constants can be independent, but must satisfy Eq. (5.7.1), an expression of the *principle of microscopic reversibility*, which states that at equilibrium of the whole system each reaction must be separately at equilibrium. From Eqs. (5.7.1) and (5.7.2) we can express the equilibrium concentrations in the following symmetric form (see Exercise 5.7.1):

$$a_e = (1 + K_1 + K_1 K_2)^{-1}, \quad b_e = (1 + K_2 + K_2 K_3)^{-1},$$
$$c_e = (1 + K_3 + K_3 K_1)^{-1}. \tag{5.7.3}$$

Figure 5.6 shows a triangular diagram in which the whole path of the reaction can be shown. By the stoichiometry of the reaction, the sum of the three concentrations is constant and, by suitably choosing the unit of con-

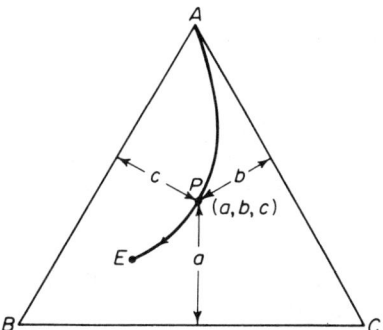

Fig. 5.6 Reaction path in the concentration triangle.

centration, $a + b + c = 1$. The point P in the triangle ABC can represent the composition (a, b, c), if the perpendicular distances to the three sides BC, CA, and AB are taken to be a fraction a, b, and c of the height, respectively. Thus the vertices represent pure A, B, and C respectively, and E might be the equilibrium point. The course of the reaction starting from pure A would be represented by a curve such as APE, and time would be a parameter along such a curve.

The equations governing the three concentrations are

$$\frac{da}{dt} = -(k_1 + k_3')a + k_1'b + k_3 c,$$

$$\frac{db}{dt} = k_1 a - (k_2 + k_1')b + k_2' c, \tag{5.7.4}$$

$$\frac{dc}{dt} = k_3' a + k_2 b - (k_3 + k_2')c.$$

Of course, only two of these are independent since $a + b + c = 1$. We could treat these equations exactly as we treated those for the reversible concurrent reactions in Sec. 5.5, by setting $x = a - a_e$, $y = b - b_e$, and $z = c - c_e$; we would obtain a similar set of equations but with x, y, and z replacing a, b, and c. However, since $x + y + z = (a + b + c) - (a_e + b_e + c_e) = 0$, it is now possible to eliminate x and obtain a pair of first order equations with constant coefficients for y and z. We know that the solution of this is the linear combination of two exponentials. This of course destroys the symmetry that we have claimed to preserve, but all we need here is the form of the solution; the reader can arrive at the same result more elegantly by doing Exercise 5.7.2. It is clear then that the solutions will be of the form

$$a(t) = a_e + A_1 e^{-\lambda_1 t} + A_2 e^{-\lambda_2 t},$$
$$b(t) = b_e + B_1 e^{-\lambda_1 t} + B_2 e^{-\lambda_2 t}, \quad (5.7.5)$$
$$c(t) = c_e + C_1 e^{-\lambda_1 t} + C_2 e^{-\lambda_2 t}.$$

The λ_1 and λ_2 must be positive if the equilibrium is approached as $t \longrightarrow \infty$. The constants $A_1, \ldots, C_2$ are not independent, for $A_1 + B_1 + C_1 = 0$ and $A_2 + B_2 + C_2 = 0$ if $a + b + c$ is to be constant. There are in fact two other conditions that the six constants satisfy if Eqs. (5.7.5) are to give a solution of Eqs. (5.7.4); these fix the ratios $A_1 : B_1 : C_1$ and $A_2 : B_2 : C_2$ so that only two (say A_1 and A_2) have to be chosen to fit the initial conditions. This is appropriate since we have agreed that $a + b + c = 1$ and so only two of a_0, b_0, and c_0 are independent.

Figure 5.7 shows some typical reaction paths in the triangular diagram. It is noticeable that all the paths except PE and QE approach the equilibrium point E tangentially to the line LEM. This also is to be expected from Eqs. (5.7.5). Suppose $\lambda_2 > \lambda_1$, then for sufficiently large times the terms proportional to $e^{-\lambda_2 t}$ will be negligible in comparison with those proportional to

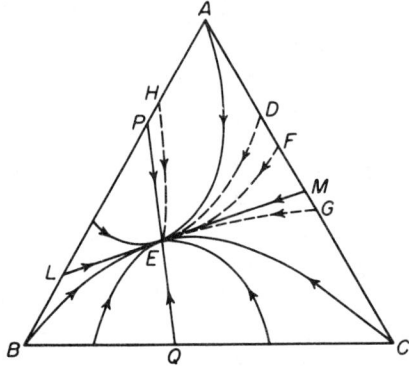

Fig. 5.7 Reaction paths for the first order system $A \rightleftharpoons B \rightleftharpoons C \rightleftharpoons A$.

$e^{-\lambda_1 t}$. But if these last terms are dropped

$$(a - a_e) : (b - b_e) : (c - c_e) = A_1 : B_1 : C_1, \qquad (5.7.6)$$

and we have mentioned that these last ratios are characteristic of the equations and so fixed. They define the direction of the line *LEM* and Eq. (5.7.6) shows that the paths become tangent to it. The exception to this rule is the solution for which $A_1 = B_1 = C_1 = 0$. This corresponds to an initial composition somewhere on the line *PEQ*. Similarly, a starting point on the line *LEM* gives $A_2 = B_2 = C_2 = 0$. In both these cases, the reaction path is a straight line and this suggests a rather sensitive way of determining the kinetics. Suppose experiments are done first with pure *A* and pure *C*. Plotting the results and drawing a tangent near equilibrium would suggest where to find the straight line path. However, the drawing of tangents is never very accurate; suppose the first attempt intersected the side *AC* at *D*. This represents an initial mixture of *A* and *C* in the ratio of about 7 : 3. With this as starting composition, the reaction path is still notably curved but a better guess can be made, say *F* or *G*. Thus it is possible by repeated trials to find the point *M*, and by plotting $(a - a_e)$ semilogarithmically against *t* to determine λ_1. A similar series of trials will lead to *P*, the starting composition giving a straight path and allowing λ_2 to be determined. From the values of λ_1 and λ_2 and the position of these paths, all the rate constants can be determined. The details of all this are beyond our present interests but there are valuable and sophisticated tools available for this work which can be grasped once the basic picture is understood.

Exercise 5.7.1. Obtain the equilibrium composition Eq. (5.7.3) as the solution of Eqs. (5.7.1) and (5.7.2).

Exercise 5.7.2. Suppose that a linear combination $g = \alpha a + \beta b + \gamma c$ can be found such that when *a*, *b*, and *c* are governed by Eqs. (5.7.4), the variable *g* satisfies

$$\frac{dg}{dt} = -\lambda g.$$

Show that λ must satisfy the cubic equation

$$\lambda(\lambda^2 - K\lambda + M) = 0$$

and find *K* and *M* in terms of the rate constants. Show further that the roots λ_1 and λ_2 are real and positive.

5.8 Numerical Methods

All the problems considered in this chapter have centered on the solution of differential equations. We have had occasion to remark at several points that solutions are not available in closed form and that it is necessary to use a direct numerical method of calculation. Although it is commonly possible to put differential equations of this kind straight onto a computer, it is worthwhile to describe a numerical method for two reasons. In the first place, contrary to public opinion, the computer does not "think," and it is therefore never safe to give it a job to do without any idea of how it would do it. In the second place, it is sometimes possible, and even desirable, to do calculations by hand. The method to be described applies to any set of ordinary differential equations including those for nonisothermal reactions. It will be illustrated for the cases of one and of two simultaneous equations.

Consider the differential equation

$$\frac{dy}{dx} = f(x, y) \tag{5.8.1}$$

with the initial condition

$$y = b \quad \text{when} \quad x = a. \tag{5.8.2}$$

In the neighborhood of this initial point

$$y(x) = y(a) + (x - a)y'(a) + \tfrac{1}{2}(x - a)^2 y''(a) + \cdots$$
$$= b + (x - a)f(a, b) + \tfrac{1}{2}(x - a)^2 (f_x + f f_y) + \cdots.$$

The crudest approximation to $y(x + h)$ would be $b + h f(a, b)$. The quadratic approximation involves finding the partial derivatives of f and evaluating them at (a, b). Since this is inconvenient, the method known as the Runge-Kutta method uses derivatives calculated at certain other points to give an accuracy equivalent to terms of the Taylor series up to and including $h^4 y^{\text{iv}}(a)/4!$. Denote by k_1 the first approximation to $y(a + h) - y(a)$; then

$$k_1 = hf(a, b). \tag{5.8.3}$$

The midpoint of the interval to this degree or approximation will be $x = a + \tfrac{1}{2}h$, $y = b + \tfrac{1}{2}k_1$. The derivative is calculated here to give

$$k_2 = hf(a + \tfrac{1}{2}h, b + \tfrac{1}{2}k_1). \tag{5.8.4}$$

This is done again,

$$k_3 = hf(a + \tfrac{1}{2}h, b + \tfrac{1}{2}k_2), \tag{5.8.5}$$

and finally an evaluation is made for the end of the interval

$$k_4 = hf(a + h, b + k_3). \tag{5.8.6}$$

Then

$$y(a + h) = b + k,$$

where

$$k = \tfrac{1}{6}(k_1 + 2k_2 + 2k_3 + k_4) \tag{5.8.7}$$

with an error proportional to h^5. If the solution is required over an interval (a, c), it is often convenient to divide it into N equal parts with $h = (c - a)/$

N. Let $x_0 = a$, $y_0 = b$, $x_n = a + nh$, and $y_n = y(a + nh)$; then the formulae given above apply to each interval, and a tabulation such as is illustrated below may be made.

Since the error at each step is of the order of h^5 and since the number of steps required is of the order of $1/h$, the total error over the interval (a, c) will be of the order of h^4. Thus we may write

$$y(c) = y_{N,h} + Fh^4, \qquad (5.8.8)$$

where $y_{N,h}$ is $y(c)$ calculated by N steps of length h, and F is a constant related to the average value of the higher derivatives in the interval. If the same calculation is done using twice the number of steps and half the interval, we can write

$$y(c) = y_{2N, h/2} + \frac{F'h^4}{16}. \qquad (5.8.9)$$

If it is reasonable to assume that the new proportionality constant F' is not too different from the old, and if we set them equal, then

$$y(c) = \frac{16 y_{2N, h/2} - y_{N,h}}{15}. \qquad (5.8.10)$$

Thus, comparison of the two calculations gives a measure of the accuracy that has been attained and shows whether h has been chosen sufficiently small.

The calculation can be conveniently set down in tabulated form, and this is shown in Table 5.1 for the typical step in the integration of two simultaneous equations

$$\frac{dy}{dx} = f(x, y, z), \qquad \frac{dz}{dx} = g(x, y, z).$$

The Runge-Kutta method is of course not the only method of numerical solution but it illustrates the notions that are fundamental to all. For much fuller details any good treatise on numerical methods may be consulted. A few are given in the references at the end of the chapter.

In this chapter we have delegated to the exercises the important question of fitting values of the rate constants to experimental data. This was done because the subject belongs to the area of pure kinetics rather than to applied kinetics, and is no reflection on its crucial importance for the whole art of reaction engineering. When data can be plotted in straight line form, as for example in the logarithmic plots implied in Exercise 5.2.2 for a first order reaction, a best straight line can be fitted to give some estimate of the accuracy of the data. When the differential equations of the reaction system have to be integrated numerically from the start, the estimation of constants is more difficult. In the simpler cases, an analog computer can sometimes be made to display the solution of the equations, the rate constants being certain potentiometer settings. Estimates of these can be made by manual adjust-

TABLE 5.1
SCHEMA FOR THE RUNGE-KUTTA METHOD OF INTEGRATING A PAIR OF FIRST ORDER DIFFERENTIAL EQUATIONS

x	y	$hf(x, y, z)$	z	$hg(x, y, z)$
x_n	y_n	$k_1 = hf(x_n, y_n, z_n)$	z_n	$m_1 = hg(x_n, y_n, z_n)$
$x_n + \tfrac{1}{2}h$	$y_n + \tfrac{1}{2}k_1$	$k_2 = hf(x_n + \tfrac{1}{2}h, y_n + \tfrac{1}{2}k_1, z_n + \tfrac{1}{2}m_1)$	$z_n + \tfrac{1}{2}m_1$	$m_2 = hg(x_n + \tfrac{1}{2}h, y_n + \tfrac{1}{2}k_1, z_n + \tfrac{1}{2}m_1)$
$x_n + \tfrac{1}{2}h$	$y_n + \tfrac{1}{2}k_2$	$k_3 = hf(x_n + \tfrac{1}{2}h, y_n + \tfrac{1}{2}k_2, z_n + \tfrac{1}{2}m_2)$	$z_n + \tfrac{1}{2}m_2$	$m_3 = hg(x_n + \tfrac{1}{2}h, y_n + \tfrac{1}{2}k_2, z_n + \tfrac{1}{2}m_2)$
$x_n + h$	$y_n + k_3$	$k_4 = hf(x_n + h, y_n + k_3, z_n + m_3)$	$z_n + m_3$	$m_4 = hg(x_n + h, y_n + k_3, z_n + m_3)$
		$k = \dfrac{k_1 + 2k_2 + 2k_3 + k_4}{6}$		$m = \dfrac{m_1 + 2m_2 + 2m_3 + m_4}{6}$
$x_{n+1} = x_n + h$	$y_{n+1} = y_n + k$	etc.	$z_{n+1} = z_n + m$	etc.

ment until the displayed solution fits the data as well as possible. When the integration is done on a digital computer, a trial and error procedure has to be used. Assume that in the reaction model there are two independent reactions and four rate constants, with the differential equations being

$$\frac{d\xi_i}{dt} = f_i(\xi_1, \xi_2; k_1, k_2, k_3, k_4) \qquad i = 1, 2$$

for the two extents ξ_1 and ξ_2. For any assumed set of constants, say $\bar{k}_1, \ldots, \bar{k}_4$, the solution can be computed, say $\bar{\xi}_1(t)$ and $\bar{\xi}_2(t)$. Suppose the available data give the values of ξ_1 and ξ_2 at times $t = t_n$, where $n = 1, \ldots, N$, as say ξ_{1n} and ξ_{2n}. Then it would be reasonable to regard the "best" estimates of $k_1, \ldots, k_4$ as those that give the least value of

$$S = \sum_{1}^{N} \{W_{1n}[\xi_{1n} - \bar{\xi}_1(t_n)]^2 + W_{2n}[\xi_{2n} - \bar{\xi}_2(t_n)]^2\},$$

where W_{1n} and W_{2n} are weighting factors that might reflect the relative accuracy of the observations. Various search methods have been suggested including steepest ascents and directed random search. Details will be found in the references below.

REFERENCES

5.1. Any good introduction to ordinary differential equations will serve as background for this chapter or for fuller study; e.g.:

E. L. Ince, *Integration of Ordinary Differential Equations*. Edinburgh: Oliver and Boyd (University Mathematical Texts), 1946.

E. A. Coddington, *An Introduction to Ordinary Differential Equations*. Englewood Cliffs, N. J.: Prentice-Hall, Inc., 1961.

Most texts on pure and applied kinetics have a section devoted to this subject and may be consulted with profit. Particularly to be commended for its range and appositeness of illustration is:

O. Levenspiel, *Chemical Reaction Engineering*. New York: John Wiley & Sons, Inc., 1962.

5.6. The concurrent-consecutive system of Exercise 5.6.5 is taken from K. G. Denbigh's paper in:

Chemical Reaction Engineering, ed. K. Rietema, pp. 125–132. London: Pergamon Press, 1957.

For a full discussion of consecutive reactions $A \to B \to C$ of all combinations of zeroth, first, and second order see:

H. A. G. Chermin and D. W. van Krevelen, "Selectivity in Consecutive Reactions," *Chem. Eng. Sci.*, **14**, 58–71 (1961).

The calculations for Fig. 5.5 are taken, with permission, from:

D. D. Konowalow, J. E. Blair, J. O. Hirschfelder, and F. Daniels, *Solutions for Complex Systems of Chemical Reactions*, Part I. Wright Air Development Center, Technical Note, 59–143, August 1959.

A large number of formulae for many special cases of chains of first order reactions can be found in:

N. M. Rodiguin and E. N. Rodiguina, *Consecutive Chemical Reactions* (trans. and ed. R. F. Schneider). Princeton, N. J.: D. Van Nostrand Co., Inc., 1964.

5.7. A full treatment of this whole area with many interesting sidelights on other questions is given in:

J. Wei and C. D. Prater, *The Structure and Analysis of Complex Reaction Systems*. Advances in Catalysis and Related Subjects series, Vol. 13. New York: Academic Press Inc., 1962.

5.8. The numerical treatment of differential equations is now a commonplace in elementary texts of numerical analysis. A comprehensive reference work is:

L. Collatz, *Numerical Treatment of Differential Equations* (3rd ed.). Berlin: Springer-Verlag, 1966.

The problems associated with fitting values of rate constants to the solution of nonlinear equations and the available methods are well described in:

H. H. Rosenbrock and C. E. Storey, *Computational Techniques for Chemical Engineers*. Oxford: Pergamon Press, 1966.

NOTATION

The notation of this chapter is quite parochial in its ambit and each symbol should be identifiable in its own context. In general, lower case letters have been used to designate the concentrations of the species denoted by the corresponding capitals, except that, as usual, c_j denotes the concentration of A_j. Lower case k's with various affixes are used for rate constants and the suffixed capitals of this letter represent equilibrium constants. Time is denoted by t and dimensionless time by τ.

The Interaction of Chemical and Physical Rate Processes

6

The subject of this chapter holds a key position in the structure of applied chemical kinetics. On the one hand, we have the development of an expression for the rate of reaction and on the other, the need to use it in the design and analysis of the reactor. But the chemical engineer cannot overlook the fact that there are physical as well as chemical processes governing the overall rate of reaction and he tries to incorporate them into his calculations as simply as possible. Thus the interaction of heat and mass transfer with reaction has been the peculiar province of the chemical engineer, because the chemist tries to arrange his experiments to eliminate or control the physical processes and, until the recent work of mechanical engineers on some of the problems of space flight, other engineers have been concerned with the physical processes only.

We have been working up to now with an expression for the rate of reaction per unit reactor volume. Now we have to recognize that the reactor volume may be packed with a stationary bed of porous catalyst particles within which the reaction is taking place and that something more than a mere chemical reaction throughout a uniform homogeneous phase is involved. In spite of this, we would like to use an equivalent reaction rate per unit reactor volume and in Sec. 6.1 we shall see what is needed to do this. Section 6.2 considers the steps of the reaction in some detail and shows how simplifications may be made when one step is much slower than the others. External mass transfer is considered in Sec. 6.3 and internal diffusion in Sec. 6.4; a simple case where both processes are of comparable importance is given in Sec. 6.5. The importance of temperature effects is described in Sec. 6.6 and the application of all this is discussed in Sec. 6.7. The scheme of the whole chapter is shown in Fig. 6.1.

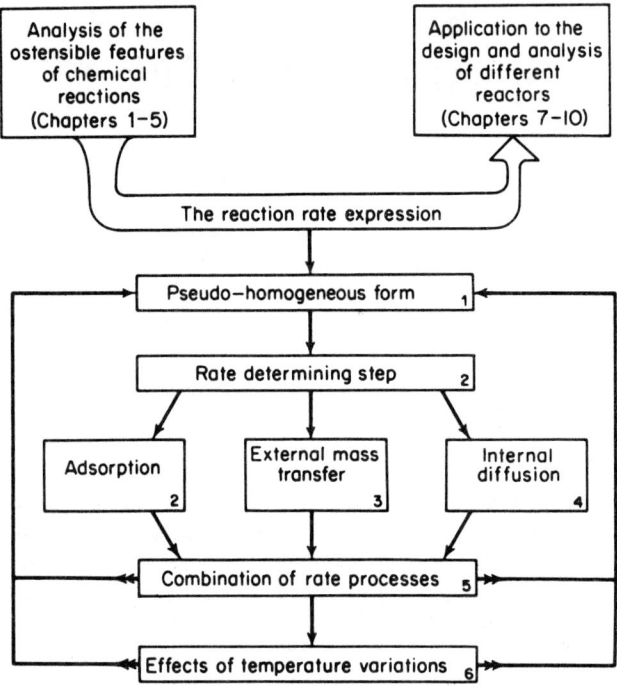

Fig. 6.1 The structure of the chapter.

6.1 Effective Reaction Rate Expressions

If a reaction is taking place in a homogeneous phase, we can readily calculate the rate of appearance or disappearance of a particular species at any point of the reactor. Suppose the reaction is $\sum \alpha_j A_j = 0$ and that in an element of volume dV, the concentration of A_j is c_j and the temperature is T. Then the rate of reaction is $r(c_1, \ldots, c_S, T)$ and the rate of production of A_j within the element is $\alpha_j r(c_1, \ldots, c_S, T) dV$ moles per unit time. If the concentrations

Sec. 6.1 Effective Reaction Rate Expressions

can be expressed in the form $c_j = c_{j0} + \alpha_i \xi$ then we have the kind of expression which we have been using up to now, namely $\alpha_j r(\xi, T)\, dV = \alpha_j r(c_{10} + \alpha_1 \xi, \ldots, T)\, dV$.

Let us take the simplest of possible examples, a reversible first order reaction which, to avoid too many subscripts later on, we shall write as $A \rightleftarrows B$. If a and b are the concentrations of A and B in the volume element dV and the reaction is a homogeneous one, taking place in the bulk phase of the volume, the reaction rate may be written $r = ka - k'b$. Thus the rates of production of A and B are $-r\, dV$ and $+r\, dV$ moles per unit time, respectively. Since the sum of the concentrations, $a + b$, is constant we can write $a = a_0 - \xi$ and $b = b_0 + \xi$ and the reaction rate is the rate of change of ξ. Now suppose that the reaction actually takes place on a solid surface of area dA within the volume element and that on the surface the concentrations of A and B are $\hat{a}$ and $\hat{b}$ moles per unit area respectively. Since the reaction is taking place on a surface it is proper to use a reaction rate per unit area, $\hat{r}$. The reaction is still $A \rightleftarrows B$ and one molecule (or mole) of A disappears for each molecule (or mole) of B that is formed; thus the rates of production of A and B within the element are $-\hat{r}\, dA$ and $\hat{r}\, dA$, respectively. Since the sum of these is zero, we can again write $a = a_0 - \xi$ and $b = b_0 + \xi$. If we set $r\, dV = \hat{r}\, dA$ (i.e., the effective bulk reaction rate $r = \hat{r}\, dA/dV$) we can regard r as the rate of change of ξ, since from it we can calculate the rates of production of each species by the standard formula, $\alpha_j r\, dV$. But $\hat{r}$, being a reaction rate on the surface, will presumably be a function of the concentrations on the surface, $\hat{a}$ and $\hat{b}$, say, $\hat{r} = \hat{k}\hat{a} - \hat{k}'\hat{b}$. If then we can relate these surface concentrations to the bulk phase concentrations, we will be able to get the pseudo-homogeneous reaction rate expression, r, in terms of the bulk phase concentrations, a and b. Let us suppose for simplicity that $\hat{a} = \alpha a$ and $\hat{b} = \beta b$ and that the temperature on the surface, T, of which $\hat{k}$ and $\hat{k}'$ are functions, is the same as the bulk phase temperature, T; then

$$\hat{r} = \hat{r}(\hat{a}, \hat{b}, T) = \hat{k}\hat{a} - \hat{k}'\hat{b} = \alpha\hat{k}a - \beta\hat{k}'b = \hat{r}^*(a, b, T)$$

and

$$r(a, b, T) = (\alpha\hat{k}a - \beta\hat{k}'b)\frac{dA}{dV},$$

or

$$r(\xi, T) = [(\alpha\hat{k}a_0 - \beta\hat{k}'b_0) - (\alpha\hat{k} + \beta\hat{k}')\xi]\frac{dA}{dV}.$$

Before returning to the general situation, let us note two things. First, the expression for r has two factors: $\hat{r}$, the reaction rate per unit area on the surface, and dA/dV, the ratio of the surface area to the volume. Second, for clarity $\hat{r}$ will denote the surface reaction rate when this is expressed as a function of the surface concentrations, but, as a reminder that the functional form may change when the relationship between the surface and bulk concentrations is substituted, the surface reaction rate as a function of the bulk concentrations is written $\hat{r}^*$. In this example, $\hat{r} = \hat{k}\hat{a} - \hat{k}'\hat{b}$, while $\hat{r}^* = \alpha \hat{k} a - \alpha \hat{k}' b$. The relationships $\hat{a} = \alpha a$ and $\hat{b} = \beta b$ were so simple that the functional form could scarcely be said to have changed, but suppose that we had the relations

$$\hat{a} = \frac{\alpha a}{1 + A'a + B'b}, \qquad \hat{b} = \frac{\beta b}{1 + A'a + B'b};$$

then we would have

$$\hat{r}^* = \frac{\alpha \hat{k} a - \beta \hat{k}' b}{1 + A'a + B'b},$$

which has a totally different form from that of $\hat{r}$. Part of our concern in this chapter is with the kind of relation between a, b and $\hat{a}$, $\hat{b}$ that may express the physical processes that are taking place.

Returning now to the case of the general reaction $\sum \alpha_j A_j = 0$, we shall suppose in the first instance that within an element of volume dV there is a catalytic area dA and that conditions are uniform over it. This will often be the surface area within a porous particle or pellet and it is frequently specified by

S_g = catalytic area per unit mass of particle or pellet.

In addition we need to know

ρ_b = bulk density of the pellet (i.e., its mass divided by its total volume)

and

ϵ = the voidage or fractional free space in the reactor.

Thus a volume dV will consist of $\epsilon \, dV$ free space and $(1 - \epsilon) \, dV$ of catalyst. The latter will contain a mass $\rho_b(1 - \epsilon) \, dV$ of catalyst and hence an area $S_g \rho_b (1 - \epsilon) \, dV = dA$. Thus the factor

$$\frac{dA}{dV} = (1 - \epsilon)\rho_b S_g. \tag{6.1.1}$$

We are still assuming that within the element of volume, the conditions on the surface dA are constant. Let the concentration of A_j on the surface be denoted by $\hat{c}_j$, moles per unit area. Then the rate of reaction on the surface, $\hat{r}$, is expressed in terms of these concentrations and the temperature of the surface, $\hat{T}$, as

$$\hat{r} = \hat{r}(\hat{c}_1, \ldots, \hat{c}_S, \hat{T}).$$

If we can find a relation between the surface conditions, $\hat{c}_j$, $\hat{T}$, and those in the bulk phase, c_j, T, then on substitution we have

$$\hat{r}^* = \hat{r}^*(c_1, \ldots, c_S, T),$$

Sec. 6.1   Effective Reaction Rate Expressions   117

and the effective or pseudo-homogeneous reaction rate is

$$r = \frac{dA}{dV}\hat{r}^* = (1 - \epsilon)\rho_b S_g \hat{r}^*(c_1, \ldots, c_S, T). \tag{6.1.2}$$

If we can write the concentrations c_j in the form $c_{j0} + \alpha_j \xi$, then we can achieve the final form of

$$r(\xi, T) = (1 - \epsilon)\rho_b S_g \, \hat{r}^*(c_{10} + \alpha_1 \xi, \ldots, T). \tag{6.1.3}$$

To see that this is possible, we must look a little more closely at what physical processes may be involved.

We are considering a situation in which the several processes have come to a steady relationship. Thus the concentration $\hat{c}_j$ may differ from c_j because the molecules of A_j are being adsorbed and desorbed at the surface. In fact, the rate of these processes will depend on the $\hat{c}_j$ and the concentration of A_j immediately above the surface. Let this be denoted by c_{js}, then c_{js} may not be the same as c_j because there is a resistance to mass transfer between the bulk phase and the surface. If A_j is a reactant we would expect c_j to be greater than c_{js} to provide the potential for driving it to the surface where it is adsorbed and it reacts; if A_j is a product, it is formed at the surface and so c_{js} would tend to be greater than c_j. However, the only place that there is any change from one species to another is by reaction at the surface and here the rate of production of A_j is proportional to its stoichiometric coefficient α_j. Now when the different physical processes have reached a steady relationship with one another, the rate of transfer of a reactant species from the bulk phase to the reaction surface and the rate of adsorption must equal the rate at which it reacts, otherwise there would be a continual accumulation or depletion at the surface. Similarly, for a product species the rate of transfer or desorption must be the rate of formation. Thus in the bulk phase, in the volume element as a whole, the rate of production of A_j must be proportional to its stoichiometric coefficient α_j; we can write $c_j = c_{j0} + \alpha_j \xi$, and calculate a pseudo-homogeneous reaction rate.

Finally, we must remove the objection that we have assumed that conditions are uniform over the catalytic reaction area. This will not necessarily be the case because diffusion into the pores of the pellet is a slow process —we would not expect the concentration of the reactant at the center of a pellet to be as great as at the outside surface. We shall overcome this difficulty by calculating the average reaction rate throughout a pellet of a given shape when the concentrations at its outside surface are specified. Let the ratio of this average reaction rate to the reaction rate at the outside surface conditions be called the *effectiveness factor* of the pellet and be denoted by η. Then, calculating $\hat{r}$ as if the conditions were those at the surface and relating them to bulk phase conditions to give $\hat{r}^*$, we have the actual rate of reaction as $\eta \hat{r}^*$ and so

$$r(\xi, T) = (1 - \epsilon)\rho_b S_g \eta \hat{r}^*. \tag{6.1.4}$$

In general, of course, η will be a complicated function of conditions, but there are circumstances under which it can be simplified to a remarkable degree. Even though it may not be possible to calculate η with great accuracy, an understanding of the way in which it depends on physical parameters is of prime importance.

6.2 The Concept of the Rate Determining Step

The complete picture of what happens in a catalytic reaction involves the diffusion of the reactants from the flowing stream to the catalytic sites within the catalyst particle. There they may be adsorbed and react and the products must then be desorbed and transferred back into the flowing stream. Again, it will be useful to start with a simplified picture of part of this process in a particularly simple case. We shall therefore ignore the transfer and diffusion to the catalytic sites and consider only what may be happening in adsorption, reaction, and desorption. Again, we shall take the simple reversible reaction $A \rightleftharpoons B$ and let a, b denote concentrations immediately above the surface and $\hat{a}, \hat{b}$ be the adsorbed concentrations. The former might be in units of moles per unit volume, the latter in moles per unit area.

The simplest picture of adsorption is that the surface has a number of sites at which molecules of A and B may be adsorbed and that if they were completely filled, the total surface concentration, $\hat{a} + \hat{b}$, would be $\hat{c}_0$. Thus when $\hat{a}$ and $\hat{b}$ are the surface concentrations of adsorbed A and B, the concentration of vacant sites is

$$\hat{c}_v = \hat{c}_0 - \hat{a} - \hat{b}. \tag{6.2.1}$$

The process of adsorption is a dynamic one in which molecules of A are constantly bombarding the surface and a fraction of them are adsorbed at active sites, while those already adsorbed may be spontaneously desorbed. The rate at which adsorption takes place is clearly going to be proportional to the concentration of the species above the surface, namely a, and also to the concentration of vacant sites. We may write it as $k_a a \hat{c}_v$, where k_a is a constant depending on temperature and having the dimensions of volume per mole per unit time. The rate at which molecules of A will desorb may be taken to be proportional to the concentration already adsorbed, $k_d \hat{a}$, where k_d has the dimensions of the reciprocal of time. Thus the net rate of adsorption, which is the difference between these two, is

$$\hat{r}_{ad} = k_a a \hat{c}_v - k_d \hat{a}. \tag{6.2.2}$$

Using Eq. (6.2.1) we can write this as

$$\begin{aligned}\hat{r}_{ad} &= k_a a(\hat{c}_0 - \hat{a} - \hat{b}) - k_d \hat{a}, \\ &= k_d[K_A a \hat{c}_0 - (K_A a + 1)\hat{a} - K_A a \hat{b}],\end{aligned} \tag{6.2.3}$$

Sec. 6.2 The Concept of the Rate Determining Step

where

$$K_A = \frac{k_a}{k_d}. \tag{6.2.4}$$

The adsorption-desorption of B follows exactly the same pattern. Its rate of desorption is proportional to $\hat{b}$, the adsorbed concentration, and its rate of adsorption to the product $b\hat{c}_v$. Since we are interested in following through the process in the sequence in which A is adsorbed, reacts to form B, and B is desorbed, we shall talk about the net rate of *de*sorption of B (rather than its rate of *ad*sorption), and this is

$$\begin{aligned}\hat{r}_{de} &= k'_d\hat{b} - k'_a b\hat{c}_v, \\ &= k'_d\hat{b} - k'_a b(\hat{c}_0 - \hat{a} - \hat{b}), \\ &= k'_d[(1 + K_B b)\hat{b} + K_B b\hat{a} - K_B b\hat{c}_0],\end{aligned} \tag{6.2.5}$$

where

$$K_B = \frac{k'_a}{k'_d}. \tag{6.2.6}$$

The reaction is supposed to take place on the surface between the adsorbed species A and B and to be a first order reversible one. Its rate may therefore be written as

$$\hat{r} = \hat{k}\hat{a} - \hat{k}'\hat{b}. \tag{6.2.7}$$

Now the concentration of adsorbed A is increasing at a rate $\hat{r}_{ad}$ but decreasing at a rate $\hat{r}$ as it reacts to form B. Therefore

$$\frac{d\hat{a}}{dt} = \hat{r}_{ad} - \hat{r}. \tag{6.2.8}$$

Similarly,

$$\frac{d\hat{b}}{dt} = \hat{r} - \hat{r}_{de}. \tag{6.2.9}$$

Now we have assumed the process to be in a steady state in the sense that there is no accumulation of adsorbed species; thus $\hat{a}$ and $\hat{b}$ are constant and

$$\hat{r}_{ad} = \hat{r} = \hat{r}_{de}. \tag{6.2.10}$$

We notice that all three rates can be written with a rate constant (k_d, $\hat{k}'$, or k'_a) outside the bracket and an equilibrium constant (K_A, K, or K_B), within, namely

$$\hat{r}_{ad} = k_d[K_A a\hat{c}_0 - (1 + K_A a)\hat{a} - K_A a\hat{b}], \tag{6.2.11}$$

$$\hat{r} = \hat{k}'(K\hat{a} - \hat{b}), \tag{6.2.12}$$

$$\hat{r}_{de} = k'_d[-K_B b\hat{c}_0 + K_B b\hat{a} + (1 + K_B b)\hat{b}], \tag{6.2.13}$$

where $K = \hat{k}/\hat{k}'$ is clearly the equilibrium constant for the reaction. All the rate constants (lower case k's) and the equilibrium constants (capital K's) are of course functions of the surface temperature, $\hat{T}$.

We could combine Eqs. (6.2.10) through (6.2.13) and by eliminating $\hat{a}$ and $\hat{b}$ calculate $\hat{r}$ as a function of a and b. This we shall do shortly, but first let us introduce the notion of a rate determining step. It is conceivable that the reaction is rather slow compared with the process of adsorption and desorption, that is, the rate constants $\hat{k}$ and $\hat{k}'$ are very much smaller, in some sense, than k_a, k_d, k'_a, and k'_d. However, although these four constants are very large, the ratios K_A and K_B may have any finite value. Now if k_d and k'_d become exceedingly large the only way in which $\hat{r}_{ad}$ and $\hat{r}_{de}$ (which are both equal to $\hat{r}$) can remain finite is for the two expressions in the brackets of Eqs. (6.2.11) and (6.2.13) to become exceedingly small. In the limit then as k_d and k'_d tend to infinity they will be zero and we shall have the two equations

$$(1 + K_A a)\hat{a} + K_A a \hat{b} = K_A a \hat{c}_0$$
$$K_B b \hat{a} + (1 + K_B b)\hat{b} = K_B b \hat{c}_0. \quad (6.2.14)$$

These two equations can be solved for $\hat{a}$ and $\hat{b}$, giving

$$\hat{a} = \hat{c}_0 \frac{K_A a}{1 + K_A a + K_B b}$$
$$\hat{b} = \hat{c}_0 \frac{K_B b}{1 + K_A a + K_B b}. \quad (6.2.15)$$

Then $\hat{r} = \hat{k}\hat{a} - \hat{k}'\hat{b}$ can be expressed in terms of a and b by

$$\hat{r}^* = \hat{c}_0 \frac{\hat{k} K_A a - \hat{k}' K_B b}{1 + K_A a + K_B b}. \quad (6.2.16)$$

We have called this $\hat{r}^*$, since this is precisely what we mean by that symbol, namely $\hat{r}$, the superficial reaction rate, expressed in terms of a and b, the bulk phase concentrations, rather than the surface concentrations $\hat{a}$ and $\hat{b}$.

Another way of saying this is that the rate of these processes can be expressed as a product of a rate constant and a term which represents a distance from equilibrium (see Exercise 6.2.1). When the rate constant becomes large, the distance from equilibrium must become small. The concentrations a and b approach the equilibrium values given by Eq. (6.2.15) very rapidly when k_d and k'_d are large (see Exercise 6.2.2) and we are entitled to say that the slow, or rate determining, step is that of the reaction on the surface. The reader should do Exercise 6.2.3 at this point to confirm his understanding of the rate determining step.

It is interesting to observe that we get the same kind of expression if none of the steps is notably faster than the others. For if we put $\hat{r}_{ad} = \hat{r}_{de} = \hat{r}$ in Eqs. (6.2.11) though (6.2.13), and rearrange them, we have the following simultaneous linear equations:

Sec. 6.2 The Concept of the Rate Determining Step

$$(1 + K_A a)\hat{a} + K_A a \hat{b} + \frac{\hat{r}}{k_d} - K_A a \hat{c}_0 = 0,$$

$$K\hat{a} - \hat{b} - \frac{\hat{r}}{\hat{k}'} = 0,$$

$$K_B b \hat{a} + (1 + K_B b)\hat{b} - \frac{\hat{r}}{k'_d} - K_B b \hat{c}_0 = 0.$$

To solve any pair of these and to substitute in the remaining equation is to eliminate $\hat{a}$ and $\hat{b}$ between the three equations. This is the same as setting to zero the determinant of the coefficients:

$$\begin{vmatrix} 1 + K_A a & K_A a & \frac{\hat{r}}{k_d} - K_A a \hat{c}_0 \\ K & -1 & -\frac{\hat{r}}{\hat{k}'} \\ K_B b & 1 + K_B b & -\frac{\hat{r}}{k'_d} - K_B b \hat{c}_0 \end{vmatrix} \quad (6.2.17)$$

The determinant can easily be evaluated to give

$$\hat{r}^* = \frac{\hat{c}_0 (KK_A a - K_B b)}{\left(\frac{1}{\hat{k}'} + \frac{K}{k_d} + \frac{1}{k'_d}\right) + \left(\frac{1}{\hat{k}'} + \frac{1+K}{k'_d}\right) K_A a + \left(\frac{1}{\hat{k}'} + \frac{1+K}{k_d}\right) K_B b}.$$

(6.2.18)

All the special cases considered so far may be obtained by letting any two of k_d, $\hat{k}'$, and k'_d tend to infinity.

As a further illustration of these ideas, we may consider one of the suggestions which have been put forward as the mechanism for the hydrogenation of ethylene. Some experimental results show that the reaction is first order with respect to the partial pressure of hydrogen but independent of the partial pressure of ethylene. The overall reaction is $C_2H_4 + H_2 \longrightarrow C_2H_6$, which, for simplicity, we shall write as $A + H \longrightarrow B$, letting p_A and p_H denote the partial pressures of ethylene and hydrogen and $\hat{a}$ and $\hat{h}$ their adsorbed concentrations. Then, if the adsorption of ethylene and hydrogen is at equilibrium,

$$K_A p_A (\hat{c}_0 - \hat{a} - \hat{h}) - \hat{a} = 0$$
$$K_H p_H (\hat{c}_0 - \hat{a} - \hat{h}) - \hat{h} = 0,$$

where K_A and K_H are the ratios of the adsorption to desorption rate constants. The assumption is now made that ethylene is much more strongly adsorbed than hydrogen and that the reaction is between gaseous ethylene and adsorbed hydrogen. Now from the adsorption equilibrium

$$\hat{h} = \frac{\hat{c}_0 K_H p_H}{1 + K_A p_A + K_H p_H},$$

and from the assumptions about the reaction, its rate can be written as

$$k\hat{h}p_A = k\hat{c}_0 \frac{K_H p_H p_A}{1 + K_A p_A + K_H p_H}.$$

But the assumption that ethylene is much more strongly adsorbed than hydrogen implies that the term $K_A p_A$ dominates the denominator. If then we neglect $1 + K_H p_H$ in comparison to $K_A p_A$, the rate becomes

$$\hat{c}_0 \frac{kK_H}{K_A} p_H,$$

which is first order with respect to hydrogen and zeroth order with respect to ethylene. Of course, this is far from being a complete discussion of the ethylene hydrogenation reaction, but it serves to show the way in which these ideas are used and the kind of assumptions that are made in trying to get a rate expression which accords with experiment.

So far in this section we have considered only the adsorption, desorption, and surface reaction. Let us next set these in the context of the transfer and diffusion steps for the same reaction. As has been mentioned above, the catalyst will often be held on porous particles so that the reactants and products have to diffuse to and from the surface of the particle and also within it. For external transfer from the flowing reaction mixture to the exterior of the particle we shall reserve the term *mass transfer* (more particularly, external mass transfer), and for the diffusion within the particle the term *diffusion* (or internal diffusion). Both are, of course, concurrently examples of mass transfer and of diffusion so that the choice of language is an arbitrary one; but it draws a useful distinction.

There are seven steps that have to occur before a molecule of A reacts to form B and returns to the bulk phase as such (see Fig. 6.2).

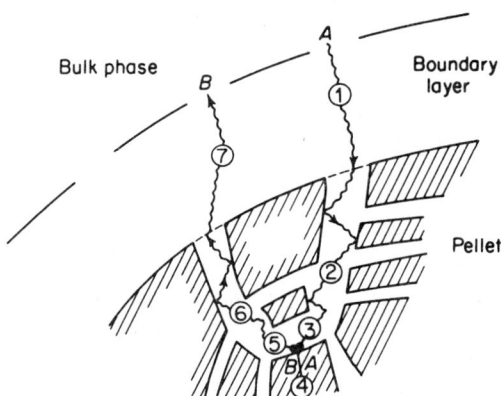

Fig. 6.2 Physical and chemical steps in heterogeneous catalysis.

Sec. 6.2   The Concept of the Rate Determining Step   123

1. The molecule of A must be transferred through a boundary layer from the flowing stream to the exterior surface of the particle. If the concentration of A in the bulk phase is denoted by a, it is reasonable to use a_s for that at the surface of the particle. The rate of transfer will then be of the form $k_{cA}(a - a_s)$ in moles per unit time per unit external area; k_{cA} is a mass transfer coefficient to be discussed in Sec. 6.3.

2. Diffusion of A within the porous pellet takes place. If the pores are very large this may be the normal type of molecular diffusion, but if the pore radius is smaller than the mean free path, a molecule will hit the pore wall more often than it hits its fellows, and this is the Knudsen regime of diffusion. Both types of diffusion can be described by Fick's law in which the flux is proportional to the concentration gradient, and if the diffusion coefficient is not in some sense large there may be large variations in the concentration of A within the pellet. Let $\mathbf{r}$ denote position within the catalyst particle; then the concentration of A within the particle is $a(\mathbf{r})$, a function of that position, and obeys the partial differential equation for diffusion with $a(\mathbf{r}) = a_s$ when $\mathbf{r}$ is a position on the exterior surface of the particle. Clearly, this is a complicated matter and we shall seek ways of simplifying it in Sec. 6.4.

3. Although there may have been some adsorption and desorption on the pore walls during the molecule's passage into the interior of the pellet, we do not consider this the third stage until adsorption occurs at a reactive catalytic site. This, and the next two, are the steps we have been considering above, so that within the pellet the concentrations of A and B, which were denoted by a and b above, will be $a(\mathbf{r})$ and $b(\mathbf{r})$. All that we have done above has local validity and only if the internal diffusion is very rapid, so that $a(\mathbf{r})$ and $b(\mathbf{r})$ are effectively uniform, will the reaction rate expression be valid for the pellet as a whole. The rate of adsorption is given by $\hat{r}_{ad}$, Eq. (6.2.11).

4. Following the adsorption of A, reaction takes place on the surface. We have already considered in detail the reversible first order reaction. The kinetics might be quite different (as illustrated in Exercise 6.2.4) and similar expressions might be derived, but the algebraic labor would be much greater. If the adsorption and desorption are rapid in comparison with reaction we can always substitute the equilibrium adsorbed concentrations in any rate law. Thus if the reaction $A \rightleftharpoons B$ were really second order in both directions we would have

$$\hat{r} = \hat{k}\hat{a}^2 - \hat{k}'\hat{b}^2,$$

and if the reaction were the rate determining step we would have

$$\hat{r} = \hat{c}_0^2 \frac{\hat{k}K_A^2 a^2 - \hat{k}'K_B^2 b^2}{(1 + K_A a + K_B b)^2}.$$

The remaining steps are symmetrical about the reaction step; we give them a positive sense that goes with the overall process $A \longrightarrow B$. They are:

5. Desorption of the product from the catalyst into the porous interior of the pellet [rate $\hat{r}_{de}$ given by Eq. (6.2.13)];
6. Diffusion of product within the pellet to the external surface;
7. Transfer of product from the external surface back to the main stream of the reaction mixture at a rate $k_{cB}(b_s - b)$.

Internal diffusion (steps 2 and 6) is the hardest feature to deal with precisely, because it is distributed throughout the volume of the pellet. It is also the most interesting from a theoretical point of view. If it is not rate determining, then we can take the concentrations of A and B to be uniform throughout the pellet; they must therefore be the same as the concentrations at the exterior surface, namely a_s and b_s.

If the mass transfer to the surface is limiting, we have to consider the rate of disappearance of A and of appearance of B in a unit volume of bed. If ϵ is the voidage of the bed, V_p the volume of a pellet, and S_x its external surface area, then the total external surface area per unit volume of the bed is $(1 - \epsilon)(S_x/V_p)$. If the rate of mass transfer is $k_{cA}(a - a_s)$ per unit area, then the rate of disappearance of A is

$$r_A = (1 - \epsilon) \frac{S_x}{V_p} k_{cA}(a - a_s) \qquad (6.2.19)$$

moles per unit reactor volume per unit time. Similarly, the rate of transfer of B from pellet to the stream is

$$r_B = (1 - \epsilon) \frac{S_x}{V_p} k_{cB}(b_s - b). \qquad (6.2.20)$$

Now if the internal diffusion is rapid and the adsorption-desorption and reaction are all very fast, the concentrations throughout the pellet are a_s and b_s, and from the equilibrium of Eq. (6.2.18)

$$b_s = a_s \frac{KK_A}{K_B}. \qquad (6.2.21)$$

Since there is no accumulation of A or B within the particle, the rates r_A and r_B are equal to one another and to the pseudo-homogeneous rate of reaction, r. Thus

$$a - a_s = \frac{k_{cB}}{k_{cA}} \left(\frac{KK_A}{K_B} a_s - b \right)$$

or

$$a_s = \frac{a + K_c b}{1 + (KK_A K_c/K_B)},$$

where $K_c = k_{cB}/k_{cA}$. Then, if $K_e = KK_A/K_B$,

$$r = (1 - \epsilon) \frac{S_x}{V_p} (K_e a - b) \left(\frac{1}{k_{cB}} + \frac{K_e}{k_{cA}} \right)^{-1}. \qquad (6.2.22)$$

This might be written as

$$r = (1 - \epsilon)\frac{S_x}{V_p} k(a - a_e),$$

where

$$\frac{1}{k} = \frac{1}{k_{cA}} + \frac{1}{K_e k_{cB}} \quad (6.2.23)$$

and $a_e = b/K_e$ is the equilibrium concentration of a for a given b. Again we see the rate expressed as the product of a rate constant and a driving force, the distance of a from equilibrium. Moreover it is evident that the reciprocal of such a rate constant is, like a resistance, the sum of the two transfer resistances, namely the reciprocals of k_{cA} and $K_e k_{cB}$. When the equations are linear we shall find this pattern of added resistances emerging constantly. In more complicated, nonlinear situations it may not apply, but can often be used to give a feel for the situation. We shall return later to the problem of deciding which step is the rate determining one.

Figure 6.3 illustrates several concentrations that are important for the

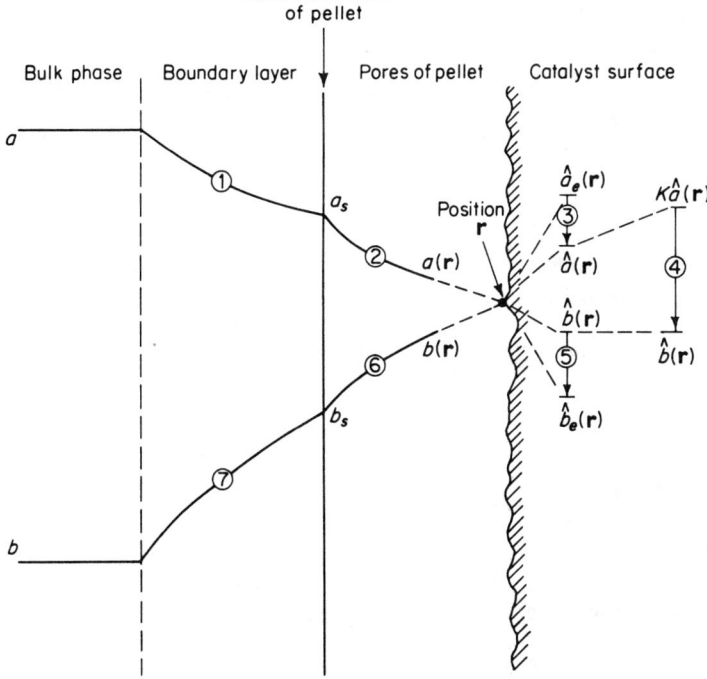

Fig. 6.3 Diagram to illustrate the relationship of the various concentrations in the heterogeneous reaction $A \rightarrow B$.

different steps. The driving force in each case is a concentration difference either between two concentrations as in steps 1 and 7, or between an equilibrium concentration and the concentration prevailing at a particular point, as in steps 3, 4, and 5. In steps 2 and 6 we have a distributed process, the driving force being a gradient rather than a difference.

All that we have done in this section carries over to the more general case of a reaction $\sum \alpha_j A_j = 0$ taking place between S different species. The situation is of course complicated by multicomponent diffusion and perhaps by a change in the total number of moles. It lies rather beyond an elementary treatment.

Exercise 6.2.1. Show that the rates of adsorption and desorption given by Eqs. (6.2.11) and (6.2.13) can be written as

$$\hat{r}_{ad} = k_d[(1 + K_A a)(\hat{a}_e - \hat{a}) + (K_A a)(\hat{b}_e - \hat{b})]$$

$$\hat{r}_{de} = -k_d'[(K_B b)(\hat{a}_e - \hat{a}) + (1 + K_B b)(\hat{b}_e - \hat{b})],$$

where $\hat{a}_e$ and $\hat{b}_e$ are the adsorbed concentrations of A and B in equilibrium with a and b, as given by Eqs. (6.2.15).

Exercise 6.2.2. If $\lambda = k_d'/k_d$, $\mu = 1/k_d$, $x = \hat{a}_e - \hat{a}$, $y = \hat{b}_e - \hat{b}$, $\tau = k_d t$, and $\hat{r}_e$ denote the right-hand side of Eq. (6.2.16), show that Eqs. (6.2.8) and (6.2.9) can be written as

$$\frac{dx}{d\tau} = -(1 + K_A a)x - (K_A a)y + \mu(\hat{r}_e - \hat{k}x + \hat{k}'y)$$

$$\frac{dy}{d\tau} = -\lambda(K_B b)x - \lambda(1 + K_B b)y - \mu(\hat{r}_e - \hat{k}x + \hat{k}'y).$$

Hence, or otherwise, show that as $\mu \longrightarrow 0$, x and y decay exponentially to zero with time, the one as $\exp - k_d t$ and the other as $\exp - k_d' t$.

Exercise 6.2.3. If k_d' and $\hat{k}'$ both tend to infinity, the rate determining step is the relatively slow adsorption of A. Show that then

$$\hat{r}^* = \hat{c}_0 \frac{k_a K a - k_d K_B b}{K + (1 + K)K_B b}.$$

Exercise 6.2.4. The hydrogenation of A to $B + C$ takes place by the irreversible reaction of gaseous hydrogen with adsorbed A. The products B and C are desorbed instantly and the rate of reaction on the surface is $\hat{k} p \hat{a}$, where p is the partial pressure of hydrogen. Find a rate expression in terms of a, p, $\hat{c}_0$, $\hat{k}$, K_A, and $K_H = \hat{k}/k_d$. Study the limiting cases $k_d \longrightarrow \infty$, K_A finite, and $\hat{k} \longrightarrow \infty$.

Exercise 6.2.5. Confirm the limiting cases of Eq. (6.2.18) and discuss their reasonableness. How is the equilibrium affected by k_d, $\hat{k}'$, and k_d'?

6.3 External Mass Transfer

Having seen the influence of external mass transfer on the rate of reaction we should now ask what is known about a mass transfer coefficient such as k_c. This coefficient is an attempt to wrap up in a single coefficient an extremely complicated situation in which mass transfer is taking place through a boundary layer around each particle at a rate varying with position. A correlation of experimental data based on so gross a simplification is bound to be approximate in the face of the vagaries of flow and the randomness of geometry in a packed bed. Nevertheless, it can provide an estimate of the mass transfer rate in the following fashion. Figure 6.3 shows a schematic view of the several concentrations involved and the external transfer of A is at a rate $k_{cA}(a - a_s)$ in moles per unit external area of the pellet per unit time. A mass transfer coefficient k_c (we shall drop the suffix A or B) is usually correlated in dimensionless form as a Sherwood number

$$\text{Sh} = \frac{k_c d_p}{D}, \quad (6.3.1)$$

where D is the molecular diffusion coefficient of the species and d_p the effective particle diameter. In this case, the effective particle is that of the sphere having the same area

$$d_p = \left(\frac{S_x}{\pi}\right)^{1/2}. \quad (6.3.2)$$

The Sherwood number is blended with the Reynolds and Schmidt numbers to form a j_D factor,

$$j_D = \frac{\text{Sh}}{(\text{Sc})^{1/3} \text{Re}}, \quad (6.3.3)$$

where

$$\text{Re} = \frac{d_p G}{\mu}, \quad \text{Sc} = \frac{\mu}{\rho D}, \quad (6.3.4)$$

and G = mass flow rate of reactants (mass per unit time per unit area),
μ = viscosity of bulk phase,
ρ = density of bulk phase.

Thus

$$k_c = \frac{j_D G}{\rho (\text{Sc})^{2/3}}. \quad (6.3.5)$$

One of the correlations between the j_D factor and the Reynolds number is based on the work of Thodos et al. and can be expressed as

$$j_D = \frac{0.725}{(\text{Re})^{0.41} - 1.5}. \quad (6.3.6)$$

It is illustrated in Fig. 6.4.

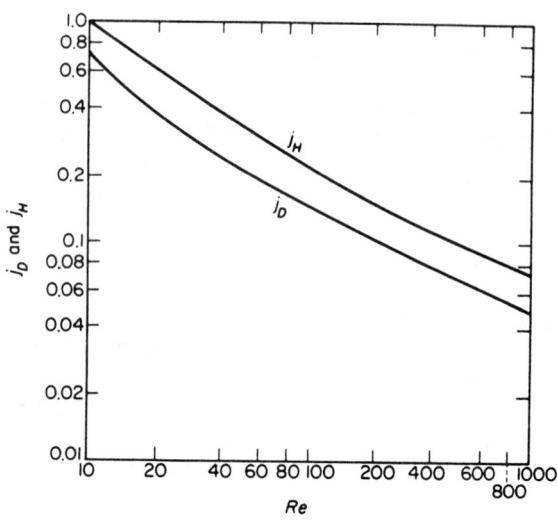

Fig. 6.4 External mass and heat transfer factors.

For ideal gases, the concentration and partial pressure are related and the mass transfer formula is often written

$$k_g(p - p_s).$$

The j_D factor for which the same curve gives a reasonable correlation is now

$$j_D = \frac{k_g P \bar{m} (\text{Sc})^{2/3}}{G}, \tag{6.3.7}$$

where P = total pressure
$\bar{m}$ = mean molecular weight.

Thus the rate of transfer per unit reactor volume can be written either as

$$(1 - \epsilon) \frac{S_x}{V_p} \frac{G}{\rho} \frac{0.725(\text{Sc})^{-2/3}}{(\text{Re})^{0.41} - 1.5} (c - c_s) \tag{6.3.8}$$

or as

$$(1 - \epsilon) \frac{S_x}{V_p} \frac{G}{P\bar{m}} \frac{0.725(\text{Sc})^{-2/3}}{(\text{Re})^{0.41} - 1.5} (p - p_s), \tag{6.3.9}$$

where c and c_s are the concentrations in moles per unit volume in the bulk phase and at the pellet surface and p and p_s are the partial pressures at those locations.

6.4 Diffusion Within the Catalyst Pellet

The most difficult step to include within the scheme of physical and chemical steps is that of diffusion within the catalyst pellet. To show what kind of

Sec. 6.4 Diffusion Within the Catalyst Pellet

effect this is going to be we shall first consider the very special case of diffusion and reaction in a single pore. After looking at the physical structure of the porous catalyst and seeing how to estimate the diffusion coefficient, we shall return to a more general problem of diffusion and reaction and take note of the effects of pellet shape and reaction kinetics.

Figure 6.5 shows a diagram of a single pore of length ℓ and diameter d. Its curved wall, but not its flat end, is capable of catalyzing the reaction $A \longrightarrow B$ so that the rate of disappearance of A is $\hat{k} \times$ (the local concentration of A) in moles per unit time per unit catalyst area. We let the blank end be $x = 0$, the

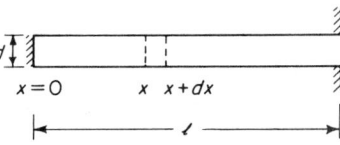

Fig. 6.5 The single pore.

pore mouth $x = \ell$ and, imagining this to open on the external surface of the pellet, we denote by a_s the concentration of A at $x = \ell$. If $a(x)$ is the concentration of A at a general position x, the rate of disappearance of A in the section between x and $x + dx$ is $\pi d \hat{k}\, a(x)\, dx$, since the wall area in this section is $\pi d\, dx$. This must be balanced by the net rate of diffusion into the section. Now the cross-sectional area of the pore is $\pi d^2/4$ so that, if D is the diffusion coefficient of A in a mixture of A, B, and any inert gases that may be present, then the flux from left to right at x is $-(\pi d^2/4)D\,(da/dx)$. A similar expression holds at $x + dx$ but here the derivative must be evaluated at $x + dx$. Hence balancing the net influx against the rate of reaction gives

$$\frac{\pi d^2}{4} D \left[\left(\frac{da}{dx}\right)_{x+dx} - \left(\frac{da}{dx}\right)_x\right] = \pi d \hat{k}\, a(x)\, dx.$$

Dividing through by dx and letting $dx \longrightarrow 0$ gives

$$\frac{\pi d^2}{4} D \frac{d^2 a}{dx^2} = \pi d \hat{k} a$$

or

$$\frac{d^2 a}{dx^2} = \frac{4\hat{k}}{dD} a = \lambda^2 a. \tag{6.4.1}$$

Since there is no flux at the unreactive end $x = 0$, we have the boundary condition

$$\frac{da}{dx} = 0 \quad \text{at} \quad x = 0. \tag{6.4.2}$$

At the mouth the concentration is given so

$$a = a_s \quad \text{at} \quad x = \ell. \tag{6.4.3}$$

Equation (6.4.1) should be one of the better known equations, but if it is not, recall that it is linear, of the second order, and amenable to the solution outlined in Sec. 5.1. The solution is

$$a(x) = L \cosh \lambda x + M \sinh \lambda x, \tag{6.4.4}$$

and by the boundary condition of Eq. (6.4.2), $M = 0$, while to satisfy Eq. (6.4.3) L must be $a_s/\cosh \lambda \ell$. Hence,

$$a(x) = a_s \frac{\cosh \lambda x}{\cosh \lambda \ell} = a_s \frac{\cosh h(x/\ell)}{\cosh h}, \qquad (6.4.5)$$

where

$$h = \lambda \ell = \ell \sqrt{\frac{4\hat{k}}{dD}}. \qquad (6.4.6)$$

Now, if the rate of diffusion were very large compared with the rate of reaction the concentration of A would be a_s throughout the whole pore and the total rate of reaction would be $\pi d\ell \hat{k} a_s$. With the concentration of a varying along the length of the pore, we have to calculate the total rate of reaction with finite D by evaluating the integral

$$\int_0^\ell \pi d\hat{k} a(x)\, dx = \pi d\ell \hat{k} a_s \frac{\tanh \lambda \ell}{\lambda}. \qquad (6.4.7)$$

A useful measure of the effectiveness of the pore will be the ratio of this rate of reaction to the rate, $\pi d\ell \hat{k} a_s$, when diffusion is no limitation; the ratio is

$$\eta = \frac{\tanh \lambda \ell}{\lambda \ell} = \frac{\tanh h}{h}. \qquad (6.4.8)$$

This is a very interesting and reasonable result. It is interesting because it tells us that the complicated interaction of diffusion and reaction in this model can be expressed, at least so far as the first order reaction $A \longrightarrow B$ is concerned, in a single effectiveness factor by which the reaction rate, when diffusion is not limiting, may be multiplied. Since the latter is easily calculated, it is a very efficient way of summing up the solution of a differential equation. It is also interesting that this effectiveness factor depends only on one parameter, h, though it is partly an artifact of the special simplicity of this first case. This parameter is often known as the *Thiele modulus*, in recognition of the importance of Thiele's work published in 1939: Damköhler had discussed diffusion and reaction in 1936 and the Thiele modulus is related to one of several Damköhler numbers. It clearly is a measure of the rate of reaction relative to the rate of diffusion, for its square could be written

$$h^2 = \ell^2 \frac{4\hat{k}}{dD} = \frac{\pi d\ell \hat{k} a_s}{(\pi d^2/4) D(a_s/\ell)}. \qquad (6.4.9)$$

The numerator is then the rate of disappearance of A when the concentration in the whole pore is a_s, while the denominator is the rate of disappearance of A into the pore when the gradient at the mouth is (a_s/ℓ).

The result is also reasonable, for $\eta = 1$ when $h = 0$ and decreases as h becomes larger. Since h becomes large when the rate of diffusion is small compared with that of reaction, this shows clearly just how diffusion becomes limiting. When h is greater than 5, $\tanh h$ is greater than 0.9999 so that for all practical purposes $\eta = 1/h$, and the actual rate of reaction is the geometric mean of the rates of reaction and diffusion which are the numerator and denominator in Eq. (6.4.9). Although this is a particularly simple case, we shall show that many of its features are retained in much more complicated situations. To do this we must look in more detail at the internal structure of the catalyst pellet.

The system of pores in a catalyst pellet consists of a tremendous number of very small and tortuous tubes of irregular cross section, some terminating in dead ends and others intersecting each other. Naturally, a structure as complex as this can only be described in statistical terms and then it must be idealized in some model for computational purposes. The first overall parameter is the internal voidage $\hat{e}$, the ratio of the internal pore volume to the total volume of the pellet, or the ratio of the bulk density of the pellet to the density of the solid material of which it is composed. In a random structure $\hat{e}$ is also the ratio of the free area in any cross-sectional plane to the area occupied by the solid. The value of $\hat{e}$ can vary between about 0.3 for Vycor glass to as much as 0.8 for alumina pellets. The second overall parameter is the total internal surface area usually specified as S_g in m²/gm. Some typical silica alumina catalysts have more than 300 m²/gm, whereas an ammonia synthesis catalyst may have 10–15 m²/gm. If all the pores were straight, nonintersecting, and of the same diameter, the ratio of the surface area to the pore volume would be the same as that of a cylinder—namely $4/d$. It is therefore not unreasonable to estimate the mean pore diameter by equating this to the known ratio of area to free volume, $\rho_b S_g / \hat{e}$. This gives an effective mean diameter of

$$d_e = \frac{4\hat{e}}{\rho_b S_g}. \qquad (6.4.10)$$

Diameters calculated in this way have been reported for a number of catalysts and vary from 20 A (10^{-8} cm = 1 A) to several microns (10^{-4} cm = 1 micron).

These two overall parameters can be measured fairly easily but there are more sophisticated techniques which measure pore size distribution. Pellets are often made by taking a powder of smaller particles and compressing it into pellets of a useful size. If these smaller particles are themselves porous we might expect a system of macropores between the particles and micropores within them with a vast difference in pore size. In fact, Rothfeld and Watson found that in an alumina they had studied, 65% of the internal free space and 99% of the internal surface area could be attributed to micropores some 0.012 microns in diameter leading out to a group of macropores with a mean diameter of 1.25 microns. It is not very meaningful to talk about an effective mean diameter with a factor of 100 separating the two sizes. Nevertheless, it is sometimes necessary to do this in the absence of any more precise information.

An admittedly crude but useful model of the porous catalyst would picture it as a system of tortuous cylinders of varying diameter whose intersections could be neglected. If D is the diffusion coefficient of the reacting species in the free space of such a pore we could obtain an effective diffusion coefficient D_s within the porous solid as follows. If we take two planes of unit area separated by a distance dx and have concentrations c_1 and c_2 we would like to express the flux as the product of the diffusion coefficient, D_s, and the concentration gradient, $(c_1 - c_2)/dx$. But if the diffusion is through the pores, only a fraction $\hat{e}$ of the unit area is available for diffusion. Moreover the diffusion is along tortuous passages of an average length $\tau\, dx$ (say) and the gradient over such tubes is only $(c_1 - c_2)/\tau\, dx$. Thus the flux is

$$\hat{e} D \frac{c_1 - c_2}{\tau\, dx} = \frac{\hat{e} D}{\tau} \frac{c_1 - c_2}{dx},$$

so that

$$D_s = \frac{\hat{e} D}{\tau}. \tag{6.4.11}$$

It could be argued that the cross-sectional area of the pore is less than that exposed by the slice (since this will be oblique) by a factor of $1/\tau$ and hence that $D_s = \epsilon D/\tau^2$. Since, however, τ is so ambiguous a quantity, it does not matter much whether it comes in as the first or second power. The principal implication of Eq. (6.4.11) is that the effective diffusion coefficient D_s is the product of the molecular diffusion coefficient and a factor depending solely on the geometry of the porous structure.

The tortuosity factor τ is thus a plausible fudge factor and could be imagined to account also for the effect of the constrictions in pore diameter. Values of τ ranging from 0.3 to 6 have been reported so that the ratio $\hat{e}/\tau$ may vary widely from 0.05 to 1 or more. However values of τ less than 1 do not make good sense physically and arise from using too simplified a model to account for diffusion in much more complex structures. More complicated models are beyond our scope here, but references are given at the end of the chapter.

We must now ask what value should be given to the diffusion coefficient D. If the diffusing substances are gases and the pores are relatively large then we may take D to be the ordinary bulk diffusion coefficient. It may be estimated by methods given in the references at the end of the chapter (see particularly the articles of Bird and the monograph of Sherwood and Satterfield) and we will not go into this in detail here. If D_{ij} is the binary diffusion coefficient of A_i in A_j, the diffusion coefficient D_j of A_j in a mixture in which x_i is the mole fraction of A_i, where $i = 1, 2, \ldots, S$, may be estimated by

$$\frac{1-x_j}{D_j} = \sum_{\substack{i=1 \\ i \neq j}}^{S} \frac{x_i}{D_{ij}}. \qquad (6.4.12)$$

The diffusion coefficient in a gas is proportional to $T^{3/2}/P$, the constant of proportionality being a rather slowly increasing function of temperature. The estimation of the diffusion coefficient in liquids is discussed briefly by Sherwood and Satterfield; it is proportional to T/μ, where μ is the viscosity. At atmospheric pressure and ordinary temperatures, the order of magnitude of D for a gas is 0.1–1 cm²/sec and for liquids it is smaller by a factor of about 10^4.

When the pore diameter is small compared with the mean free path, the mode of gaseous diffusion takes on quite another aspect. Instead of colliding with its own type, the molecule will collide much more frequently with the wall. The molecule's progress down its pore will thus depend on pore geometry as well as on the physical characteristics of the diffusing substance. For a straight cylindrical pore of radius d, Knudsen showed that Fick's law could be used with a diffusion coefficient

$$D_K = \frac{2d}{3}\left(\frac{2RT}{\pi m}\right)^{1/2}, \qquad (6.4.13)$$

where m is the molecular weight of the diffusing species. In this so-called Knudsen regime, the different species diffuse quite independently of one another, the diffusion coefficient of each being inversely proportional to the square root of its molecular weight. Though the pores are far from being straight cylinders, we can use the effective mean diameter to obtain, *faute de mieux*, the estimated Knudsen diffusion coefficient

$$D_K = \frac{8\hat{e}}{3\rho_b S_g}\left\{\frac{2RT}{\pi m}\right\}^{1/2}. \qquad (6.4.14)$$

Obviously, if one of the molecular or Knudsen diffusion coefficients is vastly greater than the other it may be ignored. Several formulae have been given for the transition between the two, but perhaps the simplest of them derives from thinking of the reciprocal of the diffusion coefficient as a resistance and the two modes of diffusion as being in parallel. Then, allowing for the area and tortuosity, the effective diffusion coefficient in the porous solid can be taken to be

$$D_s = \frac{\hat{e}}{\tau[(1/D) + (1/D_K)]}. \qquad (6.4.15)$$

This allows us to make a homogeneous model of the porous catalyst pellet in which we now have a diffusive flux given as the product of an effective diffusion coefficient and a concentration gradient, and a rate of reaction given by the product of the catalytic area and the reaction rate per unit area. Let us consider again the irreversible first order reaction $A \longrightarrow B$ taking place in a slab of thickness 2ℓ with sealed edges. We can take the external faces of

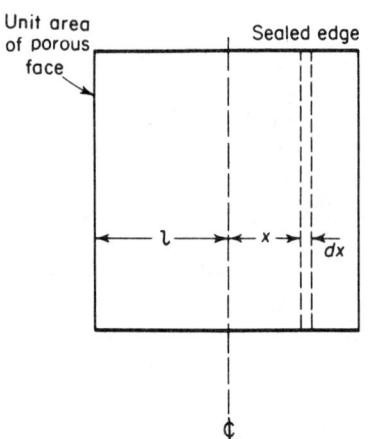

Fig. 6.6 The geometry of the rectangular slab.

the slab to be of unit area (as shown in Fig. 6.6) and do a balance on a section of the pellet between planes distant x and $x + dx$ from the center plane.

If $a(x)$ is the concentration of A at the plane x, we have

$$\begin{pmatrix}\text{Rate of}\\ \text{diffusion in}\end{pmatrix} - \begin{pmatrix}\text{Rate of}\\ \text{diffusion out}\end{pmatrix}$$
$$= \begin{pmatrix}\text{Rate of}\\ \text{reaction}\end{pmatrix},$$
$$-D_s\left(\frac{da}{dx}\right)_x + D_s\left(\frac{da}{dx}\right)_{x+dx}$$
$$= dx\, \rho_b S_g \hat{k}\, a(x),$$

or

$$\frac{d^2 a}{dx^2} = \lambda^2 a,$$

where

$$\lambda^2 = \frac{\rho_b S_g \hat{k}}{D_s}. \tag{6.4.16}$$

This equation, with the appropriate modification of λ, is exactly the same as that for a straight pore, Eq. (6.4.1). Moreover, the boundary conditions are the same, for at the surface $x = \ell$, $a = a_s$, and at the central plane $da/dx = 0$ by symmetry. It follows that the effectiveness factor for the slab, defined in the same way as the ratio of the total rate of reaction under diffusion limitation to the rate when diffusion is infinitely rapid, is again

$$\eta = \frac{\tanh h}{h}$$

where (6.4.17)

$$h = \ell \left(\frac{\rho_b S_g \hat{k}}{D_s}\right)^{1/2}$$

Notice that if we use Eq. (6.4.11) for D_s and Eq. (6.4.10) for $\rho_b S_g$ we recover the same Thiele modulus Eq. (6.4.9) for the straight tube. The geometry of the situation for a slab is identical with that for a single pore, and both give the same expression for the effectiveness factor. The assumption that the edges are sealed forces the problem into a one-dimensional form by making it possible to take the concentration constant over any plane parallel to the face. The graph of η versus h is shown as the left-hand curve in Fig. 6.7.

There are two other forms of pellet that allow a simple solution in one-dimensional terms. The first is the cylinder with sealed ends which has axial symmetry. Taking ℓ to be its radius, the effectiveness factor is given by the ratio of two modified Bessel functions (see Appendix)

$$\eta = \frac{2}{h}\frac{I_1(h)}{I_0(h)}. \tag{6.4.18}$$

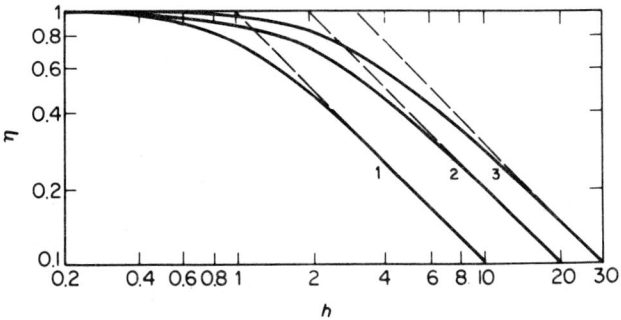

Fig. 6.7 Effectiveness factors for (1) slab, (2) cylinder, and (3) sphere.

This is shown as the middle curve in Fig. 6.7. It has the reasonable property that $\eta \longrightarrow 1$ as $h \longrightarrow 0$ and the interesting property that $\eta \sim 2/h$ as $h \longrightarrow \infty$. The sphere is the third case where symmetry about the center allows us to reduce the problem to one dimension (see Exercise 6.4.5) and to solve it easily. The right-hand curve in Fig. 6.7 shows the effectiveness factor for the sphere

$$\eta = \frac{3}{h} \frac{h \coth h - 1}{h}, \qquad (6.4.19)$$

where ℓ is taken to be the radius in calculating the modulus h. Again $\eta \longrightarrow 1$ as $h \longrightarrow 0$ and this time $\eta \sim 3/h$ as $h \longrightarrow \infty$.

Now the numbers 1, 2, and 3 associated respectively with the asymptotes for the slab, cylinder, and sphere are very suggestive, as the following table shows:

Shape	ℓ	V_p	S_x	V_p/S_x
slab of unit facial area	half thickness	2ℓ	2	$\ell/1$
cylinder of unit length	radius	$\pi\ell^2$	$2\pi\ell$	$\ell/2$
sphere	radius	$4\pi\ell^3/3$	$4\pi\ell^2$	$\ell/3$

It follows that if we were to take

$$h' = \frac{\lambda V_p}{S_x} = \frac{V_p}{S_x} \left(\frac{\rho_b S_g \hat{k}}{D_s} \right)^{1/2}, \qquad (6.4.20)$$

we would have in the three cases

$$\eta = \frac{\tanh h'}{h'} \quad \text{(slab)},$$

$$\eta = \frac{1}{h'} \frac{I_1(2h')}{I_0(2h')} \quad \text{(cylinder)}, \qquad (6.4.21)$$

$$\eta = \frac{1}{h'} \frac{3h' \coth 3h' - 1}{3h'} \quad \text{(sphere)},$$

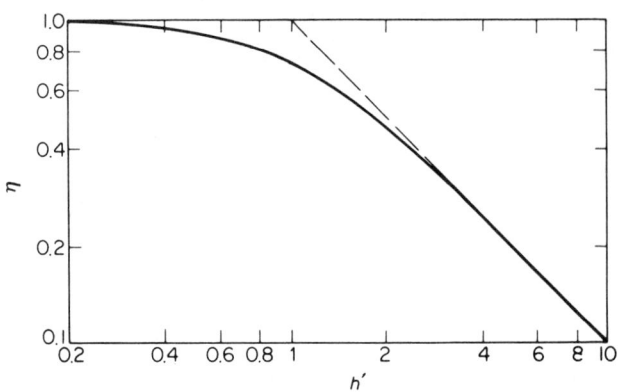

Fig. 6.8 Mean effectiveness factor for all shapes.

and in each case $\eta \sim 1/h'$ as $h' \longrightarrow \infty$. Indeed when the three curves of Eq. (6.4.21) are plotted on the same graph they lie within a few percent of one another and the mean curve shown in Fig. 6.8 is tolerably accurate.

So far we have only considered the irreversible first order reaction $A \longrightarrow B$. But we have seen that when adsorption effects are considered even a first order reaction can have more complicated kinetics. For any form of kinetics the equation for diffusion and reaction can be solved for the slab geometry. We shall give the general method and illustrate it by the already familiar case of the first order reaction.

For the general reaction we shall write $c(x)$ for the concentration of a reactant that disappears at a rate $\hat{r}(c)$ per unit volume of catalyst particle. By using $a(x)$ to denote the concentration of A in the first order, irreversible case, the example we are using to illustrate the method may be worked simultaneously without confusion; for example $\hat{k}a$ plays the role of $\hat{r}(c)$. A balance on the concentration over a differential element between x and $x + dx$ now gives

$$D_s \frac{dc^2}{dx^2} = \rho_b S_g \hat{r}(c), \qquad (6.4.22)$$

with the boundary conditions

$$c = c_s \quad \text{at} \quad x = \ell$$

$$\frac{dc}{dx} = 0 \quad \text{at} \quad x = 0.$$

Let us make the problem dimensionless by putting

$$\xi = \frac{x}{\ell}, \qquad \gamma = \frac{c}{c_s}, \qquad \hat{R}(\gamma) = \frac{\hat{r}(c_s\gamma)}{\hat{r}(c_s)}, \qquad h^2 = \frac{\ell^2 \rho_b S_g \hat{r}(c_s)}{D_s c_s}.$$

Then

$$\frac{d^2\gamma}{d\xi^2} = h^2 \hat{R}(\gamma) \qquad (6.4.23)$$

and

$$\gamma = 1 \quad \text{at} \quad \xi = 1 \qquad (6.4.24)$$

$$\frac{d\gamma}{d\xi} = 0 \quad \text{at} \quad \xi = 0.$$

For the first order case with $\alpha = a/a_s$ we have $\hat{R}(\alpha) = \alpha$. Multiply both sides of Eq. (6.4.23) by $2(d\gamma/d\xi)$ and integrate between 0 and ξ, by using γ_0 to denote the value of γ at $\xi = 0$; then

$$\int_0^\xi 2 \frac{d\gamma}{d\xi} \frac{d^2\gamma}{d\xi^2} d\xi = \left(\frac{d\gamma}{d\xi}\right)^2 - 0 = 2h^2 \int_0^\xi \hat{R}(\gamma) \frac{d\gamma}{d\xi} d\xi$$
$$= 2h^2 \int_{\gamma_0}^{\gamma(\xi)} \hat{R}(\gamma') d\gamma'. \qquad (6.4.25)$$

In our special, case,

$$\left(\frac{d\alpha}{d\xi}\right)^2 = h^2(\alpha^2(\xi) - \alpha_0^2).$$

The value of the concentration at the center of the slab is not yet known, but we can find an equation for it by a further integration. Let γ_e be the equilibrium value of the concentration so that $\hat{R}(\gamma_e) = 0$, and define the function $P(\gamma)$ by

$$P(\gamma) = 2 \int_{\gamma_e}^{\gamma} \hat{R}(\gamma) d\gamma'. \qquad (6.4.26)$$

In our example, $\alpha_e = 0$ so that $P(\alpha) = \alpha^2$.
Thus,

$$\frac{d\gamma}{d\xi} = h [P(\gamma) - P(\gamma_0)]^{1/2} \qquad (6.4.27)$$

and

$$h = h \int_0^1 d\xi = \int_{\gamma_0}^1 [P(\gamma) - P(\gamma_0)]^{-1/2} d\gamma. \qquad (6.4.28)$$

In our example,

$$h = \int_{\alpha_0}^1 \frac{d\gamma}{(\alpha^2 - \alpha_0^2)^{1/2}} = \cosh^{-1} \frac{1}{\alpha_0}$$

and

$$\alpha_0 = \frac{1}{\cosh h}.$$

The effectiveness factor can best be calculated using γ_0 as a parameter. For when $\gamma_0 = 1$ the concentration is everywhere the same as at the surface and there is no diffusion limitation, i.e., $h = 0$; whereas when $\gamma_0 \longrightarrow \gamma_e$, the reaction is near to equilibrium at the center and corresponds to the case of a large h. The effectiveness factor is the ratio of the rate of disappearance of the reactant, namely

$$D_s \left(\frac{dc}{dx}\right)_{x=\ell} = \frac{D_s c_s}{\ell} \left(\frac{d\gamma}{d\xi}\right)_{\xi=1},$$

to the rate if there were no diffusion limitation, namely $\rho_b S_g \ell \hat{r}(c_s)$. Thus,

$$\eta = \frac{D_s}{\rho_b S_g \hat{r}(c_s) \ell^2} \left(\frac{d\gamma}{d\xi}\right)_{\xi=1} = \frac{1}{h^2} \left(\frac{d\gamma}{d\xi}\right)_{\xi=1} = \frac{[P(1) - P(\gamma_0)]^{1/2}}{h}. \quad (6.4.29)$$

The two Eqs. (6.4.28) and (6.4.29) give a way of calculating both h and η as functions of γ_0, and by letting γ_0 run from 1 to γ_e we can trace out the whole curve from $h = 0$ to $h \longrightarrow \infty$. In our example, we can actually eliminate α_0 and

$$\eta = \frac{1}{h}(1 - \alpha_0^2)^{1/2} = \frac{\tanh h}{h}$$

as before.

Since $P(\gamma_e) = 0$ and $\gamma_0 \longrightarrow \gamma_e$ as h becomes large, Eq. (6.4.29) shows that

$$\eta \sim \frac{[P(1)]^{1/2}}{h} \quad (6.4.30)$$

for a large value of h. Hence by taking $h' = h[P(1)]^{-1/2}$ we could bring together the asymptotes for any kinetics and have $\eta \sim (h')^{-1}$. The same arguments as before (see Exercise 6.4.4) suggest that ℓ should be taken as V_p/S_x for the general shape of a particle.

This section has been a lengthy one and it will be useful to recapitulate what has been done. The simplest case of a single straight pore with a first order irreversible reaction occurring on its curved wall was first examined in detail and we saw how its effectiveness factor could be expressed as a function of the Thiele modulus. The question of how the complex structure of a porous solid might be more simply envisaged was then raised and methods of finding a pseudo-homogeneous diffusion coefficient were discussed. Using this model, it was found that the behavior of flat plates of catalyst was the same as that of a catalyst with straight pores. By suitably defining the Thiele modulus for other pellet shapes and more complex kinetics, the asymptotes of the effectiveness factor for large and small h' could be brought together and all curves be made to lie tolerably close to one another. We thus have in the formula $\eta = \tanh h'/h'$ a valuable estimate of the extent of diffusion limitation for isothermal reaction. It should be remarked, however, that some of the features to be noticed later for a nonisothermal reaction can be observed with more complicated kinetics.

Exercise 6.4.1. Since $\hat{k}$ has the Arrhenius dependence on temperature and D only increases some small power of T we should expect h to increase with temperature. Actual measurements of the reaction rate for a catalyst pellet would lead to an estimated rate constant $k' = \eta \hat{k}$. Show that if $\ln k'$ is plotted against T^{-1} we might expect the slope to vary between $-E/2R$ and $-E/R$, where E is the activation energy for $\hat{k}$.

Exercise 6.4.2. A pore of length ℓ and diameter d is poisoned near its mouth so that the part between $x = p\ell$ and $x = \ell$ ($0 \le p \le 1$) does not allow reaction. Using the notation of the early part of this section, find an expression for the effectiveness of the pore and discuss its reasonableness.

Exercise 6.4.3. Are the figures for the macro-micropore distribution mentioned on page 131 reasonably consistent?

Exercise 6.4.4. Show that when λ is large, the solution given by Eq. (6.4.5) is approximately

$$a(x) = a_s \exp - \lambda(\ell - x).$$

Why would you think it reasonable that for any shape of pellet, the concentration of A should fall off exponentially from the surface? How does this justify the use of V_p/S_x as the dimension ℓ in getting a universal Thiele modulus?

Exercise 6.4.5. By making a balance over a spherical annulus between the radii of x and $x + dx$, show that the first order irreversible reaction $A \longrightarrow B$ yields the differential equation

$$\frac{1}{x^2}\frac{d}{dx}\left(x^2 \frac{da}{dx}\right) = \lambda^2 a.$$

Find the appropriate boundary conditions and, using the transformation $a(x) = A(x)/x$, show that

$$a(x) = \frac{a_s \ell \sinh \lambda x}{x \sinh \lambda \ell}.$$

Hence obtain Eq. (6.4.19) and satisfy yourself that it has the asymptotic behavior claimed for it.

Exercise 6.4.6. If the reaction is an irreversible one of the nth order with $\hat{r} = \hat{k}c^n$, show that

$$h' = \ell \left(\frac{n+1}{2} \frac{\rho_b S_g \hat{k} c_s^{n-1}}{D_s}\right)^{1/2}$$

Exercise 6.4.7. Consider the case of a zeroth order reaction and show that $\eta = 1$ if $h' \le 1$ and $\eta = (h')^{-1}$ if $h' > 1$.

Exercise 6.4.8. For the reversible reaction $A_1 = A_2$ we cannot use the normal extent of reaction unless the diffusion coefficients of A_1 and A_2 are equal. If they are D_{1s} and D_{2s} respectively and $\hat{r} = \hat{k}_1 c_1 - \hat{k}_2 c_2$ show that

$$D_{1s}\frac{d^2c_1}{dx^2} = -D_{2s}\frac{d^2c_2}{dx^2} = \rho_b S_g(\hat{k}_1 c_1 - \hat{k}_2 c_2)$$

for the flat plate. By integrating the left-hand term and by using the boundary conditions, show that $D_{1s}[c_{1s} - c_1(x)] = D_{2s}[c_2(x) - c_{2s}]$ and hence solve the equation for c_1. Deduce that

$$\eta = \frac{\tanh h}{h} \quad \text{where} \quad h = \ell\left[\rho_b S_g\left(\frac{\hat{k}_1}{D_{1s}} + \frac{\hat{k}_2}{D_{2s}}\right)\right]^{1/2}.$$

Exercise 6.4.9. Show how to set up a single equation for the general reaction $\sum \alpha_j A_j = 0$ assuming that the diffusion coefficients of the A_j may all be different and that the species diffuse independently. (*Note:* This implies Knudsen diffusion.)

6.5 The Combination of External Mass Transfer and Internal Diffusion

We have seen in Sec. 6.3 that the rate of transport of A to the outside surface of a pellet can be expressed as $k_c(a - a_s)$. If A diffuses into the pellet at the same rate, the appropriate boundary condition for the diffusion and reaction equation will be

$$k_c(a - a_s) = D_s\left(\frac{da}{dx}\right)_{x=\ell}. \tag{6.5.1}$$

Again, using the flat plate geometry we have the same Eq. (6.4.16) with $da/dx = 0$ at $x = 0$ and the boundary condition Eq. (6.5.1) at $x = \ell$. Once more we have the solution Eq. (6.4.4) and $M = 0$ by the condition at the center. Thus

$$a(x) = L \cosh \lambda x,$$
$$a_s = L \cosh \lambda \ell,$$

and

$$a = L\left(\cosh \lambda \ell + \frac{D_s \lambda}{k_c}\sinh \lambda \ell\right)$$

or

$$a(x) = a\frac{\cosh \lambda x}{\cosh \lambda \ell + (D_s \lambda / k_c)\sinh \lambda \ell}. \tag{6.5.2}$$

It must be remembered that a with no suffix denotes the concentration of A in the free stream. If there were no resistance to mass transfer of any kind we would have $a(x) = a_s = a$ and the total rate of reaction $\rho_b S_g \hat{k} \ell a$. Hence the effectiveness factor

$$\eta = \frac{1}{\lambda^2 \ell a}\left(\frac{da}{dx}\right)_{x=\ell} = \frac{\tanh \lambda \ell / \lambda \ell}{1 + (D_s \lambda / k_c)\tanh \lambda \ell}. \tag{6.5.3}$$

It is best to write this in the form

$$\frac{1}{\eta} = \frac{h}{\tanh h} + \frac{\rho_b S_g \hat{k} \ell}{k_c}. \qquad (6.5.4)$$

If we regard the reciprocal of the effectiveness factor as a resistance we see that this is the sum of two resistances, the first for internal diffusion and the second for external mass transfer (see Exercise 6.4.2). Written in the form

$$\frac{V_p \rho_b S_g \hat{k}}{S_x k_c},$$

this second term is clearly the ratio of the rate of reaction to the rate of mass transport. If the first term in Eq. (6.5.4) is much larger than the second, we may refer to the reaction as diffusion limited; if the second is much larger than the first, it is limited by external mass transfer. We notice in the first case that if h is large, η is inversely proportional to $(\rho_b S_g)^{1/2}$; in the second to $(\rho_b S_g)$. It follows that the measured reaction rate which is proportional to $\rho_b S_g \eta$ is proportional to $(\rho_b S_g)^{1/2}$ when diffusion is limiting, but independent of $\rho_b S_g$ when mass transfer is limiting. When the reaction is slow, so that η is close to 1, we can say that the kinetics are limiting and in this case the measured reaction rate is proportional to $\rho_b S_g$. Koros and Nowak have suggested using this as a diagnostic test to see which regime is dominant. Its usefulness depends on being able to prepare catalyst samples with varying amounts of active catalyst, that is with varying $\rho_b S_g$.

Exercise 6.5.1. A first order reaction $A \longrightarrow B$ is taking place in a packed bed, through which the reacting fluid flows with velocity v_0. (This velocity is the volume flow rate of the reaction mixture divided by the total cross-sectional area. The fluid is assumed to be incompressible.) Set up an equation for $a(z)$, the concentration of A at a point distant z from the inlet, and show that it can be written as

$$H \frac{da}{dz} = -a.$$

H has the dimensions of a length and has been called the height of a reactor unit, by analogy with heights of transfer units and equivalent theoretical plates. Interpret Eq. (6.5.4) to show that H is the sum of a height for external mass transfer (HTU) and a term dependent on the reaction, the so-called height of a catalytic unit (HCU). Examine the contribution of these terms when the mass transfer, diffusion, and kinetic regimes are dominant.

6.6 The Effect of Temperature Variations

Up to this point, we have overlooked entirely the important effects of temperature variations, both between the pellet and stream and within the pellet itself. We shall not attempt to give a complete account of the problems thus raised, but shall describe these effects in qualitative terms.

The transport of heat to the external surface of a catalyst particle can be described in the same way as the mass transport, by a transfer coefficient. Thus if the stream temperature is T and the surface temperature of the particle T_s, the rate of heat transport is

$$q = U(T - T_s), \quad (6.6.1)$$

where $U = c_p G j_H (\text{Pr})^{-2/3}$ = heat transfer coefficient,
c_p = specific heat of the bulk phase,
$\text{Pr} = \frac{c_p \mu}{k}$ = Prandtl number,
μ = viscosity of the bulk phase,
k = thermal conductivity of the bulk phase.

The factor j_H is similar to the mass transfer factor j_D and has been correlated with the Reynolds number. It is shown in Fig. 6.4 and expressed by the equation

$$j_H = \frac{1.10}{(\text{Re})^{0.41} - 1.5}. \quad (6.6.2)$$

Just as we were able to estimate an effective diffusion coefficient for the porous pellet, so we can obtain an effective thermal conductivity, k_e, such that the heat flux across a unit area within the pellet is k_e times the gradient of temperature normal to that area. We shall not go into the relation between k_e and the other physical properties of the porous material—this is well covered by Petersen in his fifth chapter—but it is worth pointing out that whereas the diffusion of matter is largely through the pores, that of heat is rather through the solid material. Values of k_e for porous materials are of the order of 10^{-3}–10^{-2} cal/sec cm°C.

If we again consider the reaction $A \longrightarrow B$, we have the same equation for the concentration of A but must remember that $\hat{k}$ is a function of the temperature T, say $\hat{k} = \hat{A} \exp - (E/RT)$. Thus we may write

$$D_s \frac{d^2 a}{dx^2} = \rho_b S_g \hat{A} e^{-(E/RT)} a \quad (6.6.3)$$

for the flat plate with sealed edges. By doing a heat balance we have a similar equation for T, namely

$$k_e \frac{d^2 T}{dx^2} = \Delta H \rho_b S_g \hat{A} e^{-(E/RT)} a, \quad (6.6.4)$$

where ΔH is the heat of reaction. The boundary conditions for this equation are similar to those for the diffusion of A, namely

$$\frac{dT}{dx} = 0 \quad \text{at} \quad x = 0$$

and
$$T = T_s \quad \text{at} \quad x = \ell. \tag{6.6.5}$$

But dividing Eq. (6.6.4) by ΔH and subtracting from Eq. (6.6.3) we see that
$$\frac{d^2}{dx^2}\left(D_s a + \frac{k_e}{-\Delta H} T\right) = 0,$$

and so by two integrations and the use of boundary conditions
$$a = a_s - \frac{k_e}{D_s(-\Delta H)}(T - T_s). \tag{6.6.6}$$

Prater observed that this gives a useful expression for the greatest possible temperature within a particle when the reaction is exothermic (or the least when it is endothermic), for the concentration a cannot be negative. Thus
$$T \leq T_s + \frac{(-\Delta H) D_s a_s}{k_e}$$

or
$$T \geq T_s - \frac{\Delta H D_s a_s}{k_e}, \tag{6.6.7}$$

according as ΔH is negative or positive.

If we let $\xi = x/\ell$ and $\theta = T/T_s$ we can use Eq. (6.6.6) to eliminate a from Eq. (6.6.4) and obtain
$$\frac{d^2\theta}{d\xi^2} = -h^2(1 + \beta - \theta)\exp\gamma\left(\frac{\theta - 1}{\theta}\right), \tag{6.6.8}$$

where
$$h^2 = \frac{\ell^2 \rho_b S_g \hat{A} e^{-E/RT_s}}{D_s},$$
$$\beta = \frac{(-\Delta H) D_s a_s}{k_e T_s}, \tag{6.6.9}$$

and
$$\gamma = \frac{E}{RT_s}.$$

This is a nonlinear equation which we do not propose to solve, though its solution can be expressed in integrals by the same method as was used for the general isothermal case. The effectiveness factor can be computed in precisely the same way but this time it evidently depends on three parameters, h, β, and γ. The first is evidently the Thiele modulus evaluated for surface conditions. The second is the ratio of the maximum temperature rise [given by Eq. (6.6.7)] to the surface temperature. It is large for a highly exothermic reaction when the conduction of heat is poor as compared with the diffusion of reactant. The third parameter is proportional to the activation energy and is large when temperature has a large influence on the reaction rate.

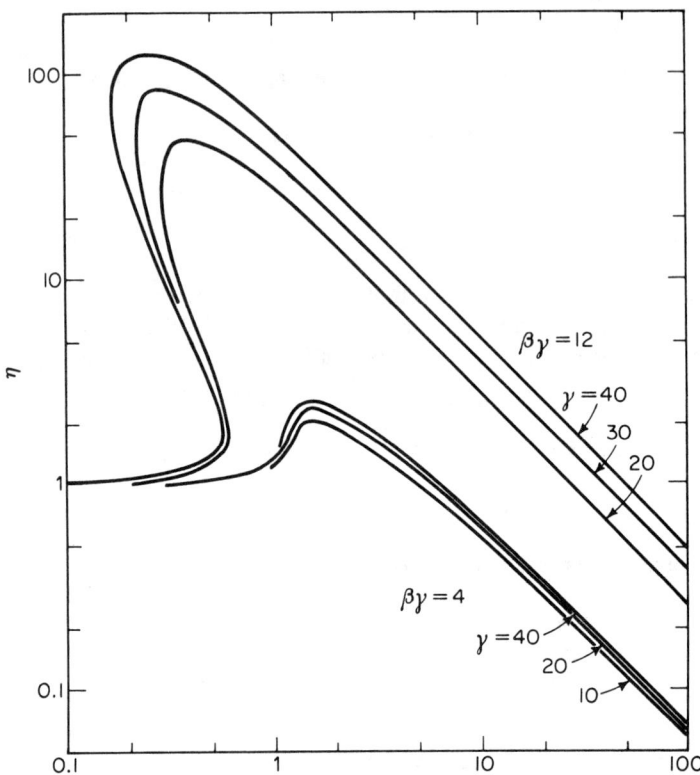

Fig. 6.9 Effectiveness factors for the sphere for several values of h, β, and γ. From *Ind. Eng. Chem.*, **58**, 32 (Sept. 1966). Copyright 1966 by the ACS. Reprinted by permission of the copyright owner.

In Fig. 6.9 several curves of η versus h are shown. Two points may be noticed before discussing the interesting way in which some of the curves double back on themselves. The figure is based on the calculations of Weisz and Hicks for spheres of radius ℓ, but these curves have, as before, precisely the same general shape as for the flat plate or cylinder. First, it may be noted that the asymptotes could be brought together by a suitable modification of the Thiele modulus h. Second, it is seen that the curves for the same value of $\beta\gamma$ tend to be close to one another. This, in conjunction with an approximation to the Arrhenius form of temperature dependence, has suggested the use of a single parameter

$$\delta = \beta\gamma = \frac{(-\Delta H)ED_s a_s}{k_e RT_s^2} \tag{6.6.10}$$

to characterize the family of curves. Although this cannot be rigorously

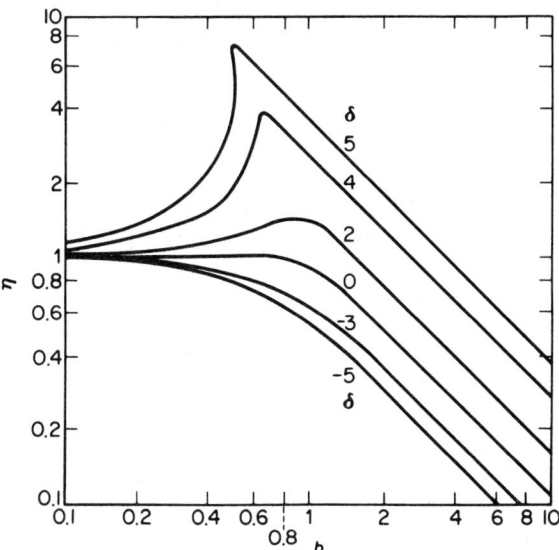

Fig. 6.10 Effectiveness factors for nonisothermal first order reaction in the slab. [Adapted from Petersen (see references) with permission.]

justified it is tolerably accurate for smaller values of β. Figure 6.10 shows a family of curves ranging from the highly exothermic ($\delta = 5$) to the endothermic ($\delta = -5$).

The striking thing about the case of the exothermic reaction is the way in which it may have a value of η which is greater than 1. This is reasonable enough when β and γ are large. A large β implies good heat of reaction and easy diffusion of matter but poor heat conduction, while a large γ implies that temperature has a marked effect on reaction rate. Under these circumstances, we can see that the heat generated by reaction can be "trapped" within the particle, leading to a marked temperature increase and an enhanced reaction rate.

Perhaps even more striking is the way in which the curves double back on themselves so that for some values of h there are three possible values of the effectiveness factor. This implies that there is more than one solution for an equation such as Eq. (6.6.8) and raises a flock of interesting problems too high-soaring to follow here. It is important however for the student to realize that such problems of nonuniqueness exist and what their physical implications are. For example, suppose we have a very active catalyst in the diffusion-controlled regime that is gradually being poisoned. Here β and γ might be unchanged and h can be gradually decreased by the effect of the poison on S_g or $\hat{A}$. If $\beta = 0.3$ and $\gamma = 40$, so that the topmost curve of Fig. 6.9 is applicable, and h were 100 at first, the effectiveness of the pellet would be about 0.5. As h decreased, the effectiveness factor would increase until for h in the neighborhood of 0.5 to 1.0 it would of the order of 100. However, when the poisoning has reduced the value of h to around 0.2, where the curve has a vertical tangent, a further decrease in h will give an effectiveness that differs little from 1, and it is as though the reaction was suddenly snuffed out. It is less easy to imagine circumstances in which h increases and β and γ are constant, but if this were possible, it is clear that as h increases past the lower point of vertical tangency (about 0.5 on this curve) the effectiveness factor takes a jump from about 2 to around 100. The reaction has been "ignited" so to speak, by reaching a rate at which heat is being generated fast enough to be trapped within the particle and to produce large internal temperatures. We notice that in neither case do we obtain an effectiveness factor given by the part of the curve between the two points of vertical tangency. It can be shown that this part of the curve corresponds to a range of unstable steady states for the catalyst pellet.

It can also be shown that, whatever values β and γ may have, there is only one steady state when h is sufficiently small or sufficiently large. On the other hand, for sufficiently small values of β and γ, the steady state, and so also the effectiveness factor, is unique for all h. A safe but imprecise condition for the latter is $\beta\gamma < 1$, but to go beyond this would lead us into the deep waters of some of the most interesting current research and the few references at the end of the chapter must suffice.

Exercise 6.6.1. Show that if $\beta = 0$ the problem reduces to an isothermal one.

Exercise 6.6.2. Show that taking $\beta\gamma$ as a single parameter has plausible accuracy if β is small.

Exercise 6.6.3. If the reaction is $\sum \alpha_j A_j = 0$, its rate (in conventional notation) is $\rho_b S_g r(c_1, \ldots, c_S, T)$ and heat of reaction is ΔH. Set up equations for the concentrations and for the temperature. Take k_e to be the effective thermal conductivity of the catalyst pellet, D_{js} to be the effective diffusion coefficient of A_j and assume the geometry of the slab. Show that these can be reduced to a single equation in temperature. Can you solve this?

Exercise 6.6.4. In the case studied in the previous problem show that the greatest possible temperature rise for an exothermic reaction is the least value of $(-\Delta H)D_{js}c_{js}/(-\alpha_j)k_e$. The choice of the least value is among all the j corresponding to reactants.

Exercise 6.6.5. Given that the external partial pressure of a gaseous reactant A in the reaction $A \longrightarrow B$ is 1 atm and that the exothermic heat of reaction is 10 kcal/mol, would you regard a temperature rise of 100°C within the catalyst pellet as possible? Assume the outside of the pellet to be at room temperature.

Exercise 6.6.6. The boundary conditions for a catalyst particle in which the reaction $A \longrightarrow B$ is taking place are given by Eq. (6.5.1) and from Eq. (6.6.1), thus allowing for external mass transfer resistance. If T_f and a_f are the temperature and concentration respectively in the reactant stream far from the pellet, show that for an exothermic reaction

$$T(x) \leq T_f + a_f \operatorname{Max}\left\{\frac{D_s(-\Delta H)}{k_e}, \frac{k_c(-\Delta H)}{U}\right\}.$$

6.7 Applications

There are many questions within the subject of diffusion and reaction that would be fascinating to pursue. Most of them lead very rapidly into deep mathematical waters and are at the forefront of research in this area. Others lead into difficult practical questions, such as the problem of knowing just what is taking place in a "trickle bed reactor," where a liquid feed is trickling over a bed of catalyst and a gaseous reactant is passing over it. For the moment however we must leave the subject with the assurance that some progress has been made in understanding the interaction of physical and chemical processes and in developing expressions which can be used for the design of packed catalytic reactors. In doing so, we have had occasion to consider the effect of the adsorption step in a catalytic reaction and the importance of internal and external mass transfer.

We might ask finally, what clues there are to show which step is the key one in the analysis of any situation. We have already mentioned some of the possibilities. If an Arrhenius plot of the apparent reaction rate constant shows that at higher temperatures the slope (i.e., the apparent activation energy) falls off to about half its value at lower temperatures, there is a clear indication of an internal diffusion limitation (see Exercise 6.4.1). If the velocity of the stream of reactants has an effect on the observed reaction rate, then external mass transfer is a controlling influence (see Sec. 6.5). If it is possible to vary the amount of active catalyst in the pellet, then the dependence of the apparent rate on the active surface area is an indication of the importance of the three regimes (see Sec. 6.6). All such evidence, together with any a priori estimates that can be made, must be used in appraising any given situation.

Certain rough criteria can be made for the importance of internal diffusion in terms of the general Thiele modulus, h. Consider first the case of the first order reaction in the slab. It is clear from the left-hand curve of Fig. 6.7 that if h is less than 1 the effect of diffusion limitation is not serious, whereas when h increases beyond 1 the asymptotic relation $\eta = h^{-1}$ is rapidly approached. When the normalized modulus h' is used for other shapes the same rough criterion holds good, as inspection of Fig. 6.8 shows. For reactions other than first order, we have shown how to normalize the modulus so that $\eta \sim (h')^{-1}$ [see Eq. (6.4.30)] and again, the value $h' = 1$ may be used as a lower limit for the region of serious diffusion limitation. Even the nonisothermal curves shown in Fig. 6.10 can be adjusted by a shift to the left (for the exothermic ones) or to the right (for the endothermic ones) so that all would be asymptotic to $\eta = (h')^{-1}$, and it is not hard to see that diffusion limitation begins to be important when $h' > 1$. Thus we can say in general terms and to a rough but useful degree of approximation that diffusion limitation is not important if

$$h' = \frac{V_p}{S_x}\left[\frac{\rho_b S_g \hat{r}(c_s)}{D_s c_s}\right]^{1/2} [P(1)]^{-1/2} < 1, \tag{6.7.1}$$

where $P(1)$ is given by Eq. (6.4.26). This formula can be modified for the nonisothermal case and when the diffusion coefficient is dependent on temperature and concentration.

However, a criterion such as this assumes that we know the reaction rate $\hat{r}$, whereas what we can actually measure is the apparent reaction rate $\eta\hat{r}$. Weisz, Prater, Petersen, and Bischoff have shown how to obtain a criterion in terms of observable quantities.

Consider first the isothermal, first order reaction for which the true reaction rate per unit volume of catalyst is $\rho_b S_g \hat{k} c$. The observed rate of reaction, r_0 (say), is a function of the observed concentration c_s and is in fact

$$r_0 = \eta \rho_b S_g \hat{k} c_s,$$

where

$$\eta = \frac{\tanh h}{h} \quad \text{and} \quad h^2 = \frac{\ell^2 \rho_b S_g \hat{k}}{D_s}.$$

Thus

$$\rho_b S_g \hat{k} = \frac{r_0}{\eta c_s} \quad \text{and} \quad \eta h^2 = \frac{\ell^2 r_0}{D_s c_s}. \tag{6.7.2}$$

Now $\eta h^2 = h \tanh h$ is an increasing function of h and, if ηh^2 is small enough, h will be less than 1. In fact, if $\eta h^2 < 1.2$, then $h < 1$, so that in terms of our rough criterion $h < 1$ we could say that diffusion is no serious limitation so long as

$$\eta h^2 = \frac{\ell^2 r_0}{D_s c_s} < 1. \tag{6.7.3}$$

The value of this criterion is that it is expressed in terms of observable quantities. Suppose that the dependence of the reaction rate on composition is known, so that the reaction rate expression in Eq. (6.4.22) might be written $\hat{r}(c) = \hat{k}g(c)$, with $\hat{k}$ an unknown rate constant and $g(c)$ a known function, with $g(c) = 0$ for some equilibrium value $c = c_e$. Further suppose that D_s is also a function of concentration c. Bischoff has shown that the generalization of the criterion Eq. (6.7.3) is

$$\frac{\ell^2 r_0 g(c_s)}{2\int_{c_e}^{c_s} D_s(c)g(c)\,dc} < 1. \tag{6.7.4}$$

For any shape of particle we should expect to be able to replace ℓ by V_p/S_x to the same degree of tolerable approximation. If D_s is constant, we see that this has the effect of replacing c_s in the first order criterion Eq. (6.7.3) by $2\int_{c_e}^{c_s} g(c)\,dc/g(c_s)$ and this correction can be most important.

As a simple example, let us consider a problem arising out of the testing of a hydrogenation catalyst in a differential reactor. Conditions are held constant in two series of runs save that the catalyst is much more finely divided for one series. The same mass of catalyst occupies the same volume in the two cases so that the external voidage of the bed is not significantly changed. There are no great differences in the shape of the particles in the two sizes, so that the normalized Thiele moduli, h_1' and h_2', will be proportional to ℓ_1 and ℓ_2, characteristic dimensions of the particles. The mean observed reaction rates in lb-mol/hr lb catalyst are r_1 and r_2 in the two series of runs and the data are as follows:

$$\ell_1 = 0.175 \text{ in.} \qquad \ell_2 = 0.025 \text{ in.}$$
$$r_1 = 0.035 \qquad r_2 = 0.070.$$

Can we say anything about the degree of diffusion limitation on the basis of this sketchy data?

Actually, the units of ℓ and r may be quite arbitrary since we shall only need the ratios. Let $h_1' = m\ell_1$ and $h_2' = m\ell_2$ where m is a constant involving catalyst properties, diffusion coefficients, rate constants, and factors for shape and kinetics. However, m will be the same for both series of runs and so

$$\frac{h_1'}{h_2'} = \frac{m\ell_1}{m\ell_2} = \frac{\ell_1}{\ell_2} = \frac{0.175}{0.025} = 7.$$

If the effectiveness factors are η_1 and η_2 respectively, we have seen that $h^2\eta/\ell^2$ is proportional to the observed reaction rate [see Eq. (6.7.2)] and so

$$\frac{r_1}{r_2} = \frac{\eta_1}{\eta_2} = \frac{0.035}{0.070} = 0.5$$

or

$$\frac{h_1\eta_1}{h_2\eta_2} = 3.5.$$

We now have to make an assumption about the reaction and the pellet.

The simplest assumption is that the kinetics are first order and the reaction is isothermal. If we further assume that the normalized curve $\eta = \tanh h'/h'$ is sufficiently accurate we have

$$\frac{h'_1 \eta_1}{h'_2 \eta_2} = \frac{\tanh h'_1}{\tanh h'_2} = \frac{\tanh 7h'_2}{\tanh h'_2} = 3.5.$$

This is a transcendental equation for h'_2 but it can be readily solved by trial and error. If h'_2 is not too small, $7h'_2$ will be large enough to make $\tanh 7h'_2$ approximately 1 so that a first guess at h'_2 might be $\tanh h'_2 = 1/3.5 = 2/7 = 0.28$. By using a table of log tanh and noting that $\log 3.5 = 0.5441$ we have:

h_2	0.28	0.281	0.282
log tanh $7h'_2$	$\bar{1}.9828$	$\bar{1}.9830$	$\bar{1}.9832$
log tanh h'_2	$\bar{1}.4360$	$\bar{1}.4374$	$\bar{1}.4389$
Difference	0.5468	0.5456	0.5443

This is certainly as close as the accuracy of the whole enterprise can justify. Thus $h'_2 = 0.282$, $h'_1 = 1.974$, and $\eta_2 = 0.974$, and $\eta_1 = 0.487$. This clearly shows that the effectiveness factor of the larger particles is slightly less than 50%, whereas the smaller particles show no sensible diffusion limitation.

Exercise 6.7.1. Let h_0 denote the Thiele modulus as calculated by using the observed reaction rate. Show that for the first order reaction $\eta h_0^2 = \tanh^2(h_0 \eta^{-1/2})$ and, hence, that $\eta \sim h_0^{-2}$ as $h_0 \longrightarrow \infty$. What would you expect the asymptotic relation to be in the general case?

Exercise 6.7.2. A first order reaction $A \longrightarrow B$ is taking place in spheres of radius $\frac{1}{4}$ in. A is a gas with partial pressure of about 1 atm and at room temperature outside the pellet. The observed reaction rate is about 10^{-2} mole/sec per unit volume of packed reactor. Would you consider diffusion to be an important limiting factor?

REFERENCES

An excellent general reference covering the whole scope of this chapter is:

E. E. Petersen, *Chemical Reaction Analysis*. Englewood Cliffs, N. J.: Prentice-Hall, Inc., 1965.

A valuable monograph, though more limited in the ground it covers, is:

C. N. Satterfield and T. K. Sherwood, *The Role of Diffusion in Catalysis*. Reading, Mass.: Addison-Wesley Publishing Co., Inc., 1963.

A large view of the whole problem by one of the pioneers in the field is to be found in:

O. A. Hougen, "Engineering Aspects of Solid Catalysis," *Ind. Eng. Chem.*, **53**, 509 (1961).

A first-rate book covering both the chemistry and chemical engineering aspects of catalysis is:

J. M. Thomas and W.J. Thomas, *Introduction to the Principles of Heterogeneous Catalysis*. New York: Academic Press Inc., 1967.

E. W. Thiele, one of the first to obtain quantitative results in this area, has written a most interesting historical review in:

E. W. Thiele, "The Effect of Grain Size on Catalyst Performance," *Am. Scientist*, **55**, 176 (1967).

6.1. Only the simplest considerations on adsorption have been mentioned here; for a thorough study see:

S. Brunauer, *The Adsorption of Gases and Vapors*. Princeton, N. J.: Princeton University Press, 1945.

J. De Boer, *The Dynamical Character of Adsorption*. London: Oxford University Press, 1953.

D. M. Young and A. D. Crowell, *Physical Adsorption of Gases*. London: Butterworths, 1962.

S. Ross and J. P. Olivier, *On Physical Adsorption*. New York: Interscience, Inc., 1964.

D. O. Hayward and B. M. W. Trapnell, *Chemisorption*, (2nd ed.) London: Butterworths, 1964.

6.3. Much work on external mass transfer is summarized in:

J. de Acetis and G. Thodos, "Flow of Gases through Spherical Packings," *Ind. Eng. Chem.*, **52**, 1003 (1960).

See also:

J. J. Carberry, "A Boundary Layer Model of Fluid Particle Mass Transfer in Fixed Beds," *A. I. Ch. E. Journal*, **6**, 460 (1960).

Chapter 6 of Petersen and Chapter 2 of Satterfield and Sherwood discuss the problems of this section in more detail.

6.4. This section is only a sketch of the field but Petersen's book may be referred to for more detail. The classic works in this area are:

G. Damköhler, "Einflüsse der Strömung, Diffusion und des Wärmeüberganges auf die Leistung von Reaktionsöfen," *Z. Elektrochem. Angew. Physik. Chem.*, **42**, 846 (1936).

E. W. Thiele, "Relation between Catalytic Activity and Size of Particle," *Ind. Eng. Chem.*, **31,** 916 (1939).

The papers of E. Wicke on this subject are most valuable; see for example: "Einfluss der Stofftransportes auf den Verlauf heterogener Gasreaktionen," in *Chemical Reaction Engineering*, ed. K. Rietema, pp. 61–72. London: Pergamon Press, 1957.

A useful survey is in:

A. Wheeler, "Reaction Rates and Selectivity in Catalyst Pores," *Advan. Catalysis*, **3,** 280 (1951).

See also:

W. C. Pollard and R. D. Present, "On Gaseous Self-diffusion in Long Capillary Tubes," *Phys. Rev.*, **73,** 762 (1948).

L. B. Rothfeld, "Gaseous Counter Diffusion in Catalyst Pellets," *A. I. Ch. E. Journal*, **9,** 19 (1963).

For an excellent survey of the theory of diffusion and the estimation of diffusion coefficients see:

R. B. Bird, "Theory of Diffusion," *Advan. Chem. Eng.*, **1,** 156–239 (1956).

An early normalization of the Thiele modulus for an isothermal pellet and arbitrary kinetics was given by R. B. Bird, W. E. Stewart, and E. N. Lightfoot on pages 335–41 of their *Notes on Transport Phenomena*, the precursor to their well-known *Transport Phenomena* (New York: John Wiley & Sons, Inc., 1960). Slightly more general forms—all of them equivalent—have been given independently and almost simultaneously in:

K. B. Bischoff, "Effectiveness factors for general reaction rate forms," *A. I. Ch. E. Journal*, **11,** 351 (1965).

E. E. Petersen, "A general criterion for diffusion influenced reaction in porous solids," *Chem. Eng. Sci.*, **20,** 587 (1965).

R. Aris, "A normalization for the Thiele modulus," *Ind. Eng. Chem. Fundamentals*, **4,** 227 (1965).

Many authors have computed effectiveness factors for different forms of kinetics and adequate references cannot be given within a reasonable compass.

6.5. The method of distinguishing regimes is described in R. M. Koros and E. J. Nowak, "A diagnostic test of the kinetic regime in a packed bed," *Chem. Eng. Sci.*, **22,** 471 (1967).

6.6. See the many references given by Petersen in his Chapters 4 and 6. A criterion for isothermal behavior of a catalyst pellet has been given by J. B. Anderson in *Chem. Eng. Sci.*, **18**, 147 (1963).

The paper of Weisz and Hicks on the results of which Fig. 6.9 is based is:

P. B. Weisz and J. S. Hicks, "The behaviour of porous catalyst particles in view of internal mass and heat diffusion effects," *Chem. Eng. Sci.*, **17**, 265 (1962).

Some further remarks on the appropriateness of the positive exponential approximation are contained in:

R. Aris, "Is sophistication really necessary?" *Ind. Eng. Chem.*, **58**, 32 (Sept. 1966).

For practical confirmation of the theory see:

F. W. Miller and H. A. Deans, "An experimental study of nonisothermal effectiveness factors in a porous catalyst," *A. I. Ch. E. Journal*, **13**, 45 (1967).

J. P. Irving and J. B. Butt, "An experimental study of the effect of intraparticle temperature gradients on catalytic activity," *Chem, Eng. Sci.*, **22**, 1859 (1967).

6.7. The criterion of Weisz and Prater is given in:

P. B. Weisz and C. D. Prater, "Interpretation of measurements in experimental catalysis," *Advan. Catalysis*, **6**, 167 (1954).

See also Petersen's paper listed above under 6.4,

K. B. Bischoff, "An extension of the general criterion of the importance of pore diffusion with chemical reaction," *Chem. Eng. Sci.*, **22**, 525 (1967),

and two communications by R. R. Hudgins and E. E. Petersen in *Chem, Eng. Sci.*, **23**, 93 and 94 (1968).

NOTATION

A	reactant in sample reaction $A \longrightarrow B$
A'	constant (p. 116 only)
A_j	chemical species in general reaction
$\hat{A}$	pre-exponential factor in $\hat{k} = \hat{A} \exp - (E/RT)$
a	concentration of A (see various suffixes)
$a(r)$	concentration of A as a function of position within the pellet
B	product in sample reaction $A \longrightarrow B$

B'	constant (p. 116 only)
b	concentration of B (see various suffixes)
$b(r)$	concentration of B as a function of position within the pellet
$c(x)$	concentration at position x in Eq. (6.4.22) and thereafter
c_j	concentration of A_j (see various suffixes)
$\hat{c}_0$	total concentration of adsorption sites
$\hat{c}_v$	concentration of vacant sites
c_p	specific heat of reactant mixture, Sec. 6.6
D	diffusion coefficient (molecular)
D_K	Knudsen diffusion coefficient
D_s	effective diffusion coefficient in porous solid
d	diameter of cylindrical pore
d_e	equivalent pore diameter, Eq. (6.4.10)
E	activation energy
G	mass flow rate of reactants through packed bed
$g(c)$	functional dependence of reaction rate, Sec. 6.7
h	Thiele modulus, Eq. (6.4.6)
h'	modified Thiele modulus
j_D	mass transfer factor
j_H	heat transfer factor
K	equilibrium constant, Sec. 6.2
K_A, K_B	adsorption constants, Sec. 6.2
k, k'	reaction rate constants
k_a, k_d, k'_a, k'_d	adsorption and desorption rate constants
k_c	mass transfer coefficient
k_e	effective thermal conductivity of catalyst pellet
ℓ	dimension of catalyst pellet or length of pore
m	molecular weight
n	reaction order
Pr	Prandtl number
R	gas constant
$\hat{R}$	dimensionless reaction rate, Eq. (6.4.23)
Re	Reynolds number
r	reaction rate per unit reactor volume
$\hat{r}$	reaction rate per unit catalyst area
$\hat{r}^*$	reaction rate per unit catalyst area expressed as a function of bulk phase composition
r_0	observed reaction rate, Sec. 6.7
Sc	Schmidt number
Sh	Sherwood number
S_g	catalytic surface area per gram
S_x	exterior surface area of particle
T	temperature

Notation

U	heat transfer coefficient, Sec. 6.6
V_p	volume of catalyst pellet
x	distance from center plane of slab or end of pore
x_j	mole fraction of A_j, Eq. (6.4.12) only
α	a/a_s, Sec. 6.4
α, β	constants, Sec. 6.1 only
α_j	stoichiometric coefficient of A_j
α_0	value of a/a_s at $\xi = 0$, Sec. 6.4
β	$(-\Delta H)D_s a_s/k_e T_s$, Sec. 6.6
γ	E/RT_s, Sec. 6.6
γ	c/c_s, Sec. 6.4
γ_0	value of c/c_s at $\xi = 0$, Sec. 6.4
ΔH	heat of reaction
δ	$(-\Delta H)ED_s a_s/k_e R T_s^2$, Sec. 6.6
ϵ	voidage of packed bed
$\hat{\epsilon}$	internal voidage of catalyst pellet
η	effectiveness factor
λ	$(\rho_b S_g \hat{k}/D_s)^{1/2}$
μ	viscosity, Sec. 6.6
ξ	x/ℓ, Eq. (6.4.23)
$P(\gamma)$	integral defined, Eq. (6.4.26)
ρ_b	bulk density of catalyst pellet
τ	tortuosity factor

AFFIXES

$\hat{}$	value appropriate to the catalytic surface
j	value appropriate to A_j
0	reference value (except r_0, α_0, and γ_0 as defined above)
s	value appropriate to the external surface of the pellet (except D_s, as defined above)
e	equilibrium value for variables or effective value for parameters

The Continuous Flow Stirred Tank Reactor 7

In these last four chapters we shall examine the design and behavior of some of the common types of reactor.* Of these there are two idealized types: the tubular reactor, in which the reactants flow through with relatively little mixing, and the stirred tank, in which mixing is deliberately promoted. Naturally, an actual reactor will depart from these idealized extremes to some degree, but their study is still worthwhile for it is the basis for design and, in spite of the simplifications of models, it will demonstrate the important features of the behavior of these reactors. Adiabatic reactors may be of either type for the word "adiabatic" refers to the way in which they are operated, namely with little or no heat exchange with the surroundings.

The stirred tank reactor is essentially just what it says—a tank into which the reactants flow and out of which the products are taken and which is stirred to keep the contents as uniform as possible. It may also be equipped with a heat exchange coil or a heating or cooling jacket. It is also known by various other names and abbreviations, such as: C.F.S.T.R., C.S.T.R., or C-star. Its virtues are its simplicity of construction and ease of temperature control, for stirring is intended to destroy any local nonuniformities such as

*When the word "reactor" is used without any qualification in any place it refers to the type of reactor under consideration in that chapter or section. If it is necessary to be very specific or a comparison is being made with another type, then the full name of the reactor will be used.

The Continuous Flow Stirred Tank Reactor

temperature hot spots. It is used extensively in the organic chemical industry for a wide range of reactions. Since mixing brings the composition of the whole reactor to that of the product, the reaction rate is low by comparison with the average reaction rate in an unmixed reactor. This means that a stirred tank reactor may have to be large to get the required conversion, but the simplicity of construction may not make this defect too serious. When, as is usually the case, there are side reactions it is not the total conversion of the reactants but the fraction which goes to the desired product that is important. There are certain reaction schemes for which the stirred tank reactor gives the best yield in this sense.

The structure of the chapter should be clear from Fig. 7.1. The design must be based on the proper mass and energy balances for the reactor. When there is only one reaction, conversion and yield are equivalent concepts, but with simultaneous reactions, the primary concern of the design

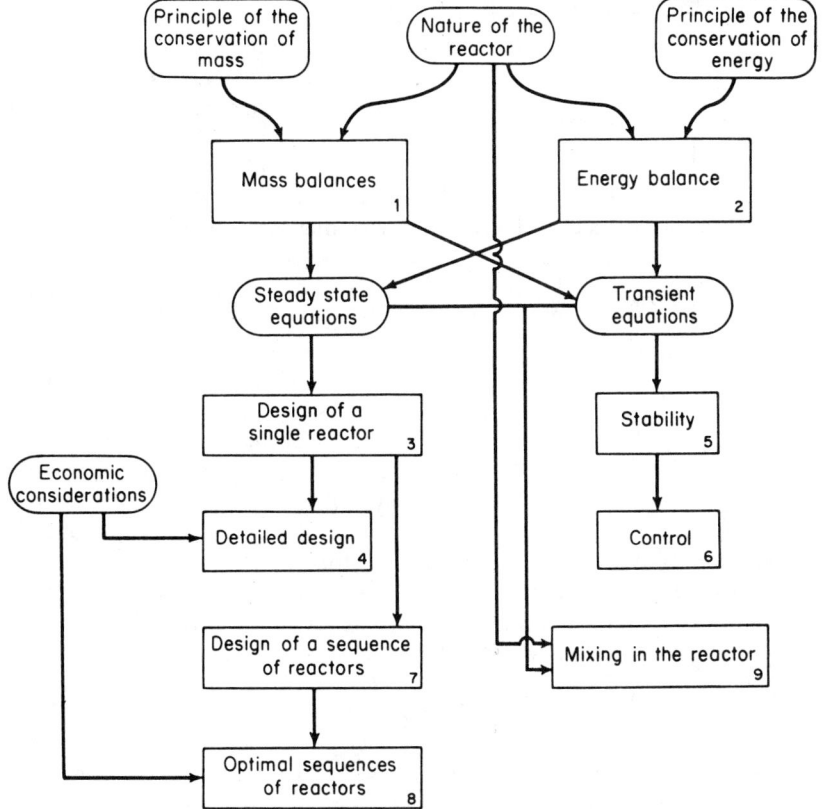

Fig. 7.1 The structure of the chapter.

must be with the yield of the desired product. Economic factors come into play in seeking a design which is in some sense optimal. All this can be done with the steady state equations, which for this reactor are algebraic rather than differential equations. Some insight into the stability of the reactor can be gained from the steady state equations, but a full treatment requires the equations for the transient state. These are ordinary differential equations and, though a full discussion of them is beyond our scope, the simplest case does provide a good insight into the general features of the behavior of this type of reactor. Moreover, they can be described in realistic terms with only the minimum of mathematics. For these reasons, the stirred tank reaction is an ideal type to consider first.

7.1 The Basic Mass Balances

Figure 7.2 shows the basic essentials of the reactor. Since it is used largely for liquid phase reactions with little change of volume, it will be appropriate to use c_j, the number of moles of A_j per unit volume, as the concentration variable with the corresponding extent variable ξ. The total volume flow rate of the feed (including both reactants and any inert gases or diluents) into the reactor is denoted by q, and the volume of the reacting mixture in the reactor by V. In normal operation, V will be constant, and so the volume flow rate out of the reactor will also be q. The quotient $V/q = \theta$ is called the holding time or residence time of the reactor. It will be shown later that θ is the expected or average time that any molecule will spend in the reactor. If the flow rate into the reactor, q_f, is not equal to the flow rate out of it, namely q, then the volume of the reaction mixture will not be constant. The rate at which V changes is just the difference between the volume flow rates into and out of the reactor, that is

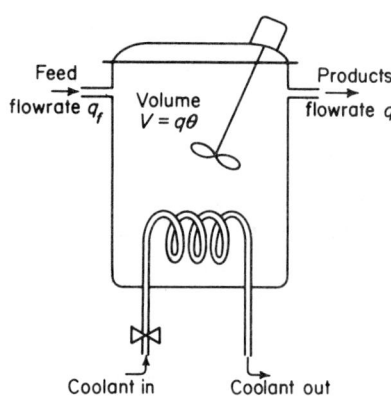

Fig. 7.2 The stirred tank reactor.

$$\frac{dV}{dt} = q_f - q. \qquad (7.1.1)$$

The reactants may be brought into the reactor by a number of different pipes for since we are assuming that there is good internal mixing, there is no need to premix the feed; q_f always refers to the total volume feed rate.

Sec. 7.1   The Basic Mass Balances

Before considering the general case, let us, as usual, look at the simplest possible example. A feed of reactant A, in concentration a_f moles per unit volume, enters the tank where it starts to react reversibly to form B. The reaction rate per unit volume at the temperature of the reactor is $r = ka - k'b$, where a and b are the concentrations of A and B in the reactor. The total amount of A in the reactor is Va and this changes because (i) A is being brought in the feed, (ii) some A is being removed with the product, and (iii) A is reacting to form B in the reactor. The rates at which these changes are taking place are:

(i) Rate of supply of A in feed $= q_f a_f$;

(ii) Rate of removal of A with product $= qa$;

(iii) Rate of disappearance of A by reaction $= V(ka - k'b)$.

Hence,

$$\frac{d}{dt} Va = q_f a_f - qa - V(ka - k'b). \qquad (7.1.2)$$

The first two terms on the right-hand side are due to advection, the third to reaction. If $q_f = q$ and, therefore, V is constant, we may divide through by q and get

$$\theta \frac{da}{dt} = a_f - a - \theta(ka - k'b). \qquad (7.1.3)$$

A similar balance on the number of moles of B in the tank would give a similar equation,

$$\theta \frac{db}{dt} = -b + \theta(ka - k'b). \qquad (7.1.4)$$

The differences between the two equations are: (1) there is no term in b_f since we assumed that the feed was pure A; (2) the sign of the reaction term is changed since the reaction uses up A but forms B.

Here then are two equations for the two concentrations a and b, but since there is only one reaction we might suspect that only one equation should be needed. This is true under certain circumstances. Suppose, for example, that at time $t = 0$ the concentrations in the tank were $a = a_0$ and $b = b_0$, and that the total concentration, $a_0 + b_0$, were equal to the total feed concentration, a_f. Then writing $c = a + b$ and adding Eqs. (7.1.3) and (7.1.4) would give

$$\theta \frac{dc}{dt} = a_f - c, \qquad (7.1.5)$$

with $c = a_f$ at $t = 0$. But the only solution to this equation is $c = a_f$ for all t (see Exercise 5.1.3), so that $a + b = a_f$, or

$$a = a_f - \xi, \qquad b = \xi. \qquad (7.1.6)$$

But now substituting into either of Eqs. (7.1.3) or (7.1.4) gives for ξ the equation

$$\theta \frac{d\xi}{dt} = -\xi + \theta[k(a_f - \xi) - k'\xi], \tag{7.1.7}$$

with $\xi = a_f - a_0 = b_0$ at $t = 0$. This is just the sort of economy we want to make in setting up equations.

The steady state equations are obtained by setting the time derivatives equal to zero,

$$\begin{aligned} 0 &= a_f - a - \theta(ka - k'b) \\ 0 &= -b + \theta(ka - k'b). \end{aligned} \tag{7.1.8}$$

In this case we notice, by addition, that $a + b = a_f$, so that we can certainly express a and b in terms of ξ by Eq. (7.1.6) and get a single equation

$$0 = -\xi + \theta[k(a_f - \xi) - k'\xi]. \tag{7.1.9}$$

This is not surprising, for the only thing that made us hesitate to write $a = a_f - \xi$ and $b = \xi$ before was the possibility that the initial composition $a = a_0$ and $b = b_0$ could not be expressed that way. But the steady state is reached after a long time of operation under steady conditions and the initial conditions are "forgotten." In fact, we can see just the way in which they are forgotten, for suppose that $a_0 + b_0 \neq a_f$ and let

$$\begin{aligned} a &= a_f - \xi + (a_0 - a_f)\zeta \\ b &= \xi + b_0\zeta. \end{aligned} \tag{7.1.10}$$

This time we still have two variables ξ and ζ and can express both the feed condition ($\xi = \zeta = 0$) and the initial conditions ($\xi = 0$ and $\zeta = 1$) in terms of them. The total concentration $c = a + b = a_f + (a_0 + b_0 - a_f)\zeta$, and substituting this into Eq. (7.1.5) gives

$$\theta \frac{d\zeta}{dt} = -\zeta. \tag{7.1.11}$$

Since $\zeta(0) = 1$ we have immediately that

$$\zeta(t) = \exp -\frac{t}{\theta}. \tag{7.1.12}$$

Substitution in the equation for b, Eq. (7.1.4), gives

$$\theta \frac{d\xi}{dt} + b_0 \theta \frac{d\zeta}{dt} = -\xi - b_0\zeta + \theta[k(a_f - \xi) - k'\xi] + \theta[k(a_0 - a_f) - k'b_0]\zeta.$$

Because of Eq. (7.1.11), the second term on the left cancels with the second term on the right and, by using Eq. (7.1.12), we have

$$\theta \frac{d\xi}{dt} = -\xi + \theta[k(a_f - \xi) - k'\xi] + Ke^{-(t/\theta)}, \tag{7.1.13}$$

Sec. 7.1 The Basic Mass Balances 161

where
$$K = \theta(ka_0 - k'b_0) - \theta ka_f$$
is the difference between the reaction rates for the initial and the feed conditions. Equation (7.1.12) shows that in three holding times of the reactor, the difference between $a + b$ and a_f has been reduced to less than 5% of its initial value (since $e^{-3} = 0.0498$) and in five holding times it is down to less than 1% ($e^{-5} = 0.0067$). Equation (7.1.13) is the same as Eq. (7.1.7) except for the exponentially varying term; how serious a difference this would make depends on the size of K. We are not suggesting that this is the best way to calculate the full behavior, for clearly it would be simpler to solve Eqs. (7.1.3) and (7.1.4) directly. What is interesting is that the incompatibility of total concentration $(a + b - a_f)$, which is initially $a_0 + b_0 - a_f$, washes out of the reactor as if it were a nonreactive substance (see Exercise 7.1.2).

All these features are seen in the case of a general reaction or a set of simultaneous reactions. Before considering this let us note that, if V is not constant, Eq. (7.1.2) becomes

$$V\frac{da}{dt} + a\frac{dV}{dt} = q_f a_f - qa - V(ka - k'b).$$

But by Eq. (7.1.1), $a(dV/dt) = (q_f - q)a$ and so

$$V\frac{da}{dt} = q_f(a_f - a) - V(ka - k'b) \qquad (7.1.14)$$

in the case of varying volume.

For the general reaction $\sum \alpha_j A_j = 0$, we use the molar concentrations c_j and suppose that the concentrations in the feed are c_{jf} and that they are constant. Then the number of moles of A_j in the tank is Vc_j and the rate of changing this has again three contributing terms:

(i) Rate of influx of A_j in feed $= q_f c_{jf}$;

(ii) Rate of removal of A_j with product $= qc_j$;

(iii) Rate of formation of A_j by reaction $= \alpha_j Vr(c_1, \ldots, c_S, T)$.

Thus we have the equation

$$\frac{d}{dt}(Vc_j) = q_f c_{jf} - qc_j + \alpha_j Vr(c_1, \ldots, c_S, T), \qquad (7.1.15)$$

which for constant volume and $q_f = q$ becomes

$$\theta \frac{dc_j}{dt} = c_{jf} - c_j + \alpha_j \theta r(c_1, \ldots, c_S, T). \qquad (7.1.16)$$

This is an ordinary differential equation and so subject to conditions describing the initial composition, say $c_j(0) = c_j^\circ$. First let us assume that the feed and initial compositions are compatible in the sense that one could be derived from the other by reaction to a certain extent. Then

$$c_j^\circ = c_{jf} + \alpha_j \xi_0 \tag{7.1.17}$$

and we can substitute

$$c_j(t) = c_{jf} + \alpha_j \xi(t). \tag{7.1.18}$$

Then substituting in Eq. (7.1.16) and dividing through by α_j gives a single equation

$$\theta \frac{d\xi}{dt} = -\xi + \theta r(\xi, T), \tag{7.1.19}$$

where the reaction rate can be expressed in terms of the extent ξ in the usual way. The initial condition for Eq. (7.1.19) is of course $\xi(0) = \xi_0$.

When there is only one reaction, it is only necessary to have a single equation provided that (1) the feed composition is constant, and (2) the initial and feed compositions are compatible in the sense of Eq. (7.1.17). If the feed composition is constant but $c_j^\circ - c_{jf} \neq \alpha_j \xi_0$ for any ξ_0, then we need only one further equation. For substituting

$$c_j(t) = c_{jf} + \alpha_j \xi(t) + (c_j^\circ - c_{jf})\zeta(t) \tag{7.1.20}$$

in Eq. (7.1.16) and dividing by α_j gives

$$\left(\theta \frac{d\xi}{dt} + \xi - \theta r\right) + \frac{c_j^\circ - c_{jf}}{\alpha_j}\left(\theta \frac{d\zeta}{dt} + \zeta\right) = 0. \tag{7.1.21}$$

Now $(c_j^\circ - c_{jf})/\alpha_j$ cannot have the same value for all j, for if it did, its common value would be ξ_0. Subtracting two equations like Eq. (7.1.21) for which this ratio has different values shows that

$$\theta \frac{d\zeta}{dt} + \zeta = 0,$$

and hence, substituting back into Eq. (7.1.21),

$$\theta \frac{d\xi}{dt} = -\xi + \theta r(\xi, \zeta, T). \tag{7.1.22}$$

It must be remembered that substituting Eq. (7.1.20) into the reaction rate expression makes it a function of both ξ and ζ. However, the initial conditions $\xi(0) = 0$ and $\zeta(0) = 1$ show that again $\zeta(t) = \exp - (t/\theta)$ and that the incompatibility again "washes out" as an inert.

In the steady state we never have to worry about the initial conditions, and Eqs. (7.1.16) and (7.1.19) become

$$0 = c_{jf} - c_j + \alpha_j \theta r(c_1, \ldots, c_S, T) \tag{7.1.23}$$

and

$$0 = -\xi + \theta r(\xi, T). \tag{7.1.24}$$

Sec. 7.1 The Basic Mass Balances

When there are R independent reactions the balance is just the same, save that now all reactions can contribute to the formation of A_j. Thus,

$$\theta \frac{dc_j}{dt} = c_{jf} - c_j + \theta \sum_{i=1}^{R} \alpha_{ij} r_i(c_1, \ldots, c_S, T). \quad (7.1.25)$$

If the initial composition c_j° is compatible with the feed composition in the sense that

$$c_j^\circ = c_{jf} + \sum_{i=1}^{R} \alpha_{ij} \xi_{i0}, \quad (7.1.26)$$

then we may substitute

$$c_j(t) = c_{jf} + \sum_{i=1}^{R} \alpha_{ij} \xi_i(t) \quad (7.1.27)$$

and obtain

$$\sum_{i=1}^{R} \alpha_{ij} \left(\theta \frac{d\xi_i}{dt} + \xi_i - \theta r_i \right) = 0. \quad (7.1.28)$$

Now it will be recalled that the very definition of independence of reactions was that the only numbers λ_i, such that $\sum \alpha_{ij} \lambda_i = 0$ for all j, were $\lambda_i = 0$ for all i [see Eq. (2.3.4)]. It therefore follows that if the reactions are independent,

$$\theta \frac{d\xi_i}{dt} = -\xi_i + \theta r_i(\xi_1, \ldots, \xi_R, T), \quad (7.1.29)$$

with $\xi_i(0) = \xi_{i0}$. This reduction from S equations for the concentrations to R equations for the extents is possible when the c_{jf} are constant and the initial composition is compatible with the feed. If this last condition is not met an extra variable ζ can be introduced exactly as before and this satisfies Eq. (7.1.11). If the feed composition is not constant, Eqs. (7.1.25) still apply but now there are time-varying terms in the equation. In the case of the steady state, it is always possible to reduce the number of equations to R, namely,

$$0 = -\xi_i + \theta r_i(\xi_1, \ldots, \xi_R, T). \quad (7.1.30)$$

Let us emphasize again, however, that the choice of concentrations or extents as the variables to work with should not be a fixed one; rather, one should always search for the simplest set.

Since the stirred tank is used mainly for liquid phase reactions we have not introduced the pressure as a second thermodynamic variable. If this were necessary, the reaction would have to be considered as a function of composition, temperature, and pressure. In any case, we shall assume the pressure to be constant.

An example of the use of these simple mass balance equations is given in the work of Eldridge and Piret [*Chem. Eng. Prog.*, **46**, 290 (1950)], who found excellent agreement between the observed and calculated extents of reaction in the hydrolysis of acetic anhydride. They carried out the reaction, which is really second order, with a large excess of one reactant so that it appeared to be first order. Thus the steady state equation would be

$$0 = -x + k\theta(1 - x),$$

or

$$x = \frac{k\theta}{1 + k\theta},$$

where $x = \xi/c_0$, the ratio of the extent to the feed concentration of acetic anhydride. The value of k is not that of the second order rate constant for the reaction but the product of that rate constant and the concentration of the reactant in excess. In testing the agreement between theory and experiment by comparing observed and calculated values of x, we are really testing two hypotheses. The first is that the kinetics of the reaction are really first order; the second that the steady state equation adequately models the reactor. The first hypothesis is amenable to an independent test, by varying the concentration of the excess reactant and by performing the reaction in a different type of reactor (for example, in a batch reactor). Granted the kinetics of the reaction, the agreement of the calculated and observed extents shown in the following extract from Eldridge and Piret's data is quite convincing.

$k\theta$	0.188	0.212	0.379	0.387	0.498	0.711	1.183	1.380	7.80
x(obs.)	0.153	0.173	0.258	0.280	0.324	0.408	0.544	0.583	0.882
x(calc.)	0.158	0.175	0.274	0.278	0.330	0.415	0.542	0.580	0.886

These are but a few of their figures and for the whole body of their work the average difference between the observed and calculated extents was 0.4%.

Exercise 7.1.1. Reactants $A_j, j = 1, 2, \ldots, S$, flow through their own pipes at volume rates q_j into the reactor. If ρ_j is the density of A_j and m_j is its molecular weight, show that

$$c_{jf} = \frac{q_j \rho_j}{q m_j} \quad \text{where} \quad q = \sum q_j.$$

Exercise 7.1.2. A stirred tank of volume V is initially filled with a solution of concentration c_0 and is washed out by a stream of pure solvent of flow rate q. Show that the concentration at time t is $c_0(t) = c_0 \exp -(t/\theta)$.

Exercise 7.1.3. A tank of acid is to be neutralized by an alkali of such a strength that an equal volume of it would be required. If the alkali is introduced as a stream into the tank, which is stirred and kept at constant volume, show that when the neutralization is very rapid some 30% less than an equal volume of acid would be needed.

Exercise 7.1.4. The first order reaction $A \longrightarrow B$ is irreversible and takes

Sec. 7.1 The Basic Mass Balances   165

place while the reactor is being filled up as well as later under steady operation. If the reactor starts by being empty, the flow rate is constantly equal to $q = V/\theta$ and the temperature is uniform. Find the concentration of A at all times. (You will need two expressions, one for $t < \theta$ and the other for $t > \theta$.)

Exercise 7.1.5. The irreversible decomposition of A is catalyzed by B. The time of half completion of the batch reaction was determined for various initial compositions. From the results estimate:

(i) The order of reaction with respect to A;
(ii) The order of reaction with respect to B;
(iii) The velocity constant at this temperature.

It is proposed to decompose A in solution, using B as catalyst at a concentration of 0.002 gm-mole/l, in a constant volume, well stirred reactor of 10,000 l. What flow rate of solution can be treated if the product is not to contain more than 5% of the original A? If the reactor is started by rapidly filling it with reagent solution and thereafter using this flow rate, how long will it be before the concentration of A in the product is 6% of that in the feed?

Initial concentration of A (gm-mole/l)	Concentration of B (gm-mole/l)	$t_{1/2}$ (min)
1.0	0.001	17.0
1.0	0.002	12.0
2.0	0.003	9.8
2.0	0.004	8.5

(C.U.)

Exercise 7.1.6. The product in the previous question cannot be sold if it contains more than 6% of A, so production during the initial period will have to be discarded. It is intended to save this loss by storing the product until the mean concentration in the storage tank is 6%. How big will this storage tank have to be?

Exercise 7.1.7. The only available storage tank is of 10,000 l capacity. It is therefore proposed to start up the reactor as in the previous questions and to start filling up the storage tank at such a time that when it is just filled, the mean concentration of A in it is 6% of the feed concentration. What volume of the process stream would have to be wasted under this plan of operation?

Exercise 7.1.8. A solution of substance A of concentration a_0 is fed at a rate q to an isothermal, constant-volume stirred tank reactor of volume V. Substance A undergoes consecutive first order reactions $A \xrightarrow{k_1} B \xrightarrow{k_2}$

C. Find the flow rate at steady state conditions for a maximum concentration of B in the product. If $k_2 = 2.25k_1$, what is the maximum yield of B? Assume complete mixing in the reactor.

7.2 The Energy Balance

It is unusual for the stirring of a reactor to contribute much to the energy balance in comparison to enthalpy flow, changes of enthalpy in the reaction mixture, and deliberate heat removal or addition. We shall therefore ignore the work term in the energy balance. As before, let h_j denote the partial molar enthalpy of A_j (a function of temperature, pressure, and composition), T_f and T, the temperatures of the feed and in the reactor respectively, and assume the pressure, volume, and flow rate to be constant. Since the reactor contains Vc_j moles of A_j, the total enthalpy of the reactor contents is $V \sum c_j h_j(T)$, where h_j is written as $h_j(T)$ as a reminder that it is evaluated for the temperature and composition within the reactor. There will again be three terms in the expression for the rate at which the total enthalpy is changing. The first two will be analogous to the advective terms in the mass balance, namely,

(i) Rate of influx of enthalpy with feed $= q \sum c_{jf} h_j(T_f)$;

(ii) Rate of outflow of enthalpy with products $= q \sum c_j h_j(T)$.

We do not introduce a term for the rate of generation of heat by reaction since this will arise naturally as an enthalpy change due to changing composition. However, if the reaction is exothermic there may well be provision for cooling it (or for heating it in case of an endothermic reaction), and this term must be included as

(iii) Rate of heat removal by deliberate cooling $= Q^*$.

The exothermic case is the one we are most concerned with; if heating is required we make Q^* negative and if the operation is adiabatic $Q^* = 0$. Then the energy balance gives

$$\frac{d}{dt} V \sum c_j h_j(T) = q \sum c_{jf} h_j(T_f) - q \sum c_j h_j(T) - Q^* \quad (7.2.1)$$

or, dividing through by q,

$$\theta \frac{d}{dt} \sum c_j h_j(T) = \sum c_{jf} h_j(T_f) - \sum c_j h_j(T) - \frac{Q^*}{q}. \quad (7.2.2)$$

Now the differentiation of the first term must take into account that it is the sum of the products of which the second factors are functions of T and of $c_1, \ldots, c_S$. In fact,

Sec. 7.2   The Energy Balance

$$\theta \frac{d}{dt}(\sum c_j h_j) = \theta \sum \frac{dc_j}{dt} h_j(T) + \theta \sum c_j \left(\frac{\partial h_j}{\partial T}\frac{dT}{dt} + \sum \frac{\partial h_j}{\partial c_k}\frac{dc_k}{dt}\right). \quad (7.2.3)$$

In this expression the last term disappears, since $\sum c_j(\partial h_j/\partial c_k) = 0$ by Eq. (3.3A). Moreover, $\partial h_j/\partial T = c_{Pj}$ is the partial molar heat capacity, so that $\sum c_j(\partial h_j/\partial T) = C_P$ is the total heat capacity per unit volume of the reaction mixture. To evaluate the first term, we take the jth mass balance equation and multiply it by $h_j(T)$. For a single reaction, Eq. (7.1.16) gives

$$\theta \frac{dc_j}{dt} h_j(T) = c_{jf} h_j(T) - c_j h_j(T) + \theta \alpha_j h_j(T) r.$$

If this is summed, the last term will give $\theta(\Delta H)r$, by definition of the heat of reaction. For simultaneous reactions, Eq. (7.1.25) shows that the last term would have a similar sum for each independent reaction:

$$\theta \sum_{j=1}^{S} \frac{dc_j}{dt} h_j(T) = \sum_{j=1}^{S} c_{jf} h_j(T) - \sum_{j=1}^{S} c_j h_j(T) + \theta \sum_{i=1}^{R} (\Delta H_i) r_i. \quad (7.2.4)$$

We can now assemble these parts of the energy balance by dropping the last term in Eq. (7.2.3), writing the second as $C_P(dT/dt)$, and substituting Eq. (7.2.4) for the first. Upon inserting this as the left-hand side of Eq. (7.2.2) and taking the terms $\sum c_{jf} h_j(T) - \sum c_j h_j(T)$ to the right, we have

$$\theta C_P \frac{dT}{dt} + \theta \sum_{i=1}^{R} (\Delta H_i) r_i = \sum_{j=1}^{S} c_{jf}[h_j(T_f) - h_j(T)] - \frac{Q^*}{q}. \quad (7.2.5)$$

If we assume that the heat capacities c_{Pj} are reasonably constant, the first term on the right of Eq. (7.2.5) is

$$\sum c_{jf} c_{Pj}(T_f - T) = C_P(T_f - T). \quad (7.2.6)$$

Then, dividing through by C_P, we have finally

$$\theta \frac{dT}{dt} = T_f - T + \theta \sum_{i=1}^{R} J_i r_i - Q, \quad (7.2.7)$$

where

$$J_i = -\frac{\Delta H_i}{C_P} \quad \text{and} \quad Q = \frac{Q^*}{qC_P}. \quad (7.2.8)$$

If we express the reaction rates r_i in terms of concentrations and temperature, Eqs. (7.1.25) and (7.2.7) give $S + 1$ equations for $c_1, \ldots, c_S, T$. If the reaction rates can be expressed in terms of extents, then Eqs. (7.1.29) and (7.2.7) provide $R + 1$ equations for $\xi_1, \ldots, \xi_R, T$.

In the steady state, the heat balance gives

$$0 = T_f - T + \theta \sum J_i r_i - Q. \quad (7.2.9)$$

The temperature enters in a highly nonlinear way into all the equations, since the reaction rates will contain Arrhenius constants. It is therefore very difficult to solve the equations for T in terms of T_f, θ, Q, and the composition. It is much easier to solve for T_f or Q given T and design problems are often turned around to be solved in this way.

7.3 The Design of a Single Reactor

The main design of a reactor rests on the steady state equations to be solved under the conditions of a given feed or a required product. After the values of the principal variables have been laid down the design must be checked for its dynamic properties. Our concern in this and the next section is with the steady state design. We shall first deal with the solution of the equations, and in the following section with some associated problems.

Let us consider first the case of a single reaction and recall that we have two equations,

$$-\xi + \theta r(c_{jf} + \alpha_j \xi, T) = 0 \qquad (7.3.1)$$

$$T_f - T + J\theta r(c_{jf} + \alpha_j \xi, T) - Q = 0. \qquad (7.3.2)$$

Normally, the feed composition will be given (the stoichiometric constants α_j, the thermostatic data J, and the parameters of the kinetic expression are, of course, known), and as these two equations contain five variables, namely T_f, ξ, T, θ, and Q, three variables must be chosen while the remaining two can be calculated. T_f is a variable associated with the feed; it may be given, or it may be necessary to ask whether the feed needs preheating or cooling. ξ and T are variables associated with the product, and though the temperature of the product may not be of critical importance, the extent certainly is, since it gives the rate of production of the product for which the whole reactor is designed. θ and Q are variables of the design and subject to relatively free choice. θ is the ratio V/q and when q is fixed by demanding some productivity, θ gives the volume of the reactor, V. Alternatively, if an existing reactor of given volume had to be used, θ would fix the flow rate q.

The heat removal rate Q is a function of other design variables, of the temperature T, and of the chosen form of cooling. Three important possibilities for the design of the temperature control may be mentioned. First, there is the jacketed pot of Fig. 7.3(a). If the flow rate of the coolant is sufficiently high for the jacket to be at a uniform temperature T_c, h is the overall heat transfer coefficient, and A is the area of the cooled wall, then

$$Q^* = hA(T - T_c). \qquad (7.3.3)$$

But the jacket itself can be considered as a stirred tank having been fed by a coolant stream of flow rate q_c, heat capacity C_{Pc}, and temperature T_{cf}. Since the heat Q^* serves to heat up the coolant from T_{cf} to T_c, we have

$$Q^* = q_c C_{Pc}(T_c - T_{cf}), \qquad (7.3.4)$$

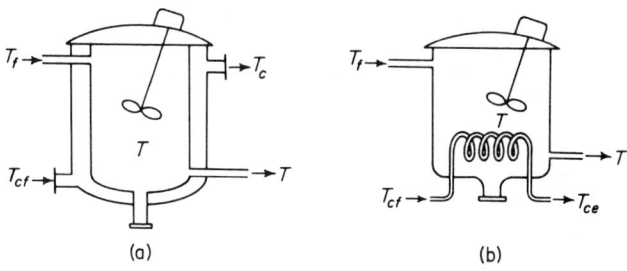

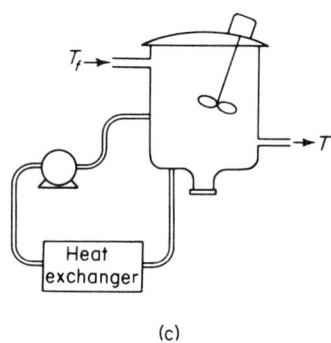

Fig. 7.3 Principal methods of cooling the stirred tank.

and we can eliminate T_c to give

$$Q^* = \frac{hAq_cC_{pc}}{hA + q_cC_{pc}}(T - T_{cf}). \tag{7.3.5}$$

The term Q in Eq. (7.3.2) is thus

$$Q = \frac{q_cC_{pc}}{qC_p}\left(1 + \frac{q_cC_{pc}}{hA}\right)^{-1}(T - T_{cf}). \tag{7.3.6}$$

The first factor is the ratio of the heat carrying capacity of the coolant stream to that of the reactant stream; the second factor depends on the ratio of the capacity of the coolant stream to the capacity of the wall to transfer heat.

The second scheme shown in Fig. 7.3(b) is that of an immersed coil. The temperature within the coil will vary along its length. If h is again an overall heat transfer coefficient so that the heat transfered per unit area is $h(T - T_c)$ at a point where the temperature within the coil is T_c, then

$$Q^* = q_cC_{pc}\left(1 - \exp-\frac{hA}{q_cC_{pc}}\right)(T - T_{cf}), \tag{7.3.7}$$

where A is the total area of the immersed coil. Thus

$$Q = \frac{q_c C_{Pc}}{qC_P}\left(1 - \exp -\frac{hA}{q_c C_{Pc}}\right)(T - T_{cf}) \qquad (7.3.8)$$

is again a function of the same dimensionless groups and, in spite of the difference of the second factor, is of similar behavior to the function in Eq. (7.3.6).

A third cooling scheme is sometimes used in which the reactants are taken out of the vessel and pumped through an external heat exchanger. This has the advantage of flexibility in design and perhaps ease of maintenance. Since the temperature in the exchanger is likely to be somewhat lower than in the reactor, it is reasonable to ignore the volume of the interchanger and assume that little reaction takes place in it.

Let us now consider some of the problems that can arise in the design of a stirred tank reactor with a single reaction. We shall speak in terms of an exothermic reaction, but most remarks will apply, *mutatis mutandis*, to the endothermic case.

A. *Product concentration and temperature specified.* If one of the products, say A_1, must be produced at a certain concentration, say γ_1, then the extent ξ is fixed at $\xi = (\gamma_1 - c_{1f})/\alpha_1$. If the temperature of operation is also given, Eqs. (7.3.1) and (7.3.2) can be written as

$$\theta = \frac{\xi}{r(\xi, T)} \qquad (7.3.9)$$

$$T_f - Q = T - J\xi. \qquad (7.3.10)$$

The first of these fixes the holding time; the second imposes a constraint between Q and T_f—the cooler the feed temperature the less the heat that would have to be removed. In all the cases we have considered, Q is proportional to $(T - T_{cf})$ so that if we write

$$Q = \kappa(T - T_{cf}), \qquad (7.3.11)$$

then

$$T_f + \kappa T_{cf} - \kappa T = T - J\xi.$$

If any two of T_f, T_{cf}, and κ are fixed, the third may be calculated from this equation. Once κ has been determined and the type of cooling system decided, Eq. (7.3.6) or (7.3.8) will provide a constraint on the two dimensionless groups $q_c C_{Pc}/qC_P$ and $hA/q_c C_{Pc}$, and the design of the cooling system can proceed from here.

B. *Production rate and temperature specified.* If the production rate of a product, say A_1, is specified rather than its concentration, we have a constraint $q(c_1 - c_{1f}) = P_1$ moles per unit time. Then $\xi = (c_1 - c_{1f})/\alpha_1 = P_1/\alpha_1 q$ is a function of q and Eq. (7.3.9) becomes

$$V = \frac{P_1}{\alpha_1 r(P_1/\alpha_1 q, T)}, \qquad (7.3.12)$$

allowing the required volume to be calculated for any chosen q. Multiplying Eq. (7.3.10) by qC_P gives

$$qC_P T_f - Q^* = qC_P T - (-\Delta H)P_1\alpha_1$$

as the constraint within which the heat exchange system must be designed.

C. *Feed conditions and size specified.* We have already remarked that the equations will be hard to solve for T on account of the nonlinearity of the Arrhenius function, but it is instructive to see how this may be done. If the feed condition, holding time, and coolant temperature are all given we have to solve

$$0 = -\xi + \theta r(\xi, T)$$
$$0 = T_f - T + J\theta r(\xi, T) - \kappa(T - T_{cf})$$

for ξ and T, with the other terms θ, T_f, J, κ, T_{cf}, and the constants in r being known. The first equation can probably be most readily solved for ξ in terms of T, since concentrations (and so the extent) appear as some small power in the reaction rate. Let $\xi = \xi_s(T)$ be the solution of this equation for any value of T, then ξ_s is the steady state extent that would be attained if the steady state temperature were T. The second equation can be rearranged and $\xi_s(T)$ substituted for $\theta r(\xi, T)$ from the first to give

$$T - T_f + \kappa(T - T_{cf}) = J\xi_s(T), \qquad (7.3.13)$$

or

$$(1 + \kappa)(T - T_c^*) = J\xi_s(T), \qquad (7.3.14)$$

where

$$T_c^* = \frac{T_f + \kappa T_{cf}}{1 + \kappa}. \qquad (7.3.15)$$

If we multiply Eq. (7.3.13) by qC_P we get

$$qC_P(T - T_f) + Q^* = (-\Delta H)Vr(\xi, T) = (-\Delta H)q\xi_s(T), \qquad (7.3.16)$$

and this is evidently a balance between the rate of heat removal, either in heating up the feed (the first term on the left) or by deliberate removal (the second term), and the rate of heat generation (given by the right-hand side). The left-hand side is linear in T; the right-hand side will certainly be small for small T, since reaction rates are always low at low temperatures, and will be bounded for all T since ξ_s is always bounded by the limiting reactant. It follows that there will certainly be a solution to this equation, as there must be if it is truly to represent a real reactor. Some of the difficulties in finding this solution can be illustrated by the very simplest case, to which we now turn.

Consider the irreversible first order reaction $A \longrightarrow B$ with rate constant $k(T)$. It is easier to use the mass balance equation on A than the equation for the extent. This is

$$0 = a_f - a - \theta k(T)a, \tag{7.3.17}$$

and since it is linear it can easily be solved to give

$$a_s(T) = \frac{a_f}{1 + \theta k(T)}. \tag{7.3.18}$$

The energy balance equation is

$$\begin{aligned} U_r &\equiv T - T_f + \kappa(T - T_{cf}) = J\theta k(T)a \\ &= Ja_f \frac{\theta k(T)}{1 + \theta k(T)} \equiv U_g, \end{aligned} \tag{7.3.19}$$

where the left-hand side, U_r, is proportional to the rate of heat removal and the right-hand side, U_g, to the rate of heat generation. It is clear that if U_r is plotted against T it gives a straight line of slope $(1 + \kappa)$ and intercept T_c^*. As T_f or T_{cf} increases, this line moves parallel to itself as shown in Fig. 7.4(b). As κ increases, the line pivots about the point with abscissa T_{cf} and ordinate $(T_{cf} - T_f)$. If $T_f > T_{cf}$ this is a point beneath the T-axis and the lines will be as shown in Fig. 7.4(c). The slope of this line can never be less

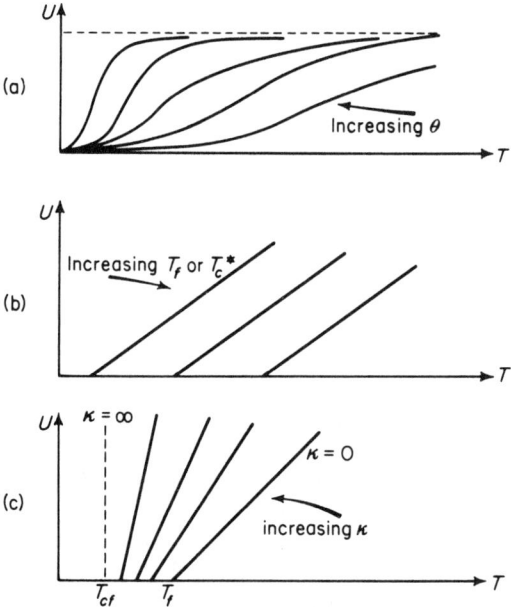

Fig. 7.4 Variation of the heat generation and rejection curves with variations of the parameters.

Sec. 7.3 The Design of a Single Reactor

than 1, corresponding to adiabatic operation $\kappa = 0$. On the other hand, the heat generation has the sigmoidal shape shown in Fig. 7.4(a), for $1 + k(T) = A \exp - (E/RT)$ is very small for small T and rises to A as T becomes very large. Thus

$$U_g = \frac{Ja_f \theta k(T)}{1 + \theta k(T)}$$

is small for small T and approaches the asymptote $Ja_f \theta A/(1 + \theta A)$, which is very close to Ja_f, as $T \longrightarrow \infty$.

The slope of the curve U_g versus T is

$$\frac{dU_g}{dT} = Ja_f \frac{E}{RT^2} \frac{\theta k(T)}{[1 + \theta k(T)]^2}, \qquad (7.3.20)$$

which is always positive and has a maximum when

$$\frac{\theta k(T)}{1 + \theta k(T)} = \frac{U_g}{Ja_f} = \left(\frac{1}{2} - \frac{RT}{E}\right). \qquad (7.3.21)$$

If this maximum slope is greater than $(1 + \kappa)$, the slope of the heat removal line, there will be the possibility of more than one intersection of the heat removal line and the heat generation curve, as shown in Fig. 7.5. As long as

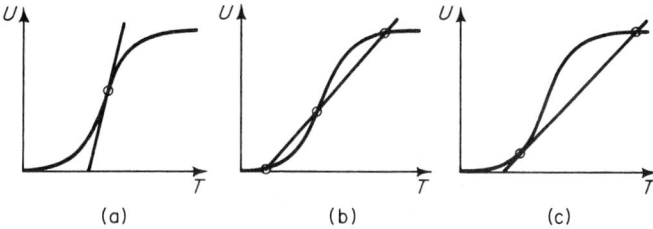

Fig. 7.5 Possible intersections of the heat generation curve with the heat removal line.

the slope of the removal line is greater than the greatest slope of the generation curve there can be only one intersection, as is clear from Fig. 7.5(a). But when the curve rises more steeply than the line there can also be positions of the line that give three intersections [Fig. 7.5(b)], or two intersections, one of which is really a point of contact [Fig. 7.5(c)]. This means that there are circumstances in which more than one steady state is compatible with a given set of feed conditions, and this raises a flock of interesting questions.

Consider what happens to a given reactor when the feed temperature is very slowly increased and everything else is held constant. This means that the heat generation curve is fixed and the heat removal line moves parallel to itself. Suppose that the maximum slope of the U_g curve is greater than $1 + \kappa$ the slope of the U_r lines. Then in Fig. 7.6 the U_g curve is fixed and A, B, C, D, and E represent five typical U_r lines for five feed temperatures.

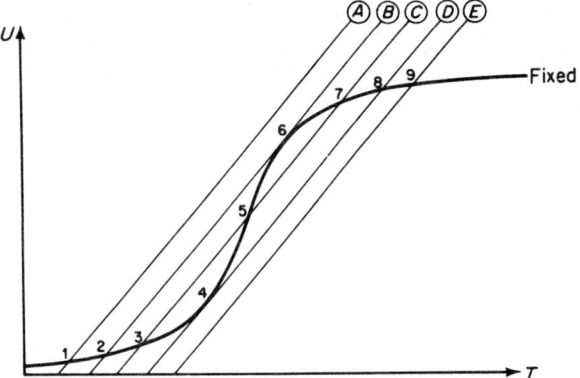

Fig. 7.6 The effect of changing the feed temperature.

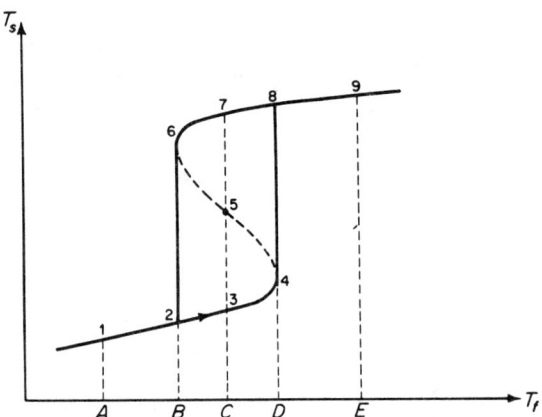

Fig. 7.7 Hysteresis of steady state with changing feed temperature.

The left-hand line corresponds to a low inlet temperature for which there can be only one steady state temperature, which we see is not much greater than T_c^*. At position B, there is the possibility of two steady states (corresponding to points 2 and 6) and at C of three steady states (corresponding to 3, 5, and 7). However, if there is no deliberate perturbation and the value of T_f is increased slowly enough for the steady state to be obtained at all times, we might expect it to follow along the bottom part of the curve until it reaches point 4 for a feed temperature corresponding to line D. This is shown by points 1 to 4 on the curve of Fig. 7.7. But here a dramatic change comes about for if the increase of T_f moves the U_r line ever so slightly to the right of D, there is only one intersection (just to the right of point 8) and this is nowhere near the previous steady state, 4. There is thus a discontinuous

Sec. 7.3 The Design of a Single Reactor

jump up from 4 to 8 as shown in Fig. 7.7. It is as though the reaction has been ignited by a sufficiently hot feed. A further increase in T_f produces no dramatic change as the line E and point 9 show.

If now T_f is slowly decreased so that the heat removal line goes through the sequence from E to A, we could again argue that in the absence of any deliberate perturbation the sequence of steady states would correspond to points 9, 8, 7, and 6. However, the slightest decrease of T_f below the value corresponding to line B gives an intersection to the left of point 2, nowhere near 6. The reaction has been suddenly "quenched" by too cold a feed and the discontinuity between points 6 and 2 in Fig. 7.7 is observed. It is noteworthy that following this very reasonable procedure never seems to give a steady state on the part of the curve between points 4 and 6. We shall see in the next section that this is because these steady states would be unstable.

The effect of changing the flow rate is similar, though less easy to show since both the slope κ and the position of the curve (governed by θ) vary with q. Figure 7.8 has been exaggerated for the sake of clarity, the lines and curves A, B, C, D, and E corresponding to an increasing sequence of flow rates and the numbered points to possible steady states. If the flow rate is increased slowly, the sequence of steady states 9, 8, 7, and 6 will be traced out, but the slightest increase of q beyond the value corresponding to D causes the steady state temperature to drop immediately to point 2. The reaction has been "blown out" by feeding the reactants so fast that the heat

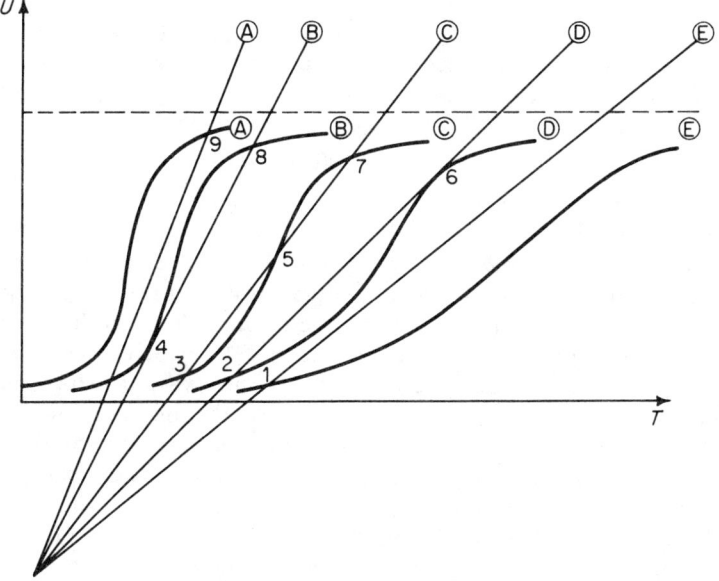

Fig. 7.8 The effect of changing the flow rate.

of reaction cannot bring them up to a sufficiently high reaction temperature. This in fact is just the sort of thing which happens if the gas supply is turned up excessively on a Bunsen burner or flame. A decreasing flow rate leads to the sequence of steady states, 1, 2, 3, 4, 8, and 9, with "ignition" at 4. Again, we notice in Fig. 7.9 that we have entirely avoided intersections on the part of the curve where the slope of U_g is greater than that of U_r. This strengthens our suspicion that such states are unstable.

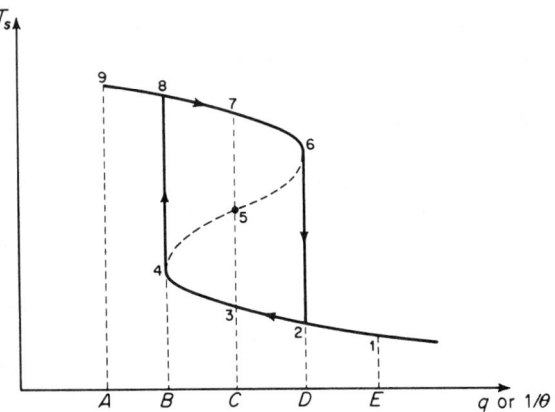

Fig. 7.9 Hysteresis of steady state with changing flow rate.

Completely analogous arguments may be applied to the most general case of a single reaction. Since $\xi_s(T)$ is obtained from the equation $-\xi + \theta r(\xi, T) = 0$, we see (by differentiating this equation with respect to T) that

$$\frac{d\xi_s}{dT} = \frac{\theta(\partial r/\partial T)}{1 - \theta(\partial r/\partial \xi)}. \tag{7.3.22}$$

Now except in the case of autocatalysis, $\partial r/\partial \xi$ is always negative, so that the denominator is always positive. Furthermore, for an irreversible or endothermic reaction $\partial r/\partial T$ is always positive. For an endothermic reactor J is negative and the heat "generation" curve is really a curve of heat absorbed by the reaction, as shown in Fig. 7.10. The U_r line still has positive slope so that there is clearly one and only one steady state; none of the interesting features of the exothermic case can arise. The U_g curves are of course bounded by the fact that ξ can never be so large that any of the concentrations become negative. For the case of a reversible exothermic reaction, we know that there is a locus, Γ_m, on which $\partial r/\partial T = 0$. It follows that the heat generation curve will have a maximum at this temperature and afterward it will go downward. In fact, at high temperatures $\xi_s(T)$ tends to approach $\xi_e(T)$, for the rate can be written as $k(T)f(\xi, T)$, where $k(T)$ is the forward rate constant and $f(\xi, T)$ vanishes for $\xi = \xi_e(T)$. If ξ satisfies the equation $\xi = \theta r(\xi, T) =$

Sec. 7.3 The Design of a Single Reactor   177

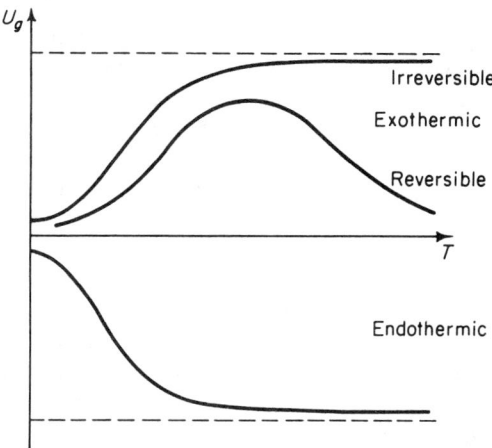

Fig. 7.10 Heat generation as a function of temperature for different types of reaction.

$\theta k(T) f(\xi, T)$, then $f(\xi, T) = \xi/\theta k(T)$, and since ξ is bounded and $k(T)$ gets very large at high temperatures, $f(\xi, T)$ must be very small and so $\xi_s(T) \longrightarrow \xi_e(T)$.

It is obviously important to bear in mind these considerations in design. For suppose that a favorable design corresponded to a point just to the right of point 6 in Fig. 7.6. At such a point, the rate of reaction would be high, as witnessed by the rate of heat generation, but the reactor temperature would not be excessive. However, if the feed temperature were to fall slightly the heat removal line might be brought to the left of line B and the reaction be virtually quenched. On the other hand, if the conditions corresponded to a steady state just to the left of point 4 and these were desirable because any temperature to the right of 5 would be dangerous, then the reactor would not be well designed, for a slight increase of feed temperature might bring the heat removal line to the right of line D and the temperature would go to a dangerous level. In such a case, it would be necessary to redesign the cooling system and make κ sufficiently large that only one intersection would be possible. With liquid phase reactions the heat capacity C_P is sufficiently high that a very large heat of reaction is needed if the greatest slope of the heat generation curve is to be greater than the least slope of the heat removal line. But the low heat capacity of gases makes it much easier for there to be multiple steady states. Some combustion chambers are similar to stirred tanks and ignition and quenching of the reaction can be readily observed. For multiple reactions such a sweeping generalization cannot safely be made; in particular, polymerization reactors are notably liable to just the kind of upset we have been describing.

As has been remarked above, the design problem when there is more than one reaction is centered around the question of getting a good yield, that is, of converting the reactants to the desired product rather than to some waste product. The waste product may just be something of no commercial value or it may be harmful to the process as, for example, when carbon forms on a catalyst and destroys its activity. If the requirements on the output are so tightly specified that the extents of all the R reactions are fixed, then for any temperature the mass and heat balance equations may be used to find the holding time and heat removal requirements. However, it is possible that too many restrictions could be put on and no consistent solution would emerge. In particular, the mass balance equations contain as variables only the extents, θ and T. If there are only two reactions and the extents are specified it *may* be possible to solve for the holding time θ and temperature T, but if there are three reactions it will clearly be impossible to specify all three extents independently.

This may be illustrated in an example which, though very simple, is often useful as a first approximation to a more complex situation. If the two irreversible reactions $A \longrightarrow B \longrightarrow C$ take place simultaneously and are first order with rate constants k_1 and k_2, we have (for a feed of pure A) the mass balance equations

$$0 = a_f - a - \theta k_1(T)a, \tag{7.3.23}$$

$$0 = -b + \theta k_1(T)a - \theta k_2(T)b, \tag{7.3.24}$$

$$0 = -c + \theta k_2(T)b. \tag{7.3.25}$$

The third equation is not necessary since by addition $a + b + c = a_f$ and so $c = a_f - a - b$. Since they are linear they can be solved to give

$$a = \frac{a_f}{1 + \theta k_1}, \tag{7.3.26}$$

$$b = \frac{a_f}{1 + \theta k_2} \frac{\theta k_1}{1 + \theta k_1}, \tag{7.3.27}$$

$$c = a_f \frac{\theta k_2}{1 + \theta k_2} \frac{\theta k_1}{1 + \theta k_1}. \tag{7.3.28}$$

If B is the useful product and C the waste material, the yield of the reactor may be defined as the fraction of reacted A that is found as B, namely

$$Y_B = \frac{b}{a_f - a} = \frac{1}{1 + \theta k_2}. \tag{7.3.29}$$

The fractional conversion of A is

$$X_A = \frac{a_f - a}{a_f} = \frac{\theta k_1}{1 + \theta k_1}. \tag{7.3.30}$$

Clearly, Y_B and X_A could be specified and θ and T determined, for

$$\frac{X_A Y_B}{(1 - X_A)(1 - Y_B)} = \frac{k_1}{k_2} = \frac{A_1}{A_2} \exp -\frac{E_1 - E_2}{RT} \qquad (7.3.31)$$

and then

$$\theta = \frac{X_A}{(1 - X_A) k_1(T)}. \qquad (7.3.32)$$

The yield of B always decreases with increasing conversion but it does so less markedly if the ratio k_1/k_2 is large. Thus if $E_1 > E_2$, it is best to have a high temperature and small holding time, while, if $E_2 > E_1$, a low temperature and large holding time are preferable.

The heat balance equation for this reactor could be written as

$$(1 + \kappa)(T - T_c^*) = J_1 \theta k_1 a + J_2 \theta k_2 b$$
$$= J_1 a_f \frac{\theta k_1}{1 + \theta k_1}\left(1 + \frac{J_2}{J_1}\frac{\theta k_2}{1 + \theta k_2}\right). \qquad (7.3.33)$$

As before the left-hand side gives the heat removal line and the right the heat generation curve, and there are now a much greater variety of shapes possible for the latter. Suppose, for example, that both reactions are exothermic but that $E_1 \ll E_2$. We have seen that the functions $\theta k_1/(1 + \theta k_1)$ and $\theta k_2/(1 + \theta k_2)$ both have sigmoidal shapes rising from zero to a value very close to A_1 or A_2. It might well be that, at the temperature at which $\theta k_1/(1 + \theta k_1)$ is rising rapidly, the second function is still very small and so the factor $1 + (J_2/J_1)[\theta k_2/(1 + \theta k_2)]$ would be little different from 1. The shape of the heat generation curve would be as shown at the left of Fig. 7.11 for low temperatures. However, at higher temperatures when the factor $J_1 a_f[\theta k_1/(1 + \theta k_1)]$ has leveled off to a constant value close to $J_1 a_f$, the second factor begins to grow in the same sigmoidal way from 1 to close to $1 + (J_2/J_1)$. The complete heat generation curve takes the form shown in Fig. 7.11 and it is clearly possible to have five steady states cor-

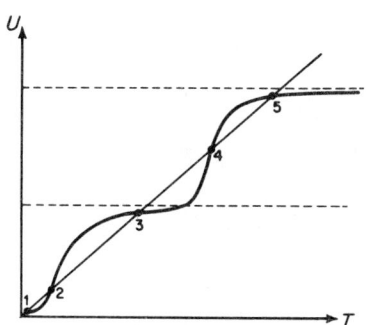

Fig. 7.11 Possible steady states with consecutive reactions.

responding to the five intersections with a suitable heat removal line. We notice that the rate of heat generation could be written

$$J_1 a_f X_A \left\{ 1 + \frac{J_2}{J_1}(1 - Y_B) \right\}.$$

If B is the desirable product, such a steady state as corresponds to point 3 is the best, for the first reaction is going well (X_A is close to 1) whereas the second reaction has scarcely started (Y_B is close to 1). Similar arguments to those that have been used above would indicate that there might be two jumps of operating temperature as the feed temperature is slowly increased. The importance of being aware of this possibility is evident for, though the temperature at point 5 might not be excessive, the yield would be extremely poor. It is again difficult to see how steady states corresponding to points 2 and 4 could ever be reached and in fact it can be proved that they are unstable.

We have chosen an extreme case to illustrate possible complexities but it does not follow that such a multiplicity of steady states will always, or even often, be obtained. With R independent reactions it is possible to have $(2R + 1)$ distinct steady states, of which R are certainly unstable. What is important is that the chemical engineer should be aware that this complexity does exist so that he can be on guard against any dangers that it may hold for his design.

Exercise 7.3.1. Establish Eq. (7.3.8) and compare it with Eq. (7.3.6).

Exercise 7.3.2. Show that the first order irreversible case discussed above can be thrown into dimensionless form with $\alpha = \theta A$ and $\tau = RT/E$ and a dimensionless heat balance

$$\nu(\tau) = \frac{\alpha e^{-1/\tau}}{1 + \alpha e^{-1/\tau}} = \beta(\tau - \tau_c^*).$$

What is β, the slope of the heat removal line? Deduce that $\sigma = d\nu/d\tau$, the slope of the heat generation curve, is given by $\nu(1 - \nu)/\tau^2$ and has its greatest value when $\nu + \tau = \frac{1}{2}$.

Show that if this greatest slope occurs at a temperature τ ($< \frac{1}{2}$), the value of α must be

$$\frac{1 - 2\tau}{1 + 2\tau} \exp \frac{1}{\tau}$$

and that the greatest slope is $(2\tau)^{-2} - 1$.

Exercise 7.3.3. Use the results of the preceding question to show that multiple steady states are only possible if

$$\alpha > \frac{(1 + \beta)^{1/2} - 1}{(1 + \beta)^{1/2} + 1} \exp 2(1 + \beta)^{1/2}.$$

For liquid phase reactions in aqueous or organic solvents, C_p varies between 0.4 and 1 cal/cm³ °K. If E is in the range 10–40 kcal/gm-mole

Sec. 7.4   Some Considerations in Detailed Design   181

and $-\Delta H$ can be between 10–80 kcal/gm-mole and a_f up to 0.25 mole/l, estimate the least value of α for multiple steady states to be possible.

Exercise 7.3.4. Discuss the possible form of the heat generation curve for the first order reactions $A \longrightarrow B \longrightarrow C$ when the second reaction is endothermic.

Exercise 7.3.5. Using the notation of Sec. 4.4 and Exercise 4.4.3, show that if $\theta A'(c_{10} + c_{20}) = \phi$ is the dimensionless holding time, then the steady state extent at dimensionless temperature τ is

$$\zeta_s(\tau) = \frac{(1 + \phi B) - [(1 + \phi B)^2 - 4\beta(1 - \beta)\phi^2 B(B - C)]^{1/2}}{2\phi(B - C)}.$$

If the heat balance is written as

$$\lambda(\tau - \tau_c^*) = \zeta_s(\tau),$$

what is the value of λ?

Exercise 7.3.6. Show that if the greatest production (i.e., greatest value of $q\xi$) is required from a given reactor for a single exothermic reaction, then the heat removal line should pass through the highest point of the heat generation curve (see Fig. 7.10).

Exercise 7.3.7. Denbigh's reaction scheme (Exercise 5.6.5) is being carried out in a stirred tank. Find the holding time that gives the greatest concentration of X in the product and the yield of X for this holding time. (Yield = production of X/consumption of A.) What must be the relationship between the four activation energies if a high temperature is to favor the yield of X?

Exercise 7.3.8. A second order reaction $A + B \longrightarrow 2C$ is taking place in the liquid phase in a continuous flow stirred flask of one l capacity. The feed concentrations of both A and B are $(20/3)$ gm-mole/l and the flow rate 100 cm³/min; their heat capacity is 650 cal/l °C; the rate constant is $3.3 = 10^9 \exp - 20,000/RT$ l/mole min; the heat of reaction $-\Delta H = 20$ kcal/gm-mole. The feed temperature is 17°C and the flask is partially immersed in a bath at 87°C (area of contact 250 cm², heat transfer coefficient 0.1 cal/cm² min). Find all the steady state conditions by plotting the fractional conversion of A against temperature. Use your graph to estimate the feed temperature at which an operating temperature jump such as at D in Fig. 7.7 might occur, all other things being left the same. What would be the magnitude of this jump in temperature?

7.4 Some Considerations in Detailed Design

To illustrate how the steady state equations lead back to the considerations that have to be faced in detailed design and cost analysis, let us sketch the design of a reactor in which two parallel, exothermic first order reactions are taking place. These are

$$A_1 \xrightarrow{k_1} A_3, \qquad A_1 + A_2 \xrightarrow{k_2} A_4,$$

with heats of reaction ΔH_1 and ΔH_2. Their rates are $k_1 c_1$ and $k_2 c_2$, respectively, the second rate being independent of c_1 provided some A_1 is present. Suppose A_4 is the desired product and is required at a rate of $P_4 = 164{,}000$ lb-moles per year. We choose to investigate the design for a particular choice of q and T and shall quote the other fixed values as we go along. By showing how the design could be done for a fixed pair of values of q and T we are in a position to find the best pair of values by repeated trial; in fact, we shall show the dependence of the economic return function on q and T over a range of both these variables. It will become clear that a computer is required if many such calculations have to be done.

If the feed consists of A_1 and A_2, then $c_4 = P_4/q$, and, by stoichiometry,

$$c_4 = \frac{P_4}{q} = c_{2f} - c_2. \tag{7.4.1}$$

The steady state equations for c_1 and c_2 are

$$c_{1f} = c_1(1 + \theta k_1) + \theta k_2 c_2 \tag{7.4.2}$$

$$c_{2f} = c_2(1 + \theta k_2), \tag{7.4.3}$$

and hence

$$V = \frac{P_4}{k_2 c_2} = \frac{P_4 q}{k_2(q c_{2f} - P_4)}. \tag{7.4.4}$$

This expresses the volume of the reactor in terms of T and q when c_{2f} and the kinetic parameters for k_2 are given. Substituting into Eq. (7.4.2) gives

$$c_1 = \frac{q c_{1f} - P_4}{q + V k_1}. \tag{7.4.5}$$

The reactor is assumed to be a cylindrical vessel of a length-to-diameter ratio of 1.75. If 5% of the reactor volume is allowed for an external pump around the exchanger, the diameter of the reactor, d (ft), is given by

$$d = \left(\frac{V}{1.70}\right)^{1/3} \tag{7.4.6}$$

The thickness of the vessel walls will depend on the design pressure. If the reagents have the physical properties of benzene their vapor pressure will be appreciable and the design pressure P_D(psia) should exceed this by 25 psia for safety. If the minimum design pressure is 50 psia, we have

$$P_D = \begin{cases} 50 & T < 196°F \\ 25 + 3.3 \times (T/100)^3 & T > 196°F. \end{cases} \tag{7.4.7}$$

With a corrosion allowance of $\frac{1}{8}$ in. and a working stress of 16,000 psi, the

Sec. 7.4 Some Considerations in Detailed Design

thicknesses of the cylinder and its ends should be $(P_D d/32{,}000) + 0.0104$ and $(P_D d/64{,}000) + 0.0104$, respectively. The weight of steel in the vessel, allowing for a skirt (10 ft long and equal in diameter to that of the vessel) to represent its supports, is

$$W = (0.0909 d^3 + 0.482 d^2) P_D + (36.6 d^2 + 160.5 d). \qquad (7.4.8)$$

This can be calculated as a function of q and T from Eqs. (7.4.4), (7.4.6), and (7.4.7). Though the exact breakpoint in design temperature is 196°F according to Eq. (7.4.7), it will be convenient to round this off to 200°F in what follows.

The vessel will also have to be insulated if its temperature is above 200°F; the thickness of insulation would be $(T/150) - 1$. The cost of insulation depends on the product of its thickness and the area to be covered, and in this case is $(17.2 + 0.0133 T) d^2$. The cost of fabricating a vessel of weight W is taken to be $3.5 W^{0.782}$ and, since the same break point has been chosen, the cost of the vessel plus insulation may be written as

$$\begin{cases} 3.5(4.56 d^3 + 60.7 d^2 + 160.5 d)^{0.782}, & T < 200°\text{F} \\ 3.5[(0.0909 d^3 + 0.482 d^2)(25 + 3.3 \times 10^{-6} T^3) + 36.6 d^2 \\ \qquad + 160.5 d]^{0.782} + [17.2 + 0.0133 T] d^2, & T > 200°\text{F}. \end{cases}$$

Arbitrary limits were imposed on d, namely that it should lie between 1 ft 3 in. and 9 ft 8 in. If d was found to exceed this, a number n_1 of parallel units, with feed equally divided between them, were to be used.

An external tube and shell heat exchanger was chosen for reasons of flexibility and the following assumptions were made:

(i) The temperature drop of reactants in the cooler is 10°F;

(ii) Constant cooling-water velocity of 3 ft/sec through standard 1-in. 16 BWG, 16 ft exchanger tubes;

(iii) Fouling factors of 0.002 and 0.005 (Btu/lb °F ft^2)$^{-1}$ on the water and reagent sides;

(iv) In evaluating the overall heat transfer coefficient for the exchanger tube–side film resistance and wall resistances were taken as constant at 0.00103 and 0.00022, respectively; the shell side resistance was taken to be temperature dependent. Then by standard calculations the overall heat transfer coefficient is

$$h = 43 + 0.0452 T. \qquad (7.4.9)$$

The coolant temperature is such that the LMTD is $(T - 94)$, so that if Q^* is the required heat removal rate, the interchanger area required is

$$a = \frac{Q^*}{(43 + 0.0452 T)(T - 94)}. \qquad (7.4.10)$$

The area per shell is restricted to lie between 50 and 4000 ft^2; anything larger than this requires n_2 coolers in parallel. With a minimum design pressure of 150 psia the cost is $193a^{0.546}$, but above 330°F this must be corrected by a pressure correction factor of

$$P_{CF} = 0.962 + 1.68 \times 10^{-9}T^3. \qquad (7.4.11)$$

Thus C_2, the cost of the exchanger, is:

$$C_2 = \begin{cases} 193n_2 a^{0.546} & T < 330°F \\ 193 P_{CF} n_2 a^{0.546} & T > 330°F. \end{cases} \qquad (7.4.12)$$

Now the heat balance gives

$$Q^* = V(-\Delta H_1)k_1 c_1 + V(-\Delta H_2)k_2 c_2 + q(T_f - T), \qquad (7.4.13)$$

and Q^* can be calculated by Eqs. (7.4.4) and (7.4.5) as a function of T and q; $T_f = 100°F$; then Eqs. (7.4.10), (7.4.11), and (7.4.12) give C_2.

The miscellaneous costs are:

(i) Foundations and platforms, estimated at

$$C_3 = 1000 + 100d; \qquad (7.4.14)$$

(ii) Mixer, at input of 0.05 hp per unit volume and cost

$$C_4 = 255(V)^{0.3}; \qquad (7.4.15)$$

(iii) Reactor feed pump: if the feed pump has to boost the pressure to 50 psi above the pressure in the reactor and is 70% efficient, it will require a horsepower of

$$H_1 = (6.95 \times 10^{-4} + 4.59 \times 10^{-11}T^3)q. \qquad (7.4.16)$$

The cost of such a pump is estimated to be

$$C_5 = 580 H_1^{0.467}; \qquad (7.4.17)$$

(iv) Circulating pump for cooler: if the temperature drop of the reactants on passing through the interchanger is to be 10°F and their specific heat is 0.5 Btu/lb, a mass flow rate of $0.2Q^*$ will be required to circulate. Let us boost this to $0.25Q^*$ for flexibility in control and assume a 40 psi pressure increase and 70% efficiency; then the required horsepower is

$$H_2 = 1.88 \times 10^{-5} Q^*. \qquad (7.4.18)$$

Above 250°F the pump design must be modified and more expensive packing glands must be used; thus

$$C_6 = \begin{cases} 580 H_2^{0.467} & T < 250°F \\ 922 H_2^{0.467} & T > 250°F. \end{cases} \qquad (7.4.19)$$

All these cost estimates can be evaluated for any choice of T and q and give the total design cost

$$C = n_1 \sum_1^6 C_p. \qquad (7.4.20)$$

Sec. 7.4 Some Considerations in Detailed Design

To allow for installation, piping, instrumentation, and contingencies, this cost is multiplied by a factor of 5 to give a total investment of

$$I = 5n_1 \sum_1^6 C_r. \qquad (7.4.21)$$

Based on such a reasonably detailed estimate of the cost of constructing this type of reactor, let us now see what the economic return would be.

The running costs are as follows:

(i) Labor costs of 4 man-hours per shift at $2.52 per hr would amount to $11,000 per year;

(ii) Plant burden for depreciation, taxes, insurance, and maintenance is reckoned at 18% of the investment I per year;

(iii) Utility costs per year per reactor are:
 (a) mixed power, $3.10\,V$;
 (b) feed pump power, $61.10 H_1$;
 (c) circulating pump power, $0.00115 Q^*$;
 (d) cooling water, $6.92 Q^*$;

(iv) Raw material cost. Assuming cheap recovery and recalling Eq. (7.4.1), the raw material cost is

$$410 q(c_{1f} - c_1) + 82{,}000$$

at prices per mole of 0.05 and 0.50 for A_1 and A_2.

The sum of these is the total operating cost, and this must be subtracted from the value of the product, which at $3/mole of A_4 is $492,000, to give the gross profit. If the net profit II is taken to be half the gross, the percent return is

$$E = \frac{II}{I} \qquad (7.4.22)$$

We have thus arrived at a single economic parameter as a basis for comparing designs, which are calculable as soon as T and q are chosen. The complexity of the design makes this calculation laborious, elementary though it is, and clearly calls for the use of a computer.

In a very careful study, on which the foregoing description is based, G. T. Westbrook examined some 7000 complete designs (including a test for stability which we give in the next section) using a computer program of more than 100 instructions. He covered the range $100 \leq T \leq 500$ and $1000 \leq q \leq 8000$, and varied the heats of reaction, activation energies, and frequency factors over wide ranges. Figures 7.12 and 7.13 show the economic parameter E as a function of q and T in two cases. The greatest economic return is obtained on the adiabatic boundary where no cooling is required. What is very significant is the fact that in any one case the return can vary by a factor of as much as 4 between the best and the worst choices of q and

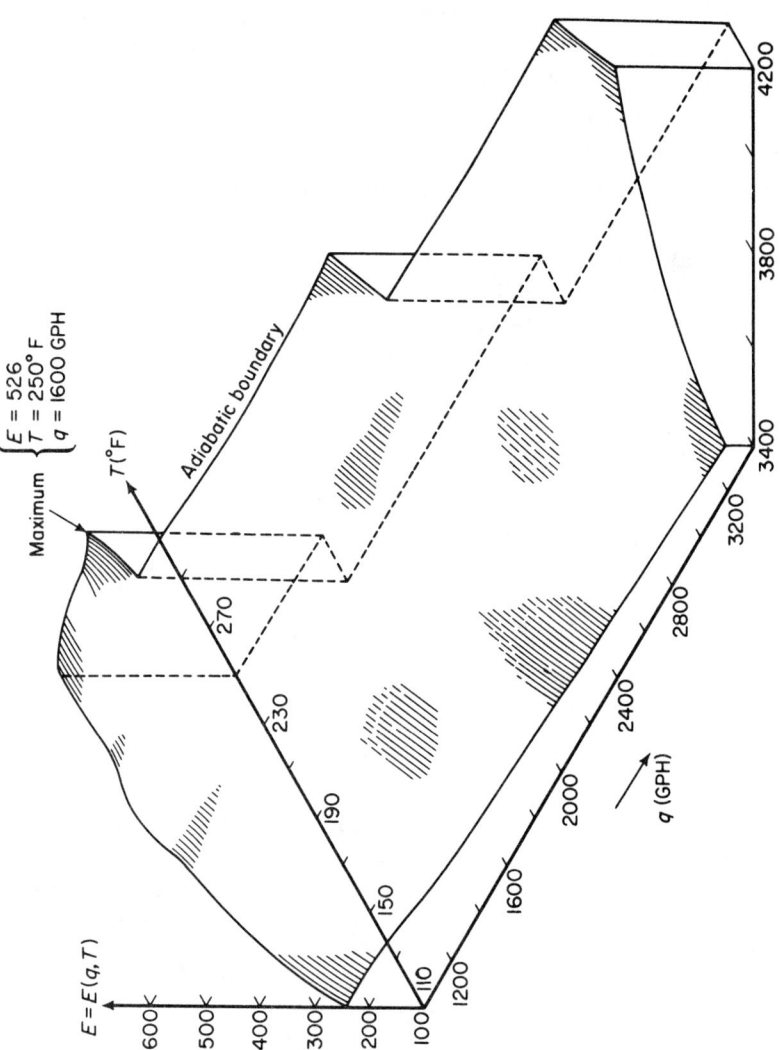

Fig. 7.12 Economic return from a single reactor ($\Delta H_1 = \Delta H_2 = -8000$). Reproduced from *Ind. Eng. Chem.*, **53**, 181 (1961). Copyright 1961 by the ACS. Reproduced by permission of the copyright owner.

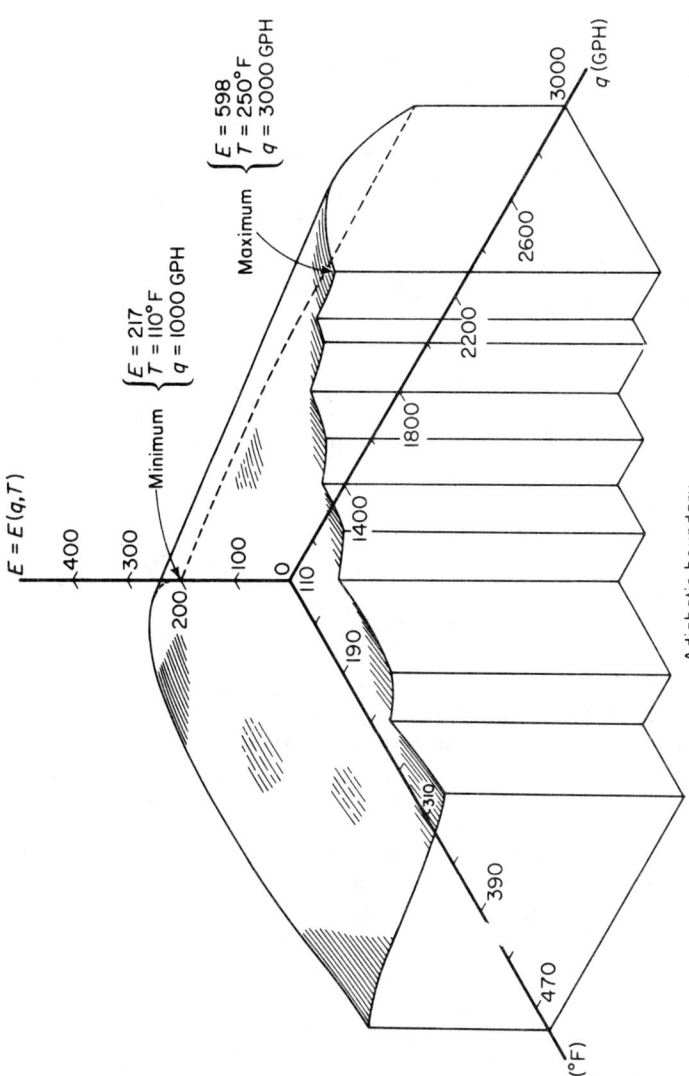

Fig. 7.13 Economic return from a single reactor ($\Delta H_1 = \Delta H_2 = -30,000$). Reproduced from *Ind. Eng. Chem.*, **53**, 181 (1961). Copyright 1961 by the ACS. Reproduced by permission of the copyright owner.

T, as shown in Table 7.1. Nevertheless, the worst choices of q and T are quite reasonable ones, so that without this detailed exploration a very inferior return might be obtained. With a big investment even the high cost of electronic computing is fully justified.

TABLE 7.1
EXAMPLES OF THE RANGE OF E

Heats of reaction (cal)		Activation energies (cal/gm-mole)		Frequency factors (min^{-1})		Economic return (%)	
$-\Delta H_1$	$-\Delta H_2$	E_1	E_2	A_1	A_2	E_{min}	E_{max}
8,192	8,192	3,000	1,000	200	100	234	771
3,000	30,000	4,000	1,000	200	100	302	884
4,000	6,000	4,000	4,000	100	10	108	397
8,000	8,000	7,800	4,000	100	10	134	421
10,000	30,000	10,000	1,000	500	1	120	432

Exercise 7.4.1. A stirred tank reactor is designed for the reaction $A + B = C + D$ under the following conditions:
  Feed: A and B each at a concentration of 1 lb-mole/cu ft
  Flow rate: $q = 50$ cu ft/min
  Heat capacity of reactants: $C_P = 50$ Btu/cu ft
  Feed temperature: $T_f = 350°$K
  Values of one mole of A, B, C, and D: \$1, \$1, \$10, and \$2, respectively
  Cost of reactor of V cu ft: $2.07V$ \$/hr
  Cost of removing Q Btu/min: $0.0324Q$ cent/min
  Operating temperature: $T = 400°$K
  The extent of reaction ξ may be taken to be the concentration of C, and the reaction rate, in lb-mole/cu ft min, is:

$$r(c, T) = (\tfrac{10}{9}) \left[\exp\left(12.5 - \frac{5000}{T}\right) \right] (1 - \xi)^2 - 2\left[\exp\left(25 - \frac{10,000}{T}\right) \right] \xi^2.$$

The rate of profit is the rate of increase in the value of the process stream less the reactor and cooling costs. Find the value of V that maximizes the profit and the corresponding maximum profit and volume of the reactor. Explain how you would check that this is truly a maximum.

7.5 Stability of the Steady State

We have already had reason to notice that it is difficult to reach the intermediate one of the three possible steady states which can exist. This is because it is unstable in the following sense. Figure 7.14 shows the heat generation

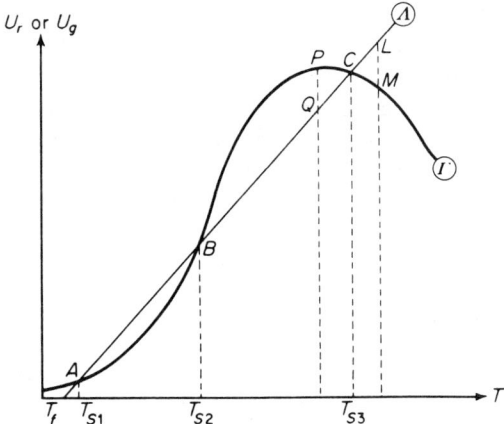

Fig. 7.14 Heat generation and rejection curves for a single exothermic reaction.

curve, Γ, for a general exothermic reaction with a superposed heat removal line, Λ, for just such a feed temperature T_f as would give three intersections. These are the points A, B, and C with corresponding steady state temperatures T_{s1}, T_{s2}, and T_{s3}. (The figure is drawn as if for an adiabatic reactor with $\kappa = 0$ and so $T_c^* = T_f$, but it could equally well stand for the general case with T_f replaced by T_c^*.) Suppose it were possible to start the reactor exactly at temperature T_{s2} and corresponding extent $\xi_s(T_{s2})$; then if some slight upset in the operating conditions were to take it away from this steady state, it would not go back to it but would finish up either at the low or at the high temperature steady state, at A or at B. By contrast, these states, A and B, are stable and if the state were perturbed away from one of them it would return there naturally.

For the moment, we shall consider the adiabatic case since the argument we wish to use is then valid both for stability and instability. Suppose the reactor were operating at the steady state corresponding to point C in Fig. 7.14. A slight increase in temperature would bring us to a region where Λ, the heat removal line, lies above Γ, the heat generation curve, as can be seen from points L and M. Thus at temperatures just above the steady state there is a net removal of heat which would tend to reduce the temperature. On the other hand, a slight decrease of temperature from C brings us to a region where the heat generation (corresponding to point P) is greater than the heat removal (corresponding to Q), so that there is a net generation of heat and a tendency for the temperature to rise. This suggests that the reactor has a built-in propensity to remain at the steady state C, due to the fact that the slope of the heat removal line is greater than that of this heat generation curve.

By contrast, at the intermediate steady state B, the slope of the heat generation curve is greater than that of the heat removal line. Hence a slight increase in temperature produces a net heat generation which would tend to drive the temperature even higher. Conversely, a slight drop of temperature induces a net removal of heat, which would cause the temperature to fall even more. In this sense, we say that the intermediate steady state is unstable.

This is the classical argument introduced by van Heerden in 1953 for the adiabatic stirred tank. It is a most important one to grasp firmly for it can be used in more complicated situations to get some insight into the stability of a system. However, its limitations must be also thoroughly understood. In particular, it can be used to establish instability, but it does not count conclusively for stability because of several reasons. First, we should be suspicious of a single condition for a system in which there are two variables. Second, the diagram for the heat generation was drawn in a rather special way, for the steady state–mass balance equation, $\xi = \theta r(\xi, T)$, was first solved for $\xi_s(T)$, the steady state extent for any given reactor temperature. This means that when we perturb the temperature from T_s to $T_s + \delta T$ and then compare the rates of heat generation and removal, we are making the comparison between states $[\xi_s(T_s), T_s]$ and $[\xi_s(T_s + \delta T), T_s + \delta T]$. Thus two rather particular states are being compared, for $\xi_s(T_s + \delta T) = \xi_s + (d\xi_s/dT)\delta T$, and the perturbation $\delta\xi$ in the extent is proportional to the perturbation δT in temperature. A completely general perturbation away from (ξ_s, T_s) would be to use $(\xi_s + \delta\xi, T_s + \delta T)$ where $\delta\xi$ is not necessarily proportional to δT. If we could show that for such a perturbation there would be a net removal of heat for $\delta T > 0$ and a net generation for $\delta T < 0$, then we might be able to show that the state was stable. It follows that the rather special nature of the perturbation implied by the heat generation and removal diagram does not allow us to conclude stability when $dU_g/dT < dU_r/dT$. However, it does allow us to say the state will be unstable when

$$\frac{dU_r}{dT} < \frac{dU_g}{dT}. \tag{7.5.1}$$

For to prove instability, we only have to show some circumstance, no matter how special, under which there will be a tendency to fly away from the steady state. To put it in another way, we can say that Eq. (7.5.1) is a sufficient condition for instability, while its converse, $dU_r/dT < dU_g/dT$ is a necessary, but not sufficient, condition for stability.

The complete set of necessary and sufficient conditions for stability, as first given by Amundson in 1955, is derived in a rather different way. The basic idea is to focus attention on small perturbations away from a given steady state. If they are sufficiently small, they can be described by linear equations and we shall be able to see just how they grow or die away. It can be proved that this establishes local stability, in the sense that sufficiently small perturbations will certainly die out. It does not say anything

Sec. 7.5   Stability of the Steady State   191

about the possibility that after a really large perturbation, the reactor might return to some other steady state; for this the nonlinear equations have to be solved. We shall show this method by examining the reaction $A \longrightarrow B$ in some detail.

For a feed of pure A, the general mass balance equations are

$$\theta \frac{da}{dt} = a_f - a - \theta k(T)a \qquad (7.5.2)$$

and

$$\theta \frac{db}{dt} = -b + \theta k(T)a. \qquad (7.5.3)$$

These can be obtained from Eqs. (7.1.3) and (7.1.4) by setting $k' = 0$ and thus making the reaction irreversible. We shall assume that the initial state is always compatible with the feed, as is possible if the feed is always pure A. Then for the extent $\xi = a_f - a = b$ we have

$$\theta \frac{d\xi}{dt} = -\xi + \theta k(T)(a_f - \xi) \qquad (7.5.4)$$

[see Eq. (7.1.7)]. The heat balance equation is

$$\theta \frac{dT}{dt} = T_f - T + J\theta k(T)(a_f - \xi) - \kappa(T - T_{cf}),$$
$$= -(1 + \kappa)(T - T_c^*) + J\theta k(T)(a_f - \xi). \qquad (7.5.5)$$

The steady state equations are

$$0 = -\xi_s + \theta k(T_s)(a_f - \xi_s) \qquad (7.5.6)$$
$$0 = -(1 + \kappa)(T_s - T_c^*) + J\theta k(T_s)(a_f - \xi_s). \qquad (7.5.7)$$

We now subtract Eq. (7.5.6) from Eq. (7.5.4) and Eq. (7.5.7) from Eq. (7.5.5). Then

$$\theta \frac{d\xi}{dt} = -(\xi - \xi_s) + \theta[k(T)(a_f - \xi) - k(T_s)(a_f - \xi_s)] \qquad (7.5.8)$$

and

$$\theta \frac{dT}{dt} = -(1 + \kappa)(T - T_s) + J\theta[k(T)(a_f - \xi) - k(T_s)(a_f - \xi_s)]. \qquad (7.5.9)$$

If

$$\xi - \xi_s = x, \quad T - T_s = y, \qquad (7.5.10)$$

and x and y are sufficiently small that their squares, product, and all higher powers can be neglected, we can expand the expression in brackets by Taylor's theorem, giving

$$k(T_s + y)(a_f - \xi_s - x) - k(T_s)(a_f - \xi_s) = -k(T_s)x + k'(T_s)(a_f - \xi_s)y,$$

where

$$k'(T_s) = \left(\frac{dk}{dT}\right)_s = \frac{E}{RT_s^2}k(T_s).$$

Substituting this into the two equations and noting that $d\xi/dt = dx/dt$ and $dT/dt = dy/dt$, we have

$$\theta\frac{dx}{dt} = -[1 + \theta k(T_s)]x + \frac{E}{RT_s^2}\theta k(T_s)(a_f - \xi_s)y$$

$$\theta\frac{dy}{dt} = -J\theta k(T_s)x - \left[1 + \kappa - J\frac{E}{RT_s^2}\theta k(T_s)(a_f - \xi_s)\right]y.$$

Now all the coefficients on the right are evaluated for the steady state we are considering and so are constants. Let us write them as

$$L = 1 + \theta k(T_s), \qquad M = 1 + \kappa, \qquad N = \frac{JE}{RT_s^2}\theta k(T_s)(a_f - \xi_s), \qquad (7.5.11)$$

so that

$$\theta\frac{dx}{dt} = -Lx + \frac{N}{J}y \qquad (7.5.12)$$

$$\theta\frac{dy}{dt} = J(1 - L)x - (M - N)y. \qquad (7.5.13)$$

Dropping the nonlinear terms is justified, provided that x and y are sufficiently small. We have here a pair of equations which could be solved to see what happens to an initial perturbation x_0, y_0.

It was just such a pair of equations as this which we reviewed in Sec. 5.1(IV). From a comparison with these equations (replacing x, y, and z of Sec. 5.1 by t/θ, x, and y; and α, β, γ, and δ by $-L$, N/J, $J(1 - L)$, and $-(M - N)$, respectively) we see that the solution will be

$$\begin{aligned}x(t) &= A_1 e^{m_1 t/\theta} + A_2 e^{m_2 t/\theta},\\ y(t) &= B_1 e^{m_1 t/\theta} + B_2 e^{m_2 t/\theta},\end{aligned} \qquad (7.5.14)$$

and m_1 and m_2 are roots of

$$m^2 + (L + M - N)m + (LM - N) = 0. \qquad (7.5.15)$$

If the roots are real, the possible types of behavior of x or y are shown in Fig. 7.15. If both m_1 and m_2 are negative, x and y will behave like one of the curves A, B, or C depending on the values of the constants A_1, A_2, B_1, and B_2. Clearly, this indicates stability, for the perturbation dies away. If either m_1 or m_2 is zero, the behavior will be similar to curve D and this is not very satisfactory since x and y do not go to zero, even though they may stay small. If either m_1 or m_2 is positive, then the perturbations will grow exponentially, as shown by curves E and F. Evidently, it is necessary for stability that, when m_1 and m_2 are real, they should both be negative. But then their sum

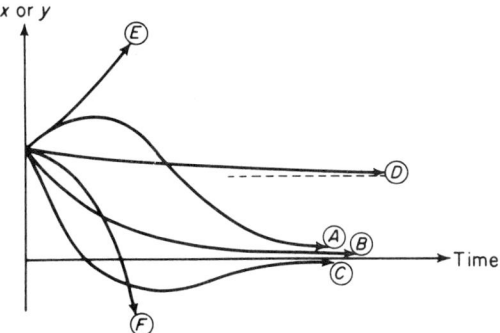

Fig. 7.15 Types of solution corresponding to real roots.

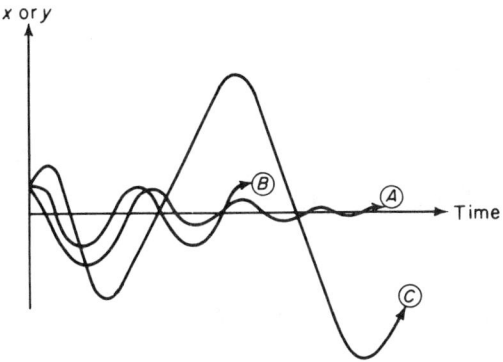

Fig. 7.16 Types of solution corresponding to complex roots.

$m_1 + m_2 = -(L + M - N)$ would be negative and their product $m_1 m_2 = LM - N$ would be positive, that is

$$L + M > N \quad \text{and} \quad LM > N. \tag{7.5.16}$$

If m_1 and m_2 are complex conjugates, the types of behavior are shown in Fig. 7.16. Curve A corresponds to a negative real part of m_1 and m_2, giving a damped oscillation in which the perturbations die out. If the real part is zero, a sustained oscillation is possible (curve B), but if the real part is positive, the amplitude of the perturbations will grow exponentially (curve C). Now for complex roots, the product $m_1 m_2$ is always positive and the sum $(m_1 + m_2)$ is twice the real part. Hence the condition for stability is again that the sum of the roots be negative and the product positive. It follows that Eq. (7.5.16) gives the necessary and sufficient conditions for a sufficiently small perturbation to die away and hence for the steady state to be stable.

In the case of the first order reaction we have from Eq. (7.5.6) that

$$a_f - \xi_s = \frac{a_f}{1 + \theta k(T_s)}$$

so that

$$N = a_f \frac{JE}{RT_s^2} \frac{\theta k(T_s)}{1 + \theta k(T_s)}.$$

Now

$$U_g = J\theta k(T_s)(a_f - \xi_s) = Ja_f \frac{\theta k(T_s)}{1 + \theta k(T_s)},$$

so

$$\frac{dU_g}{dT_s} = a_f \frac{JE}{RT_s^2} \frac{\theta k(T_s)}{[1 + \theta k(T_s)]^2} = \frac{N}{L}. \tag{7.5.17}$$

On the other hand,

$$\frac{dU_r}{dT_s} = 1 + \kappa = M, \tag{7.5.18}$$

and hence the condition $dU_r/dT > dU_g/dT$ is just $M > N/L$, that is the second of the two conditions Eq. (7.5.16). We again see that the condition on the slopes of the U_g and U_r curves is necessary [being one of the conditions (7.5.16)], but not sufficient (being only one of two conditions). On the other hand, the steady state is certainly unstable if either of the conditions is violated, and so Eq. (7.5.1) is a sufficient condition for instability.

We have done this in detail for the first order irreversible reaction, but exactly the same method can be employed with the most general reaction rate expression, $r(\xi, T)$ (see Exercise 7.5.4). In this case,

$$L = 1 - \theta\left(\frac{\partial r}{\partial \xi}\right)_s, \quad M = 1 + \left(\frac{dQ}{dT}\right)_s, \quad N = J\theta\left(\frac{\partial r}{\partial T}\right)_s, \tag{7.5.19}$$

and the necessary and sufficient conditions for stability are again given by Eq. (7.5.16). It can be shown (see Exercise 7.5.4) that the possibility of a perturbation which is incompatible with the feed does not affect the stability; indeed we have seen that the incompatibility variable ζ always decays exponentially. The physical meaning of the stability parameters L, M, and N is worth remarking. If an increment $\delta\xi$ is made in the extent, it will induce an increment $q\alpha_j \delta\xi = \alpha_j \delta P$ in the flow of any species A_j in the product stream. At the same time, it will cause an increment of $(\partial r/\partial \xi)_s \delta\xi$ in the reaction rate. The net difference in production rate is $q\alpha_j \delta\xi - V\alpha_j(\partial r/\partial \xi)_s \delta\xi = \alpha_j \delta R$. Then

$$L = 1 - \theta\left(\frac{\partial r}{\partial \xi}\right)_s = \frac{q\,\delta\xi - V(\partial r/\partial \xi)_s\,\delta\xi}{q\,\delta\xi} = \frac{\delta R}{\delta P}. \tag{7.5.20}$$

Similarly, an increment of δT in the reactor temperature will produce an increment in the enthalpy flux in the product stream of $qC_P \delta T = \delta Q_f$. It will also increase the rate of cooling by $(dQ^*/dT)_s \delta T$ and the rate of cooling due to heating the incoming feed by $qC_P \delta T$. Hence the total heat removal rate is increased by $[qC_P + (dQ^*/dT)_s] \delta T = \delta Q_r$, and

$$M = 1 + \left(\frac{dQ}{dT}\right)_s = 1 + \frac{(dQ^*/dT)\delta T}{qC_P \delta T} = \frac{\delta Q_r}{\delta Q_f}. \qquad (7.5.21)$$

Finally, an increment of δT will produce an increment in the rate of heat production of $(-\Delta H)V(\partial r/\partial T)_s \delta T = \delta Q_g$, and so

$$N = J\theta\left(\frac{\partial r}{\partial T}\right)_s = \frac{(-\Delta H)V(\partial r/\partial T)_s \delta T}{qC_P \delta T} = \frac{\delta Q_g}{\delta Q_f}. \qquad (7.5.22)$$

The two conditions $N < LM$, and $N < L + M$ are

$$\delta Q_g < \delta Q_r \frac{\delta R}{\delta P} \qquad \delta Q_g < \delta Q_r + \delta Q_f \frac{\delta R}{\delta P}, \qquad (7.5.23)$$

and they give precise form to the instinctive feeling that the increment of heat generation must not be too large relative to the increment in heat removal and the other increments.

Before leaving the question of stability, a valuable way of presenting the solution to the nonlinear equations is worth remarking:

$$\theta\frac{d\xi}{dt} = -\xi + \theta r(\xi, T)$$
$$\theta\frac{dT}{dt} = T_f - T - Q(T) + J\theta r(\xi, T). \qquad (7.5.24)$$

They will have to be solved numerically, but no matter how complex the reaction rate expression this can be certainly done (see Sec. 5.8). If we do this for any initial values $\xi(0) = \xi_0$ and $T(0) = T_0$ we can draw a curve representing the solution in the ξ, T plane by plotting points $[\xi(t), T(t)]$ for various t. The only place where two such trajectories can cross is at a steady state. The slope of the trajectory is $d\xi/dT$, which can be calculated as $(d\xi/dt)/(dT/dt)$, namely the ratio of the right-hand sides of Eqs. (7.5.24). Except at points where both of these right-hand sides vanish—and these are just the steady states—this ratio can only have one value. But the crossing of two trajectories would imply two different slopes at the intersection point.

Figure 7.17 shows just such a phase-plane plot or phase portrait, as it is called, for an irreversible first order reaction with a single steady state. The steady state is at a temperature of 337°K and an extent of 0.0021, and evidently, whatever the initial condition may be, the reactor reaches this final steady state, which is therefore globally stable. However, the phase portrait gives important information about what happens during start up. For example, if the reactor is filled at a temperature of 320°K so rapidly that no reaction takes place, the curve starting at $\xi = 0$ and $T = 320$ shows how the reactor behaves. Evidently the temperature and extent increase together and overshoot the steady state values slightly before settling down. Just how rapidly it happens cannot be judged from this diagram. For this

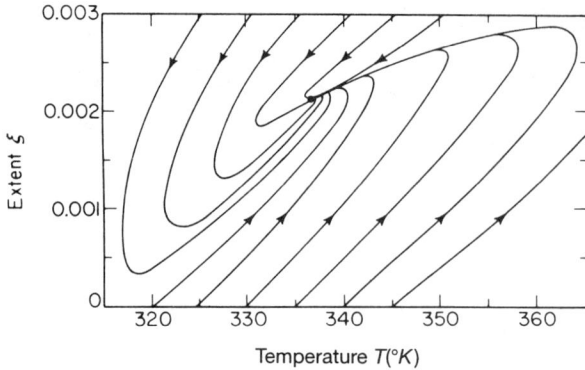

Fig. 7.17 Phase portrait of the reactor with a single steady state. [Adapted from calculations of Bilous and Amundson, with permission.]

it would be necessary to graduate the curves with the values of t. The overshoot in temperature is quite mild (about 3°K) when the start up is from 320°K. However, from $\xi = 0$ and $T = 340$, the trajectory reaches a temperature of 364°K before settling down to a steady state. This excess of 27°K might be important in some circumstances and, although the designer is concerned first with the steady state, he cannot afford to overlook the transients lest they wreck his plant before the steady state is reached.

The phase portrait when the reactor has three possible steady states is even more interesting. Figure 7.18 shows a typical case with points A, B, and C marking the three steady states. A is the low temperature steady state with poor conversion; C is the high temperature steady state with a good conversion. Both are stable in the sense that a small perturbation will die out. Suppose however that we filled up the reactor so slowly that the reaction went almost to completion by the time it was full and heated up. If our intention was to operate at C, we might want to heat it up to the same temperature as the final steady state ($T = 2.2J$). The starting point would thus be I which has an extent $\xi = 1$ and temperature $T = 2.2J$. However, the solution shows that we do not get to the desirable steady state C, but go instead to A. It is necessary to heat it to a slightly higher temperature, such as at the point J, if the trajectory is to lead to C. On the other hand, if the reactor is filled very rapidly so that ξ is negligible at the start, then any starting temperature to the right of point D would give a trajectory leading to C. Suppose, however, that the starting temperature were close to the desired steady state temperature, say at point K. Then there would be a very rapid rise of reactor temperature before it started to fall to the steady state; in fact the trajectory from K goes out of the figure and up to nearly $T/J = 3$ before returning close to the upper boundary above the broken curve BC. The broken line EBD is known as the *separatrix* and divides the plane into two regions of

Sec. 7.5   Stability of the Steady State   197

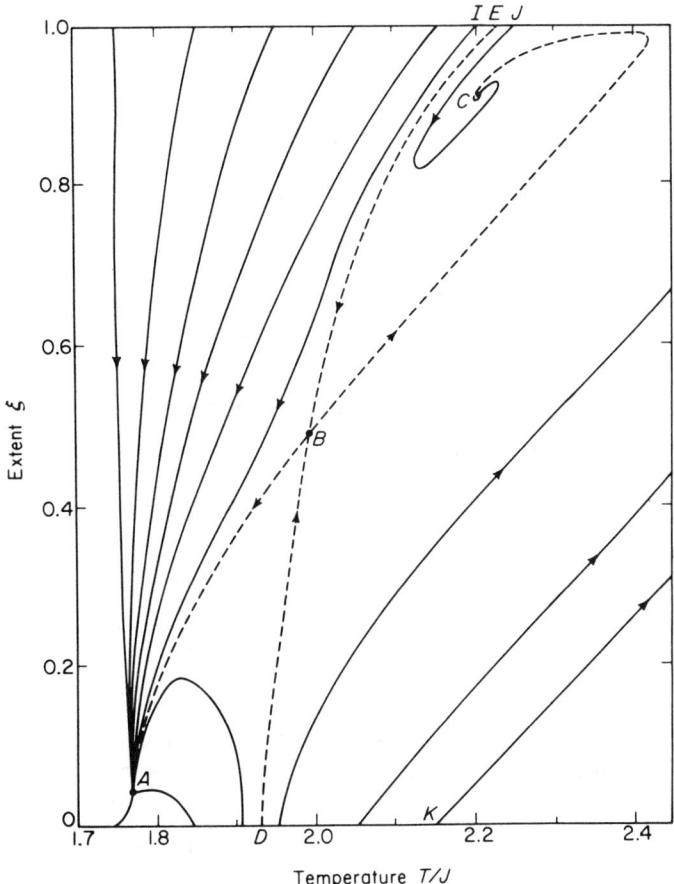

Fig. 7.18 Phase portrait for reactor with three steady states. (Adapted from Aris and Amundson, with permission.)

influence. To start to the left of it is to finish at A, and to the right, to go to C. It is theoretically possible to reach the unstable steady state by starting precisely at D or at E, but this could never happen in practice for the very least error in the starting point or the slightest perturbation on the path would make the state of the reactor swing away to A or C. It is not even possible to compute the path starting from D or E because the rounding off error in computation is likewise magnified. It is necessary to change the sign of time and go backwards, starting from a point on BE or BD close to B.

This kind of representation of transient behavior has been used for even more complicated situations. It provides a useful picture of the whole range of starting conditions and reveals possible failures or dangers.

Exercise 7.5.1. By writing the transient equations as

$$\theta \frac{d\xi}{dt} = -\xi + \theta r(\xi, T)$$

$$\theta \frac{dT}{dt} = T_f - T - Q(T) + J\theta r(\xi, T),$$

subtracting the steady state equations, and expanding by Taylor's theorem with $\xi = \xi_s + x$ and $T = T_s + y$, obtain Eqs. (7.5.12), (7.5.13), and (7.5.19).

Exercise 7.5.2. Show that for a single endothermic reaction the steady state is always stable. Show also that in the case of an adiabatic reactor the slope condition $LM > N$ is sufficient as well as necessary.

Exercise 7.5.3. Show that a reactor is stable when its productivity is greatest.

Exercise 7.5.4. If the perturbation is incompatible with the feed, we must make the rate a function of ξ, ζ, and T, i.e., $r(\xi, \zeta, T)$, and add to the equations in Exercise 7.5.1 the further equation

$$\theta \frac{d\zeta}{dt} = -\zeta.$$

Linearize these three equations and, by trying a solution in which each deviation is proportional to $\exp mt/\theta$, show that

$$(m+1)[m^2 + (L + M - N)m + (LM - N)] = 0.$$

Deduce that the criteria of Eq. (7.5.16) are both necessary and sufficient for stability under the most general perturbation in concentration and temperature.

Exercise 7.5.5. Is the state for the reactor designed in Exercise 7.4.1 stable?

7.6 Control of the Steady State

If the desirable steady state is unstable or if perturbations die away rather slowly, we may ask whether it is possible to make it stable or improve its stability. The answer is that it is always possible to improve the stability by making the heat removal line steeper. For since $L > 0$ we can increase both $L + M$ and LM by increasing M, the slope of the heat removal line. If one or both of them were less than N we would only have to make M greater than the greater of the two numbers $(N - L)$ and N/L for the stability criteria to be satisfied. Now $M = 1 + \kappa$ can be increased either in the basic design which fixes κ or by the addition of a control system. The simplest control system we could think of would measure the departure of the temperature from steady state, increasing the rate of coolant flow if the temperature deviation were positive and decreasing it if it were negative. In fact, by Eq. (7.3.6) or Eq. (7.3.8) Q is a function of both T and q_c, $Q(T, q_c)$.

Suppose that the basic design calls for a steady state temperature T_s and coolant flow rate so that, for a fixed $q_c = q_{cs}$,

$$M = 1 + \kappa = 1 + \left[\frac{d}{dT} Q(T, q_{cs})\right]_{T=T_s}. \tag{7.6.1}$$

If we make

$$q_c = q_{cs} + \mu^*(T - T_s), \tag{7.6.2}$$

where μ^* is a positive proportionality constant, then this will represent a controller which increases the coolant flow rate proportionately to the temperature deviation. But now

$$\frac{dQ}{dT} = \left(\frac{\partial Q}{\partial T}\right)_s + \left(\frac{\partial Q}{\partial q_c}\right)_s \frac{dq_c}{dT} = \kappa + \mu^* \left(\frac{\partial Q}{\partial q_c}\right)_s.$$

Since $\partial Q/\partial q_c$ is always positive we can write $\mu = \mu^*(\partial Q/\partial q_c)_s$ and the action of the controller is to increase from $M_0 = 1 + \kappa$, with no control, to $M_c = 1 + \kappa + \mu$, when control is used. A sufficiently large value of μ^*, and so of μ, will thus always achieve stability, or improve it. Of course, a large gain is not always desirable from other points of view so that a good designer will consider the steady state design (involving the choice of κ) as well as that of the imposed control (corresponding to the choice of μ). It is highly undesirable to design for an ill-understood steady state and rely on a sufficiently good controller to make it work.

Another phenomenon which these nonlinear equations can show is that of the limit cycle in which the reactor tends to take up a continual, very nonlinear oscillation. An example of this was found when the control of the unstable state B of Fig. 7.18 was studied. For this case $L = 2$ and $L + M - N = LM - N = -2.25$. Addition of the control would thus make $L + M_c - N = -2.25 + \mu$ and $LM_c - N = -2.25 + 2\mu$, and the steady state B is stable if $\mu > 2.25$. However, because of other changes, which we shall not go into here, at a value of $\mu = 2$, when the steady state B was still unstable, the steady states A and C had disappeared. The phase plane must now contain a limit cycle as shown in Fig. 7.19, and every trajectory ultimately winds itself around the limit cycle. The behavior of temperature and extent as functions of time is shown in Fig. 7.20. The point does not move uniformly around the limit cycle in the phase plane, but much more rapidly at the high temperature end. Thus the reactor gets into this oscillatory behavior in which there are periodic excursions to high temperatures. Moreover, this limit cycle is quite stable in the sense that any perturbation from it will die out and that starting at any point whatever the oscillation eventually builds up.

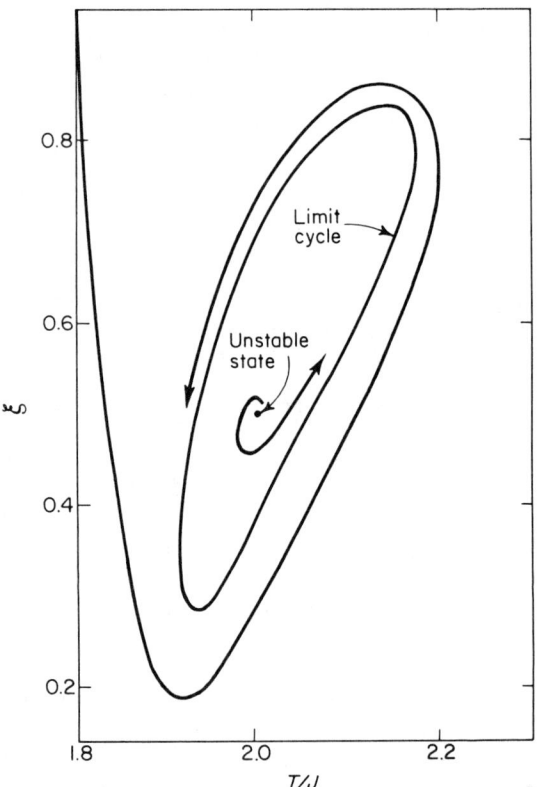

Fig. 7.19 The phase plane with a limit cycle.

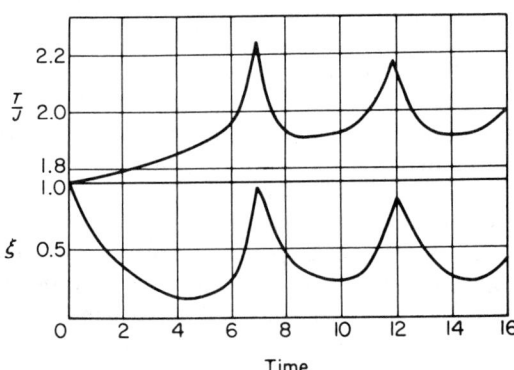

Fig. 7.20 Limit cycle behavior as a function of time.

This kind of oscillation is often very undesirable and it is important to be aware of the possibility of its arising. Douglas has shown that there are some cases in which it can be exploited; as for instance when the mean value of the production rate is greater than that of the steady state about which the limit cycle is oscillating. Though it is beyond our scope to try to describe this in detail, it is vital for the student to realize that this kind of behavior is possible.

Exercise 7.6.1. Show that if the controller acts on deviations of composition and $q_c = q_{cs} + \nu(\xi - \xi_s)$, it is not always possible to get control of an unstable steady state.

7.7 Sequences of Stirred Tank Reactors

Up to this point, we have considered only the design and behavior of a single reactor. We now have to see how one of the principal defects of the stirred tank reactor may be at least partially removed. This defect is the large holding time that is needed because the reaction is being conducted at product conditions where the reaction rate is low because the extent is high. For example, for an irreversible first order reaction with $r(\xi, T) = k(1 - \xi)$, we have a holding time

$$\theta = \frac{\xi}{r(\xi, T)} = \frac{1}{k}\frac{\xi}{1 - \xi} \qquad (7.7.1)$$

if the product extent is to be ξ. If the reaction were conducted as a batch reaction at the same temperature we would have the equation

$$\frac{d\xi}{dt} = k(1 - \xi),$$

i.e.,

$$t = \frac{1}{k} \ln \frac{1}{1 - \xi} \qquad (7.7.2)$$

is the time required at the same temperature to reach the same extent. Now, the ratio

$$\frac{\theta}{t} = \frac{\xi}{1 - \xi} \bigg/ \ln \frac{1}{1 - \xi} \qquad (7.7.3)$$

is shown in Fig. 7.21 (curve for $n = 1$), and it is seen that this ratio rises very rapidly as the reaction approaches completion. The other curves in Fig. 7.21 are for irreversible reactions of order n for which (see Exercise 7.7.1)

$$\frac{\theta}{t} = \frac{\xi}{1 - \xi}\frac{n - 1}{1 - (1 - \xi)^{n-1}}. \qquad (7.7.4)$$

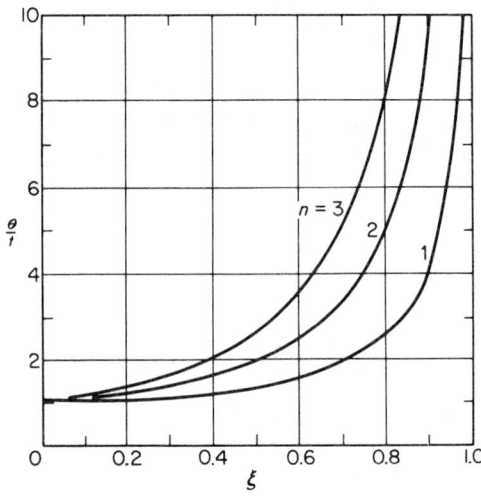

Fig. 7.21 The ratio of stirred tank holding time to batch reaction time as a function of conversion and order of reaction.

Another way of showing that the holding time requirement of the stirred tank must always be greater than that of the batch time at the same temperature is to compare the two formulae

$$t = \int_{\xi_f}^{\xi_p} \frac{d\xi}{r(\xi, T)} \tag{7.7.5}$$

and

$$\theta = \frac{\xi_p - \xi_f}{r(\xi_p, T)}. \tag{7.7.6}$$

Here we have called the extent in the feed not $\xi = 0$ but ξ_f, and denoted the extent of the product by ξ_p. If we plot $1/r(\xi, T)$ against ξ for constant T, the curve must be as shown in Fig. 7.22 for $\partial r/\partial \xi$ is always negative so that r decreases and $1/r$ increases. At $\xi = \xi_e(T)$, $r = 0$ so that the curve $1/r$ has a vertical asymptote there. Equation (7.7.5) shows that the batch reaction time t is the area under the curve between the ordinates ξ_f and ξ_p ($ABGCD$). On the other hand, the stirred tank holding time θ is the area of the rectangle $AECD$ of base $(\xi_p - \xi_f)$ and height $1/r(\xi_p, T)$. It is clear that θ is always greater than t and in fact θ/t becomes unbounded as $\xi_p \longrightarrow \xi_e$. But the same diagram suggests the way to overcome this. For if we get from extent ξ_f to extent ξ_p by way of two reactors, in the first of which ξ increases to ξ' and in the second from ξ' to ξ_p, the holding time of the first reactor will be the area $AFGH$ and that of the second $HJCD$. Clearly, the total holding time of two tanks is less than that of one. The reason, of course, is that the first tank is working under conditions where the reaction rate is greater. If we were to use a number of stirred tanks we should have a number of such rectangles with their upper right-hand corners on the curve. The greater the number of stages, the smaller the total holding time, and in the limit we reach the area under the curve, namely the batch time.

There is thus a real incentive to study the design of sequences of stirred tanks. Many graphical presentations have been given since Denbigh pointed out the advantage in 1944 (see References). Figure 7.23 shows one type of diagram which is convenient for stirred tanks of equal holding time and tem-

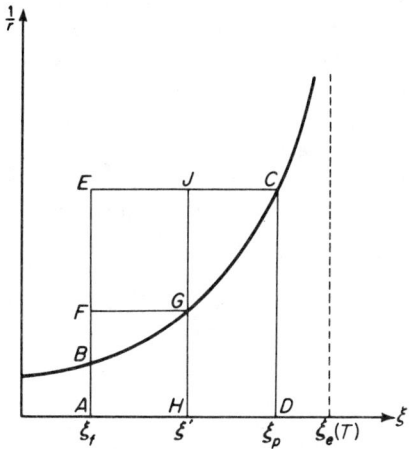

Fig. 7.22 Graphical construction for holding times of a sequence of reactors.

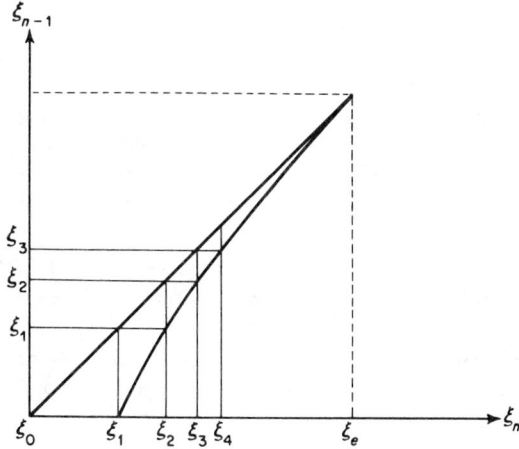

Fig. 7.23 Graphical construction for extent of reaction in a sequence of equal reactors.

perature. If ξ_{n-1} is the extent of reaction in the feed to the nth reactor and ξ_n the extent of its product, the steady state balance gives

$$\xi_{n-1} = \xi_n - \theta r(\xi_n, T). \qquad (7.7.7)$$

Figure 7.23 is a plot of ξ_{n-1} against ξ_n for constant θ and T, together with the line of unit slope. Starting with ξ_0, the feed to the whole system ξ_1 is immediately obtained from the curve. We may then step between the line and the curve, in the escalating fashion common to many graphical constructions in chemical engineering, to determine ξ_2, ξ_3, etc. Different values of θ and T would give different curves. The distance of the curve below the line is proportional to θ; the point of intersection of the curve and the diagonal is ξ_e, the equilibrium extent for that temperature.

The work of Eldridge and Piret, referred to on page 164, contains a confirmation of the validity of these equations. It will be recalled that they carried out the hydrolysis of acetic anhydride under pseudo–first order conditions. For a first order reaction Eq. (7.7.7) can be most usefully written as

$$(1 - x_{n-1}) = (1 - x_n)(1 + \theta_n k),$$

where $x_n = \xi_n/c_0$ and c_0 is the concentration of acetic anhydride in the feed. Thus

$$x_n = 1 - \prod_{r=1}^{n}(1 + \theta_r k)^{-1}.$$

In their experimental work, Eldridge and Piret had up to five reactors in sequence each of approximately the same holding time, though they used the exact formula above for calculating the extent. The following very brief extract from their extensive experimental work illustrates the good agreement with theory.

Mean $k\theta$		$n = 1$	2	3	4	5
0.187	x (obs.)	0.153	0.290	0.397	0.489	0.576
	x (calc.)	0.159	0.291	0.402	0.497	0.575
1.359	x (obs.)	0.583	0.823	0.924	0.968	0.987
	x (calc.)	0.580	0.821	0.924	0.968	0.986

Such a system in which the reaction appears to be of first order lends itself to very simple calculations. For example, if we were given that a single reactor achieves a fractional extent of 0.547, we could immediately find the number of stages required for 99% conversion. Here $x_1 = 0.547$ so $1 - x_1 = (1 + k\theta)^{-1} = 0.453$. If $x_n \geq 0.99$, then $1 - x_n \leq 0.01$ and a slide rule calculation suffices to give

$$1 - x_2 = (0.453)^2 = 0.205, \quad 1 - x_3 = 0.093,$$
$$1 - x_4 = 0.042, \quad 1 - x_5 = 0.0191, \quad 1 - x_6 = 0.0087,$$

and hence six equal stages would suffice. When we are asked how many stages are necessary to achieve a given final extent it is often best to work back from the end. To illustrate this we shall assume that the conversion of 0.547 in one stage was achieved with quite a modest excess of the other reatant and that we now want to work with the system when the feed is in stoichiometric proportions. Now the extent $x_1 = 0.547$ corresponds to a value of $k\theta = 1.21$, where k is the pseudo–first order rate constant. If x_1 was obtained using 100% stoichiometric excess in the feed and if c_0 denotes the concentration of acetic anhydride, then the concentration of the second reactant in the feed was $2c_0$, whereas in the reactor it was $(2 - 0.547)c_0 = 1.453c_0$. If k_2 is the true second order rate constant, $k = 1.453 c_0 k_2$ so that $k_2 c_0 \theta = 0.83$. Now, if the reactants are to be used in stoichiometric proportions, the concentration of each at stage n will be $c_0(1 - x_n)$, and $\theta r = k_2 c_0^2 \theta (1 - x_n)^2$. For equal stages, Eq. (7.7.7) may be written as

$$1 - x_{n-1} = 1 - x_n + \theta k_2 c_0 (1 - x_n)^2$$
$$= (1 - x_n)[1 + 0.83(1 - x_n)].$$

If an 80% conversion is required from n stages, $1 - x_n = 0.2$ and we have

$$1 - x_{n-1} = 0.2(1 + 0.83 \times 0.2) = 0.233,$$
$$1 - x_{n-2} = 0.233(1 + 0.83 \times 0.233) = 0.278,$$
$$1 - x_{n-3} = 0.278(1 + 0.83 \times 0.278) = 0.343,$$
$$1 - x_{n-4} = 0.343(1 + 0.83 \times 0.343) = 0.440,$$
$$1 - x_{n-5} = 0.440(1 + 0.83 \times 0.440) = 0.600,$$
$$1 - x_{n-6} = 0.600(1 + 0.83 \times 0.600) = 0.895.$$

Clearly, the next value of $1 - x$ will exceed 1, i.e., $x_{n-7} \leq 0$. It follows that if $n = 7$ and $x_0 = 0$, then x_7 will exceed 0.8 and seven stages will suffice.

If we ask what temperature will give the greatest extent of reaction, we would immediately suspect that the answer is $T = T_m(\xi)$, the temperature at which the reaction rate is maximum. This is indeed so, for $\xi_f - \xi + \theta r(\xi, T) = 0$; hence, by differentiation,

$$\frac{d\xi}{dT} = \frac{\theta(\partial r/\partial T)}{1 - \theta(\partial r/\partial \xi)}, \qquad (7.7.8)$$

which only vanishes when $\partial r/\partial T = 0$, that is when r is maximum. If there are practical restrictions on the range of temperature $T_* \leq T \leq T^*$, then r should still be made as large as possible within these limits. We know from the study of the reaction rate expression in Sec. 4.3 that for an endothermic or irreversible reaction this is merely a question of making the temperature as high as possible, $T = T^*$. However, for a reversible exothermic reaction, the temperature $T_m(\xi)$ can decrease from infinity, for $\xi = 0$ (when there are no product species in the reference composition), to zero as ξ approaches the limiting value which would make the concentration of one of the reactants zero. Hence there must be extents ξ_* and ξ^* such that $T_m(\xi_*) = T^*$ and $T_m(\xi^*) = T_*$. The optimal temperature is shown in the upper part of Fig. 7.24. In the lower part is the corresponding

$$r_m(\xi) = \begin{cases} r(\xi, T^*), & 0 \leq \xi \leq \xi_* \\ r[\xi, T_m(\xi)], & \xi_* \leq \xi \leq \xi^* \\ r(\xi, T_*), & \xi^* \leq \xi \leq \xi_{\max}. \end{cases} \quad (7.7.9)$$

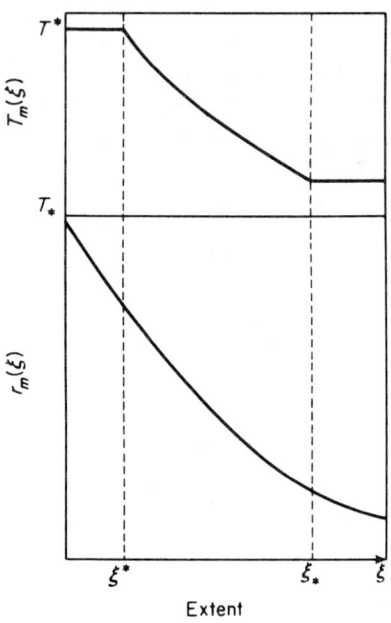

Fig. 7.24 The optimal temperature and reaction rate as functions of extent.

It is not unreasonable to expect—and we shall prove this in the next section—that in a sequence of reactors the temperature should always be chosen to maximize the reaction rate. If we had to consider a sequence of tanks all of the same holding time we could draw just such a curve as in Fig. 7.23 for the relation

$$\xi_{n-1} = \xi_n - \theta r_m(\xi_n) \quad (7.7.10)$$

and use it in a similar way. In fact, this almost proves the assertion that the reaction rate should be as large as possible for then the curve would be as far as possible beneath the diagonal at each point and so the largest possible steps would be taken in the graphical construction.

For multiple reactions, the algebra of the manipulations becomes tedious even in the simplest cases and it is necessary to use purely numerical methods at an early stage. It is therefore all the more important to get a good physical feel for what benefits can accrue from a properly designed sequence of reactors. For example, in the sequence of reactions $A \longrightarrow B \longrightarrow C$, suppose that the activation energy of the second reaction is greater than that of the first. We might to get a better yield of B by running the reaction in two reactors and having expect the temperature of the first reactor somewhat higher than that of the second. For in the first reactor the concentration of B may be smaller than in the second and a higher temperature will allow the main reaction $A \longrightarrow B$ to proceed at a healthy rate without the danger of too much $B \longrightarrow C$. To make this precise involves a lot of algebraic work, but if in doing a particular example the final answer conflicted with this general conclusion it would have to be regarded with suspicion.

Sec. 7.7 Sequences of Stirred Tank Reactors

Exercise 7.7.1. Establish Eq. (7.7.4).

Exercise 7.7.2. The first order reversible reaction rate

$$r = k_1(c_{20} - \xi) - k_2(c_{10} + \xi)$$

at constant temperature can be written as $r = (k_1 + k_2)(\xi_e - \xi)$. Show that if ξ increases from ξ_{n-1} to ξ_n in the nth reactor of a sequence, then the holding time of this reactor is θ_n, where

$$1 + (k_1 + k_2)\theta_n = \frac{\xi_e - \xi_{n-1}}{\xi_e - \xi_n}.$$

Deduce that the problem of finding the sequence of $\theta_n, n = 1, 2, \ldots, N$, with $\sum \theta_n = \Theta$ for which ξ_N is greatest, is equivalent to finding the set of $x_n, n = 1, 2, \ldots, N$, whose arithmetic mean is given and whose geometric mean is to be as large as possible. What do you know about this classical problem of elementary mathematics that will allow you to solve the problem immediately? If you don't know anything about the relationship of geometric and arithmetic means, find out what it is at the first opportunity and in the interim prove that $\theta_n = \Theta/N$ and

$$\xi_N = \xi_e - (\xi_e - \xi_0)\left[1 + \frac{(k_1 + k_2)\Theta}{N}\right]^{-N}.$$

Exercise 7.7.3. It is proposed to carry out a reaction $2A \rightleftharpoons P + Q$ in one or more stirred tank reactors at a uniform flow rate of 125 cu ft per hr. The initial concentration of A is 3 lb-mole per cu ft and that of P and Q is zero. The velocity constant of the forward reaction is 12 ft^3 (lb-mole)$^{-1}$ hr^{-1}, and the equilibrium constant has a value of 16. If the final concentrations of P and Q are to be 85% of the equilibrium values, what size of vessel will be needed if only one is used? If the available vessels are limited to 5% of the capacity of the single vessel, how many smaller vessels will be needed for the same conversion when operated in series?
(C.U.)

Exercise 7.7.4. A process for manufacturing X from A and B according to the reaction $A + B \longrightarrow X$ is being carried out in two well-stirred tanks in series. The reaction is second order, and the two reactants are fed in solution in equimolar proportions. For 90% efficiency of conversion find the ratio of tank volumes which gives the maximum output for a given total reactor volume.

Exercise 7.7.5. A continuous saponification process is carried out in two stirred tanks in series. The reaction is second order. The ester and alkali are added to the system in solutions of the same constant molar concentration and the total volume of the system remains constant. Find the ratio of the volume of the second tank to that of the first so as to give the optimum unit output of soap, when the overall degree of conversion tends to 100%.

Exercise 7.7.6. A system following Denbigh's reaction scheme (see Exercise 5.6.5) is to be carried out in a sequence of two reactors with a feed of pure A. If $E_1 > E_2$ and $E_3 < E_4$, what relationship between the two temperatures is likely to promote a good yield of Y? What can you say about the case $E_1 > E_2$ and $E_3 > E_4$?

7.8 Optimal Sequences of Stirred Tank Reactors

In this section we shall study in detail the optimal choice of holding time and temperature in a sequence of stirred tanks in which a single exothermic reversible reaction is taking place. As soon as the word optimal is used the question must be asked, "In what sense is the design to be optimal?" Here we shall only treat the simplest of optimization problems, but their interplay is instructive.

First, however, let us introduce a notation* which may look a little perverse at first sight but which helps to clarify the problem. It is to number the stages from the last to the first as shown in Fig. 7.25. Thus the feed to reactor n is from reactor $n+1$ and its extent is denoted by ξ_{n+1} (*not* ξ_{n-1} as in other sections); accordingly, the extent of reaction in the feed to the whole process of N stages is

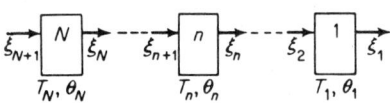

Fig. 7.25 The reactor sequence.

ξ_{N+1}. If the holding time and temperature of reactor n are θ_n and T_n, respectively, the mass balance equations are

$$\xi_{n+1} - \xi_n + \theta_n r(\xi_n, T_n) = 0 \quad \text{where} \quad n = 1, 2, \ldots, N, \quad (7.8.1)$$

and given ξ_{N+1}, θ_n, and T_n, where $n = 1, 2, \ldots, N$, we could calculate ξ_1.

We could set up three basic optimal problems. First, suppose the feed flow rate and composition are given (i.e., q and ξ_{N+1} are specified) and a certain production is required. If the important product is A_1, the production is $\alpha_1 q(\xi_1 - \xi_{N+1})$ and since α_1, q, and ξ_{N+1} are given ξ_1 is also known. We could then ask for the design that minimizes the total volume of the reactors or, equivalently, the total holding time. Second, we could specify the feed composition and the total holding time $\Theta_N = \sum \theta_n$ and ask for the design that maximizes the conversion, that is ξ_1. Note that in the first problem when we were minimizing the total holding time Θ_N, we had to specify ξ_1 to avoid the vacuous solution of finding $\Theta_N = 0$ by setting $\xi_1 = \xi_{N+1}$, which would be not to run a reaction at all. Correspondingly, in the second problem we had to fix Θ_N when seeking to maximize ξ_1 in order to avoid the impossible solution of achieving the equilibrium conversion by having an infinite holding

---
*This notation is peculiar to this section.

Sec. 7.8   Optimal Sequences of Stirred Tank Reactors   209

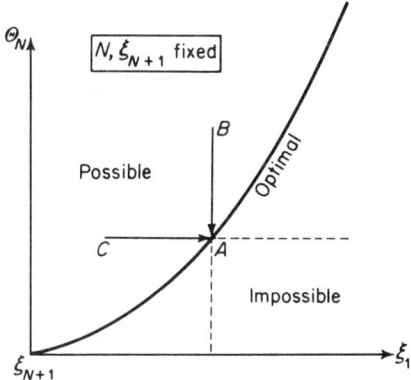

Fig. 7.26 Equivalence of optimal problems.

time. The relationship between these problems is shown in Fig. 7.26, where for the moment we suppose we have solved them. The curve is the optimal Θ_N plotted against the extent ξ_1 for fixed N and ξ_{N+1}. It naturally starts at the origin where $\Theta_N = 0$ if $\xi_1 = \xi_{N+1}$ and has a vertical asymptote at the equilibrium extent at the lowest permitted temperature, for even the best design would require an infinite Θ_N to achieve this. This optimal curve is the boundary between possible and impossible designs, for it is quite possible to be less than optimal and take a greater holding time than is necessary, but if the curve shown is truly optimal it is impossible to achieve a given conversion in less than the optimal holding time. The first problem consists in approaching the optimal design (say point A) from above (along the line BA) because ξ_1 is constant on this. The second problem asks for the maximum ξ_1 for a fixed Θ_N and so approaches the boundary point A along the horizontal CA.

A third problem could be arranged in which a very crude cost estimate would be set up. If p_j is the price of a mole of A_j, the cost of the feed stream is $q \sum p_j(c_{j0} + \alpha_j \xi_{N+1})$, while the value of the product stream is $q \sum p_j(c_{j0} + \alpha_j \xi_1)$. Hence the rate of making profit from the conversion is $q\bar{p}(\xi_1 - \xi_{N+1})$, where $\bar{p} = \sum \alpha_j p_j$. If the cost of building and operating the reactor is proportional to the total volume it can be written as $P \sum V_n = qP\Theta_N$. Then the net rate of profit is $qP[\lambda(\xi_1 - \xi_{N+1}) - \Theta_N]$, where $\lambda = \bar{p}/P$ is the value of the conversion in units of reactor cost. To maximize the value of $\lambda(\xi_1 - \xi_{N+1}) - \Theta_N$ we would move a line of slope λ in Fig. 7.26 as far as it could go to the right and still have some part of it in the "possible" region. This would of course make it the tangent to the optimal curve of slope λ. In this way, we see that all three problems are equivalent and we shall solve them by treating the first of them.

More complicated and realistic problems could be devised in which we could do a complete cost estimate for each stage, given its temperature and volume. This, however, can be solved by similar methods and adds little to the basic understanding. The weakness of the method of dynamic programming which we shall use is the difficulty of applying it when there are more than two reactions.

Our immediate problem is therefore: Given the system of Eq. (7.8.1) and the values of ξ_{N+1} and ξ_1, find the values of $\theta_1, T_1, \ldots, \theta_N, T_N$ so that $\Theta_N = \sum \theta_n$ is a minimum. The key to the method of solution is to regard ξ_1 as fixed and ξ_{N+1} as free until we have finally solved the problem and can give it its specified value. We then work back from the end, solving the problem fully for one stage and embedding this in our solution for two stages and so on. To this end we set

$$f_N(\xi_{N+1}) = \text{Min} \sum_1^N \theta_n \qquad (7.8.2)$$

and find first $f_1(\xi_2)$, then $f_2(\xi_3)$, and so on. Obviously, $f_0(\xi_1) = 0$, and in fact $f_N(\xi_1) = 0$ for then no reactor would be necessary.

Let us first observe that there are really only $(2N - 1)$ quantities that can be freely chosen. From the given ξ_{N+1} and any choice of $\theta_N, T_N, \theta_{N-1}, \ldots, \theta_2, T_2$, we can calculate successively $\xi_N, \xi_{N-1}, \ldots, \xi_2$ by the use of Eq. (7.8.1). But for the last stage

$$\theta_1 = \frac{\xi_1 - \xi_2}{r(\xi_1, T_1)}, \qquad (7.8.3)$$

so that if T_1 is chosen, θ_1 is fixed by the requirement that the extent ξ_1 is to be achieved; both cannot be chosen independently. This suggests that it would be more natural to decide on the N temperatures, T_n, and the $(N - 1)$ intermediate extents, $\xi_2, \xi_3, \ldots, \xi_N$. Doing this has two advantages. First, the ξ's are ordered, which makes it easier to find them:

$$\xi_{N+1} < \xi_N < \xi_{N-1} < \cdots < \xi_1.$$

Second, Eq. (7.8.1) is much easier to solve for θ_n in terms of ξ_n than vice versa:

$$\theta_n = \frac{\xi_n - \xi_{n+1}}{r(\xi_n, T_n)}. \qquad (7.8.4)$$

In fact, as soon as we write the total holding time

$$\Theta_N = \sum_1^N \frac{\xi_n - \xi_{n+1}}{r(\xi_n, T_n)}, \qquad (7.8.5)$$

we have immediate proof that the optimal choice of T_n (which we shall denote by T_n°) maximizes the reaction rate. For when any choice of ξ_n has been made, each term in the sum of Eq. (7.8.5) can be separately minimized by taking

Sec. 7.8 Optimal Sequences of Stirred Tank Reactors

$$T_n^\circ = T_m(\xi_n). \tag{7.8.6}$$

If this is true for any choice of ξ_n, it must also be true of the optimal choice. From now on we may incorporate this decision and replace $r(\xi_n, T_n)$ by $r_m(\xi_n)$, writing

$$\Theta_N = \sum_1^N \frac{\xi_n - \xi_{n+1}}{r_m(\xi_n)}. \tag{7.8.7}$$

We have now only to choose $\xi_2, \ldots, \xi_N$.

In case $N = 1$ there is no decision left to be made:

$$f_1(\xi_2) = \frac{\xi_1 - \xi_2}{r_m(\xi_1)}. \tag{7.8.8}$$

In Fig. 7.27, Γ is the curve with equation $z = 1/r_m(\xi)$. Since ξ_1 is fixed, we may draw the ordinate AB and through B the horizontal line Γ_1. Equation (7.8.8) then has the immediate graphical construction that if ξ_2 is given, we need only draw the vertical line through it (DC) to intersect Γ_1, and $f_1(\xi_2)$ is the area of the rectangle $ABCD$. We could graduate the curve Γ in $T_m(\xi)$ and Γ_1 in $f_1(\xi_2)$, and then θ_1° could be read from the top left-hand corner (C) and T_1° from the top right-hand corner (D) of the rectangle we have constructed.

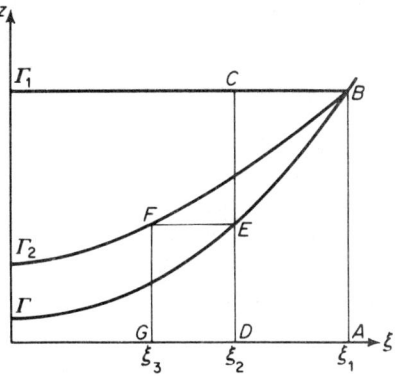

Fig. 7.27 Graphical presentation of the optimal design.

Now consider the case $N = 2$ and ξ_3 fixed (point G, Fig. 7.27). We have only to choose ξ_2 to minimize:

$$\Theta_2 = \frac{\xi_2 - \xi_3}{r_m(\xi_2)} + \frac{\xi_1 - \xi_2}{r_m(\xi_1)}. \tag{7.8.9}$$

Quite evidently, the ξ_2 that does this, $\xi_2^\circ(\xi_3)$, lies strictly between ξ_3 and ξ_1. For if $\xi_2 = \xi_1$ or ξ_3, one of the stages has vanished and we are back at a one-stage process. We know from the previous section (Fig. 7.22) that we can reduce the total holding time by having two stages instead of one, so that $\xi_3 < \xi_2 < \xi_1$ is a strict inequality. It follows that we can find ξ_2° by differentiating Θ_2 to give the equation

$$\frac{1}{r_m(\xi_2)} - \frac{r_m'(\xi_2)}{r_m^2(\xi_2)}(\xi_2 - \xi_3) - \frac{1}{r_m(\xi_1)} = 0. \tag{7.8.10}$$

As we have seen on other occasions, it often proves useful to turn equations inside out. In this case, if we try to solve for ξ_2 in terms of ξ_3, we have a complicated equation; but solving for ξ_3 in terms of ξ_2 is trivial:

$$\xi_3 = \xi_2 + \frac{r_m^2(\xi_2)}{r_m'(\xi_2)}\left[\frac{1}{r_m(\xi_1)} - \frac{1}{r_m(\xi_2)}\right]. \tag{7.8.11}$$

Let us record the result of this calculation by plotting the curve Γ_2. The typical point of Γ_2 has abscissa ξ_3 as calculated from Eq. (7.8.11) and ordinate $1/r_m(\xi_2)$. If these calculations are done for a few values of ξ_2, a smooth curve can be readily drawn. We now claim that by starting from $(\xi_3, 0)$, point G in Fig. 7.27, and drawing GF vertically to Γ_2, FE horizontally to Γ, EC vertically to Γ_1, and CB horizontally to Γ we can construct the optimal design. Since ξ_3 and ξ_2 are connected by Eq. (7.8.11), $\xi_2 = \xi_2^\circ(\xi_3)$, it follows that θ_2 is the area of $GFED$, θ_1 is the area of $DCBA$, and their sum is the least possible. If we wished, Γ_2 could be graduated with values of $f_2(\xi_3)$ which could then be read off from the top left-hand corner of the rectangle, as could T_2° from the right-hand corner.

It can be shown that

$$\frac{df_n(\xi_{n+1})}{d\xi_{n+1}} = \frac{-1}{r_m[\xi_n^\circ(\xi_{n+1})]}. \tag{7.8.12}$$

For the moment let us notice that it is true for $n = 1$ and $n = 2$. For $n = 1$ we obtain it immediately upon differentiating Eq. (7.8.8). To prove it for $n = 2$, let us differentiate Eq. (7.8.9) with respect to ξ_3, by making ξ_2 a function of ξ_3,

$$\frac{d\Theta_2}{d\xi_3} = \frac{\partial\Theta_2}{\partial\xi_3} + \frac{\partial\Theta_2}{\partial\xi_2}\frac{d\xi_2}{d\xi_3}.$$

But if $\xi_2 = \xi_2^\circ(\xi_3)$, $\partial\Theta_2/\partial\xi_2 = 0$, since that is how the minimum was determined, so that the second term is zero. When $\xi_2 = \xi_2^\circ(\xi_3)$, $\Theta_2 = f_2$ and Eq. (7.8.12) follows immediately.

Turning next to the case $n = 3$ we will invoke the so-called *principle of optimality* that Bellman first discovered for systems of this type. We may write

$$\Theta_3 = \frac{\xi_3 - \xi_4}{r_m(\xi_3)} + \left[\frac{\xi_2 - \xi_3}{r_m(\xi_2)} + \frac{\xi_1 - \xi_2}{r_m(\xi_1)}\right], \tag{7.8.13}$$

and to find $f_3(\xi_4)$ we have to choose ξ_3 and ξ_2. But suppose we have made some choice of ξ_3, perhaps not the best; we clearly cannot do better than to choose $\xi_2 = \xi_2^\circ(\xi_3)$, for any other choice of ξ_2 would make the terms in the bracket larger than they need be. With this choice the right-hand side of Eq. (7.8.13) can be written

$$\frac{\xi_3 - \xi_4}{r_m(\xi_3)} + f_2(\xi_3),$$

and now we have only to find the ξ_3 which minimizes this. If we find the best ξ_3, denoted by $\xi_3^\circ(\xi_4)$, we shall have

Sec. 7.8 Optimal Sequences of Stirred Tank Reactors

$$f_3(\xi_4) = \underset{\xi_3}{\text{Min}} \left[\frac{\xi_3 - \xi_4}{r_m(\xi_3)} + f_2(\xi_3) \right]. \quad (7.8.14)$$

The principle involved is that whatever the first decision (in this case ξ_3) may be, the subsequent decisions (here only ξ_2) must be optimal with respect to the result of the first decision.

An argument similar to the one used before shows that $\xi_4 < \xi_3 < \xi_1$, and we may find the minimum by differentiating the right-hand side of Eq. (7.8.14) with respect to ξ_3. Doing this and using Eq. (7.8.12), we obtain

$$\frac{1}{r_m(\xi_3)} - \frac{r'_m(\xi_3)}{r_m^2(\xi_3)}(\xi_3 - \xi_4) - \frac{1}{r_m[\xi_2^\circ(\xi_3)]} = 0. \quad (7.8.15)$$

Once again, this is hard to solve for ξ_3, but readily gives

$$\xi_4 = \xi_3 + \frac{r_m^2(\xi_3)}{r'_m(\xi_3)} \left[\frac{1}{r_m(\xi_2^\circ)} - \frac{1}{r_m(\xi_3)} \right]. \quad (7.8.16)$$

For various ξ_3 we can calculate ξ_4 from Eq. (7.8.16) and plot a curve Γ_3 whose typical point has abscissa ξ_4 and ordinate $1/r_m(\xi_3)$. A similar construction of vertical and horizontal lines will serve to give us the optimal three-reactor design.

In the general case of N stages we may suppose that we have already calculated $f_{N-1}(\xi_N)$. Then

$$\Theta_N = \frac{\xi_N - \xi_{N+1}}{r_m(\xi_N)} + \left[\sum_{1}^{N-1} \frac{\xi_n - \xi_{n+1}}{r_m(\xi_n)} \right], \quad (7.8.17)$$

and we argue again from Bellman's principle of optimality that whatever choice of ξ_N may be made, the second term will be least, if it is followed by the optimal choices of $\xi_{N-1}, \ldots, \xi_2$. But in this case the sum is $f_{N-1}(\xi_N)$, and if we further choose ξ_N optimally, Θ_N will be $f_N(\xi_{N+1})$, and

$$f_N(\xi_{N+1}) = \underset{\xi_N}{\text{Min}} \left[\frac{\xi_N - \xi_{N+1}}{r_m(\xi_N)} + f_{N-1}(\xi_N) \right]. \quad (7.8.18)$$

Differentiating with respect to ξ_N and using Eq. (7.8.12) with $n = N - 1$ gives the equation

$$\frac{1}{r_m(\xi_N)} - \frac{r'_m(\xi_N)}{r_m^2(\xi_N)}(\xi_N - \xi_{N+1}) - \frac{1}{r_m(\xi_{N-1}^\circ)} = 0, \quad (7.8.19)$$

which can again be rearranged to give

$$\xi_{N+1} = \xi_N + \frac{r_m^2(\xi_N)}{r'_m(\xi_N)} \left[\frac{1}{r_m(\xi_{N-1}^\circ)} - \frac{1}{r_m(\xi_N)} \right]. \quad (7.8.20)$$

It follows that a sequence of curves can be constructed, $\Gamma_1, \Gamma_2, \ldots, \Gamma_N$, and that a sequence of alternate vertical and horizontal lines from $(\xi_{N+1}, 0)$ to $\Gamma_N, \Gamma, \Gamma_{N-1}, \Gamma, \ldots, \Gamma_1, \Gamma$ will give a rapid construction of the optimal design. Figure 7.28 shows the case of $N = 4$. Notice that, since the construction involves only horizontal and vertical lines, the scales may be distorted in

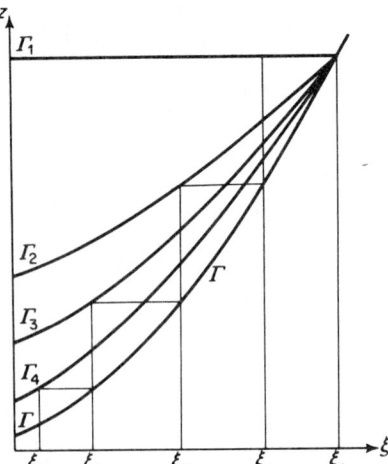

Fig. 7.28 Graphical construction for the optimal design.

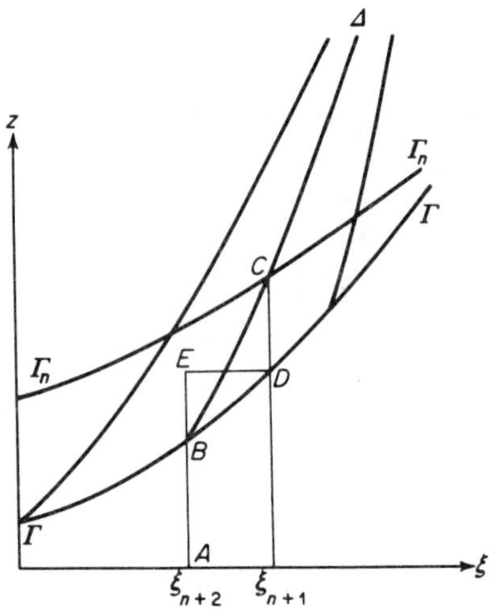

Fig. 7.29 Graphical construction for the Γ curves.

any manner we please and the construction will still be valid. The curves, however, would have to be graduated, for the areas of the rectangles would no longer give the θ_n.

The weakness of this construction lies in the fact that the curves Γ_n are all attached to Γ at the point of abscissa ξ_1. If we were to change ξ_1, we would have to begin the calculation all over again. However, we can give a graphical construction for obtaining Γ_{n+1} from Γ_n, by using a system of fixed curves that do not depend on ξ_1. Let us recall that Γ_n is the locus $z = 1/r_m[\xi_n^\circ(\xi)] = g(\xi)$. Now consider the family of curves

$$\Delta : z = \frac{1}{r_m(\xi)} - \frac{r'_m(\xi)}{r_m^2(\xi)}(\xi - \xi') = h(\xi; \xi'). \tag{7.8.21}$$

The several members of the family, Δ, are distinguished by different values of ξ', as shown in Fig. 7.29. In fact, ξ' is the abscissa of the point at which $z = h(\xi; \xi')$ meets Γ, for when $\xi = \xi'$ the second term in Eq. (7.8.21) vanishes. Now suppose Γ_n is known and we wish to construct Γ_{n+1}, i.e.,

$$z = \frac{1}{r_m[\xi_{n+1}(\xi)]}.$$

For any ξ_{n+2} the equation for $\xi_{n+1}^\circ(\xi_{n+2})$ is

$$h(\xi_{n+1}; \xi_{n+2}) = g(\xi_n^\circ) \tag{7.8.22}$$

[see Eq. (7.8.19)]. Graphically this means that we go vertically from $(\xi_{n+2}, 0)$ (point A in Fig. 7.29) to Γ (point B), pick out the curve from the Δ family that starts there (BC), and follow it up to Γ_n (point C). By Eq. (7.8.22) the abscissa of this point is $\xi_{n+1}^{\circ}(\xi_{n+2})$. But Γ_{n+1} is the curve $z = 1/r_m[\xi_{n+1}^{\circ}(\xi)]$, and if we drop vertically to Γ (point D) and proceed horizontally until meeting AB produced upward, this intersection (point E) is a point on the curve Γ_{n+1}. Now Γ_1 can be drawn immediately since it is just a horizontal straight line, and from this we can construct Γ_2, and from Γ_2, Γ_3, etc. Clearly this sequence of curves will converge on Γ.

7.9 Mixing in the Reactor

We must now examine the assumptions about the mixing in the stirred tank reactor that are implicit in the kind of equations we have been using. In saying that the feed is immediately dispersed throughout the reactor, which is itself in a uniform state of composition and temperature, we have committed ourselves to a very stringent conception of mixing and the actual state of mixing may not be nearly as perfect as this. We shall do this in three stages. First, we want to examine the concept of a residence time distribution; this idea will apply to any reactor or flow system and is not confined to the stirred tank. Secondly, we must try and see how the opposite of complete mixing might be defined. This leads to the concepts of maximum mixedness and complete segregation; again, these have a bearing beyond the stirred tank reactor. Finally, we must see what effect this might have on the design and performance of the stirred tank reactor.

In the first place, it is clear that not all molecules entering a reactor of holding time $\theta = V/q$ will reside there for exactly θ. It is because of the vigorous mixing, which causes some molecules to pass out of the reactor almost immediately and contribute so little to production, that the volume requirements of this type of reactor are so much higher. To determine the residence time distribution, a suitable experiment would be to inject a sharp pulse of some nonreacting tracer material at time $t = 0$ and measure its concentration in the effluent. If $c(t)$ were the concentration at time t, the number of molecules emerging between times t and $t + dt$ would be proportional to $c(t)\,dt$. The total number emerging for all time would be $\int_0^{\infty} c(t)\,dt$ with the same constant of proportionality. Thus

$$p(t)\,dt = \frac{c(t)\,dt}{\int_0^{\infty} c(t)\,dt} \quad (7.9.1)$$

would be the fraction of molecules emerging in the interval $(t, t + dt)$. If all the tracer material went in at $t = 0$, $p(t)\,dt$ would be the probability of a molecule residing in the reactor for a time between t and $t + dt$. Let us determine this.

If the tracer does not react, its concentration is governed by Eq. (7.1.3) with $r = 0$ and the suffix j dropped:

$$\theta \frac{dc}{dt} + c = c_f(t). \qquad (7.9.2)$$

We can solve this for an arbitrary feed concentration, and if $c(0) = 0$, the solution is

$$c(t) = \frac{1}{\theta} \int_0^t c_f(s)\, e^{-(t-s)/\theta}\, ds. \qquad (7.9.3)$$

Suppose that a unit amount of tracer is injected during a very short interval $(0, t_1)$, i.e., $c_f(t) = 0$, $t > t_1$, and that

$$\int_0^\infty c_f(t)\, dt = \int_0^{t_1} c_f(t)\, dt = 1. \qquad (7.9.4)$$

Then by the second mean value theorem

$$c(t) = \frac{1}{\theta} e^{-(t-t_2)/\theta} \int_0^{t_1} c_f(s)\, ds = \frac{1}{\theta} e^{-(t-t_2)/\theta},$$

where $0 < t_2 < t_1$. But if the injection is made instantaneously, an idealization we can attain by letting $t_1 \longrightarrow 0$, then $t_2 \longrightarrow 0$ and

$$c(t) = \frac{1}{\theta} e^{-t/\theta}. \qquad (7.9.5)$$

It can be easily verified that $\int_0^\infty c(t)\, dt = 1$ [it is a consequence of Eq. (7.9.4)], so that Eq. (7.9.5) also gives $p(t)$, the probability density of residence time t. The expected, or mean, residence time is

$$\mu = \int_0^\infty t\, p(t)\, dt = \theta \int_0^\infty \left(\frac{t}{\theta}\right) e^{-t/\theta} d\left(\frac{t}{\theta}\right) = \theta, \qquad (7.9.6)$$

since $\int_0^\infty x\, e^{-x}\, dx = 1$. Thus the nominal holding time $\theta = V/q$ is also the average time a molecule spends in the reactor. There is a wide dispersion about this mean as Fig. 7.30 shows. The variance of residence times about the mean is

$$\sigma^2 = \int_0^\infty (t - \theta)^2 p(t)\, dt$$

$$= \theta^2 \int_0^\infty (x - 1)^2 e^{-x}\, dx = \theta^2. \qquad (7.9.7)$$

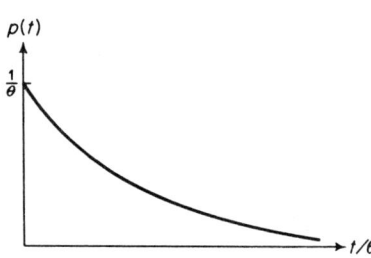

Fig. 7.30 The residence time distribution in a single reactor.

The standard deviation is thus as large as the mean itself.

This kind of tracer experiment is possible for any reactor or flow system

Sec. 7.9 Mixing in the Reactor 217

and, if a really sharp impulse can be used for $c_f(t)$, it will give a good estimate of the residence time distribution $p(t)$. However, even if the input is not a sharp impulse the mean residence time and variance of residence times can be calculated as follows. Assume that the input is given by $c_f(t)$, with $\int_0^\infty c_f(t)\,dt = 1$ and its own mean and variance:

$$\mu_f = \int_0^\infty t c_f(t)\,dt, \qquad \sigma_f^2 = \int_0^\infty (t - \mu_f)^2 c_f(t)\,dt. \tag{7.9.8}$$

The response of tracer at the outlet is $c(t)$ and again $\int_0^\infty c(t)\,dt = 1$ and the measured mean and variance are

$$\mu_t = \int_0^\infty t c(t)\,dt \quad \text{and} \quad \sigma_t^2 = \int_0^\infty (t - \mu_t)^2 c(t)\,dt. \tag{7.9.9}$$

Then the mean residence time and variance of residence times (the same as would be given by the ideal tracer experiment) are

$$\mu = \mu_t - \mu_f \quad \text{and} \quad \sigma^2 = \sigma_t^2 - \sigma_f^2. \tag{7.9.10}$$

This can be easily proved for the stirred tank (see Exercise 7.9.1) and is true in general, if the tracer is conserved and does not diffuse upstream from the point of injection. Another way of viewing this is to see that $\mu_t = \mu_f + \mu$ and $\sigma_t^2 = \sigma_f^2 + \sigma^2$, so that the mean of the input is increased by the mean residence time of the vessel and the variance is likewise additive.

It follows that if we had in sequence n stirred tanks each of holding time θ and performed the ideal tracer experiment ($\mu_f = \sigma_f^2 = 0$), the mean residence time in the sequence and variance of residence times would be

$$\mu_n = n\theta \quad \text{and} \quad \sigma_n^2 = n\theta^2. \tag{7.9.11}$$

By applying the mass balance equation to each reactor (see Exercise 7.9.2), it can be shown that just as $p_1(t)$, the residence time distribution for one stage, is $e^{-t/\theta}/\theta$, so in general

$$p_n(t) = \frac{t^{n-1}}{(n-1)!\,\theta^n} e^{-t/\theta}. \tag{7.9.12}$$

If we fix the total holding time by putting $n\theta = \Theta$ and use the dimensionless time $\tau = t/\Theta$ for the residence time, we see that

$$\Theta\, p_n(\tau) = \frac{\tau^{n-1} n^n}{(n-1)!} e^{-n\tau}. \tag{7.9.13}$$

This is shown in Fig. 7.31. It is clear that a larger number of smaller stages produces a relatively sharper peak* until an infinite number of infinitesimal stages produces the sharp spike at $\tau = 1$. This means that the probability of getting anything other than a residence time of $t = \Theta$ or $\tau = 1$ is zero;

*A measure of the sharpness of a peak is the smallness of the ratio of the variance to the square of the mean. In this case, $\sigma_n^2/\mu_n^2 = 1/n$, which tends to zero as $n \longrightarrow \infty$.

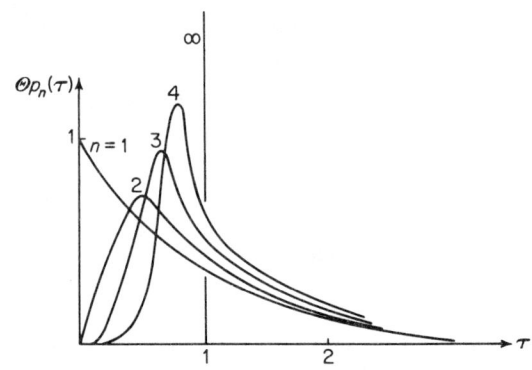

Fig. 7.31 The residence time distributions in sequences of equal reactors.

in other words all the molecules have exactly the same residence time. But this is the plug flow tubular reactor with no mixing in which each portion of the feed moves uniformly through the reactor. In fact, we shall see later that a sequence of a large number of small stirred tanks is a useful model for the tubular reactor with a certain amount of back-mixing.

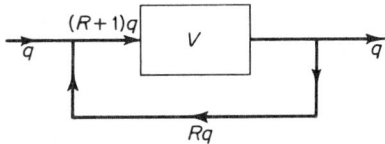

Fig. 7.32 The recycle reactor.

Another way of passing between the extremes of a single stirred tank and plug flow has been suggested by Carberry in the context of reactor design (Fig. 7.32). The flow rate through the whole system is q and the volume of the tubular element is V, so that the nominal holding time of the whole system is $\theta = V/q$. However, there is a recycle stream of rate Rq so that the actual rate through the tube is $(R + 1)q$. If $R = 0$ we have a tubular reactor and as $R \longrightarrow \infty$ the system approaches a perfectly stirred vessel of holding time θ. [In fact, it can be shown (see Exercise 7.9.3) that whatever the residence time distribution of the unit of volume V, the limit as $R \longrightarrow \infty$ gives the residence time distribution of a simple stirred tank.] Wolf and Resnick have used this model to describe the residence time distributions observed by tracer experiments.

Other combinations have been used to describe different residence time distributions in actual reactors. Cholette, Blanchet, and Cloutier have suggested that combinations of stirred tank and plug flow reactors would better represent the real state of affairs. At one extreme, the reactants might enter and flow without mixing for some time before reaching the mixing region; this would be like having a tubular reactor followed by a completely mixed one, a combination denoted by TM. At the other extreme, the reactants

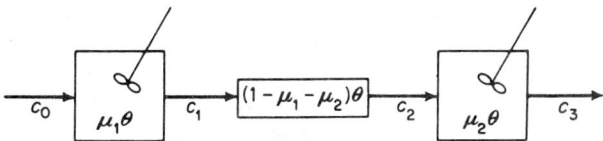

Fig. 7.33 Combined types of reactor.

might be fed into a well-agitated zone but leave from a relatively unmixed one; this would correspond to a well-mixed reactor followed by a tubular (MT). Figure 7.33 shows a combined model MTM, which gives either of the foregoing as a special case. Van de Vusse has looked at the natural circulation loops that might develop in the stirring of a reactor and has proposed a model with two parameters, which has been experimentally confirmed. Asbjørnsen has described a triangular network of stirred tanks in which two feed streams are distributed to the tanks on the sloping sides and the product streams from the tanks at the base are combined to give the product stream from the whole system. There is thus no lack of models for representing the imperfection of mixing.

However, there is more to mixing than just the residence time distribution. Another term for the residence time is "age"—the length of time that a molecule has spent in the reactor since being "born" into it by entering with feed. We can also consider the "life expectancy" of a molecule, namely the length of time before it passes out of the system. When they enter the reactor all molecules have the same age (zero) but differ in life expectancy, since in general some will leave the reactor sooner than others. When they are about to leave, they have the same life expectancy (zero) but differ in age, since some will have spent a longer time in the reactor than others. Only in the case of plug flow with no mixing do the molecules entering have the same life expectancy and those leaving have the same age, because the singular character of the residence time distribution gives them no choice. For all other systems, we can think of an "entering environment," where all ages are much the same, and a "leaving environment," where life expectancies are the same. Mixing is in some sense a transition from one environment to the other. If the transition is immediate so that the molecule spends all its time in the reactor in the company of molecules of different ages but of the same life expectancy we speak of a state of "maximum mixedness"—a concept introduced by Zwietering in 1959. If the transition is delayed until the end, so that the molecule spends all its time with those of the same age then we have a state of "complete segregation."

Danckwerts had earlier introduced similar ideas for the stirred tank, for there maximum mixedness corresponds to the complete molecular mixing which is presumed in writing the mass balance with uniform concentrations throughout the reactor. The state of complete segregation can be best imagined by supposing that the feed is broken up into drops of a dispersed phase. These drops are whirled around and have a variety of residence times, but they do not coalesce. Therefore any molecule spends all of its time in the company of those of the same age and each drop acts like a little batch reactor. For a first order reaction the two extremes give the same conversion in the reactor. If the reaction is isothermal, its rate can be written as $r = k(\xi_e - \xi)$, where $k = k_1 + k_2$ is the sum of the forward and backward rate constants. Thus in a batch the extent after time t is given by

$$\frac{d\xi}{dt} = k(\xi_e - \xi)$$

or

$$\xi(t) = \xi_e(1 - e^{-kt}).$$

The fraction of volume spending a time between t and $t + dt$ in the reactor is of course given by the residence time distribution $p(t)\, dt = e^{-t/\theta}\,(dt/\theta)$. Therefore the average extent of reaction over these segregated "drops" is

$$\bar{\xi} = \int_0^\infty p(t)\xi(t)\, dt = \xi_e\left(1 - \frac{1}{1 + k\theta}\right).$$

But for complete molecular mixing we have the equation

$$0 = -\xi + \theta k(\xi_e - \xi),$$

and

$$\xi = \xi_e \frac{\theta k}{1 + \theta k}.$$

That segregation makes no difference with a first order reaction is scarcely surprising, for a first order reaction is in a sense a spontaneous thing which happens to a molecule independently of other similar molecules. For a second order reaction we should expect a difference in conversion, for this is essentially a cooperative phenomenon between neighboring molecules. Consider the irreversible reaction with rate $k(c_0 - \xi)^2$. For complete molecular mixing

$$0 = -\xi + \theta k(c_0 - \xi)^2,$$

and this can be solved to give

$$\frac{\xi}{c_0} = 1 - \frac{(1 + 2\kappa)^{1/2} - 1}{\kappa}, \qquad (7.9.14)$$

where

$$\kappa = 2kc_0\theta. \qquad (7.9.15)$$

On the other hand, the batch equation is

$$\frac{d\xi}{dt} = k(c_0 - \xi)^2,$$

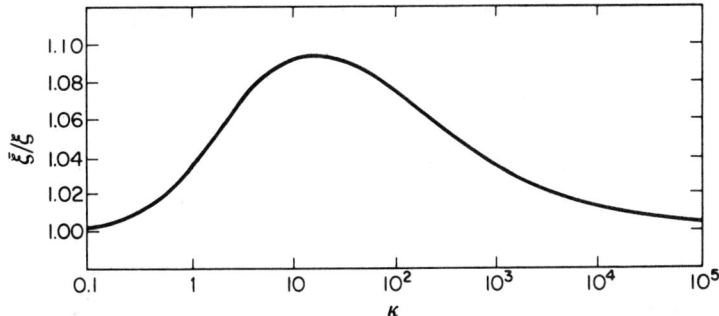

Fig. 7.34 The effect of imperfect mixing with a second order reaction.

and

$$\xi(t) = \frac{c_0(\kappa t/\theta)}{2 + (\kappa t/\theta)}. \tag{7.9.16}$$

Then

$$\begin{aligned}\frac{\bar{\xi}}{c_0} &= \int_0^\infty e^{-t/\theta}\left[1 - \frac{2/\kappa}{(2/\kappa) + (t/\theta)}\right] d\frac{t}{\theta}, \\ &= 1 - \frac{2}{\kappa} e^{2/\kappa} E_1\left(\frac{2}{\kappa}\right),\end{aligned} \tag{7.9.17}$$

where $E_1(z)$ is the exponential integral defined in the Appendix (Eq. A.14). Figure 7.34 shows the ratio $\bar{\xi}/\xi$ as a function of κ; the curve has a maximum of 1.095 when $\kappa = 16$.

For reaction orders greater than 1, the segregated conversion is greater than that for maximum mixedness, so that a conservative design is obtained by using the steady state equations. The difference, however, is at most 9.5%. To put it in this way is somewhat deceiving if the objective is to calculate the holding time necessary for a given conversion and if the cost of holding time is critically important. For the ratio of holding times required to reach an extent Xc_0 becomes very large as X approaches 1 (see Exercise 7.9.8), and it may be worthwhile to find out more about the state of segregation in case this can be exploited to give a more economical design. For reaction orders less than 1 the conversion with segregation falls below that for complete mixing. Rippin has proposed a single parameter model for incomplete mixing, which involves the concept of the two environments and a rate of transfer between them. This is perhaps the simplest bridge that has been constructed between the extremes of complete segregation and maximum mixedness. The value of Zwietering's work remains in showing how the two limiting cases can be calculated for an arbitrary residence time distribution, not just for that of the stirred tank as has been done here.

Exercise 7.9.1. By multiplying Eq. (7.9.2) by t or by t^2 and integrating both sides, show that $\mu_t = \mu_f + \theta$ and $\sigma_t^2 = \sigma_f^2 + \theta^2$.

Exercise 7.9.2. The residence time distributions for the sequence of equal stirred tanks satisfy

$$\theta \frac{dp_n}{dt} + p_n = p_{n-1}, \qquad p_1 = \frac{1}{\theta} e^{-t/\theta}.$$

Use these equations to establish Eq. (7.9.12) and confirm Eqs. (7.9.11) by direct computation.

Exercise 7.9.3. By taking the tubular element in Fig. 7.32 to have an output equal to the input at a time $\theta/(R+1)$ earlier, show that as $R \longrightarrow \infty$ the response of the system to a tracer experiment tends to obey Eq. (7.9.2). Can you generalize this to prove the parenthetic assertion on p. 218?

Exercise 7.9.4. When n is large, $n!$ is approximately $n^n e^{-n} (2\pi n)^{1/2}$. Use this approximation of Stirling's in Eq. (7.9.13) to show that as $n \longrightarrow \infty$, the probability of having a residence time different from θ tends to zero.

Exercise 7.9.5. Show that the residence time distribution for the system shown in Fig. 7.33 is

$$p(t) = \begin{cases} 0 & t \leq (1 - \mu_1 - \mu_2)\theta = \mu\theta \\ \dfrac{\exp - (t - \mu\theta)/(\mu_2\theta) - \exp - (t - \mu\theta)/(\mu_1\theta)}{(\mu_2 - \mu_1)\theta} & t \geq \mu\theta. \end{cases}$$

Exercise 7.9.6. Investigate the model in which the feed is split into fractions λ and $(1 - \lambda)$ and sent to parallel reactors whose volumes are in the ratio $\mu:(1 - \mu)$. Find the residence time distribution, and the mean and variance of residence times. For a first order reaction $A \longrightarrow B$, show that the ratio of the concentration of A in the product stream for this system to its concentration from a single tank of the same mean residence time θ is

$$\frac{(1 + \kappa)[1 + (M + N - 1)\kappa]}{(1 + \kappa M)(1 + \kappa N)},$$

where $M = \mu/\lambda$, $N = (1 - \mu)/(1 - \lambda)$, and $\kappa = k\theta$.

Exercise 7.9.7. A reactor of holding time θ is found to have the residence time distribution $p_1(t) = (t/\theta^2) \exp - (t/\theta)$. Show that for n such reactors in series

$$p_n(t) = \frac{t^{2n-1}\theta^{-2n}}{(2n - 1)!} \exp - \frac{t}{\theta}.$$

If one such reactor gives 60% conversion when used for a particular first order reaction, how many similar reactors would have to be added in series to give a final conversion of 99%? (Adapted C.U.)

Exercise 7.9.8. If the extent $\bar{\xi}$ given by Eq. (7.9.17) is written as $\bar{\xi} = c_0 X$ and X is very close to 1, show that the holding time θ will give a value of κ satisfying $1 - X = \ln(\kappa\, e^{-\gamma/2})/(\kappa/2)$, where γ is Euler's constant. (You will need the expansion of E_1 given in the Appendix.) Let θ_m be the holding time required to achieve the extent Xc_0 with perfect mixing and θ_s the holding time required with complete segregation. Show that, as $X \longrightarrow 1$ and so $\kappa = 2kc_0\theta_s \longrightarrow \infty$, the ratio

$$\frac{\theta_m}{\theta_s} \longrightarrow \frac{\kappa}{2}(\ln \kappa\, e^{-\gamma/2})^{-2}.$$

REFERENCES

All the standard texts have much to say on this important type of reactor; the following list mentions references only on specific points.

7.3. For a more detailed discussion of the transfer of heat in reaction vessels see:

J. M. Coulson and J. F. Richardson, *Chemical Engineering*, Vol. 1, pp. 200–205. London: Pergamon Press, Inc. (1954).

F. A. Holland and F. S. Chapman, *Liquid Mixing and Processing in Stirred Tanks*. New York: Reinhold Publishing Corp. (1966).

The problem of multiple reactions is well handled in Levenspiel's text. See also:

J. J. Carberry, "Yield in Chemical Reactor Engineering," *Ind. Eng. Chem.*, **58,** 40 (Oct. 1966).

7.4. The details of these examples are taken from a very careful study by G. T. Westbrook summarized in:

G. T. Westbrook and R. Aris, "Chemical Reactor Design," *Ind. Eng. Chem.*, **53,** 181 (1961).

7.5. The most accessible early work is:

C. van Heerden, "Autothermic Processes," *Ind. Eng. Chem.*, **45,** 1242 (1953).

O. Bilous and N. R. Amundson, "Chemical Reactor Stability and Sensitivity," *A. I. Ch. E. Journal*, **1,** 513 (1955).

An excellent discussion of the application of these ideas in the case of simultaneous reactions is given in:

K. R. Westerterp, "Maximum Allowable Temperature in Chemical Reactors," *Chem. Eng. Sci.*, **17,** 423 (1962).

See also:

> C. van Heerden, "The Character of the Stationary State of Exothermic Processes," in *Chemical Reaction Engineering* (1st Eur. Symp.), p. 133. London: Pergamon Press, Inc., 1957.

Some of the early work is in German and the stability parameters L, M, and N are related to Damköhler numbers; see, for example, p. 196 of the translation of Brotz' text cited above in the references to Chap. 1. Some dimensionless graphs of the first order plot are given in:

> W. Regenass and R. Aris, "Stability Estimates for the Stirred Tank Reactor," *Chem. Eng. Sci.*, **20**, 60 (1965).

The transient behavior and temperature overshoot in a stirred tank was first discussed in:

> K. G. Denbigh, M. Hicks, and F. M. Page, "The Kinetics of Open Reaction Systems," *Trans. Faraday Soc.*, **44**, 479 (1948).

See also:

> D. S. Sabo and J. S. Dranoff, "On the Stability of a Continuous Flow Stirred-Tank Reactor," *A. I. Ch. E. Journal*, **12**, 1223 (1966).

7.6. For much fuller detail of the case outlined here, see:

> N. R. Amundson and R. Aris, "An Analysis of Chemical Reactor Stability and Control," Parts I–III, *Chem. Eng. Sci.*, **7**, 121 (1958).

See also:

> W. Oppelt and E. Wicke (Ed.), *Grundlagen der chemischen Prozessregelung*. München: R. Oldenbourg, 1964.
>
> A. S. Foss, "Chemical Reaction System Dynamics," *Chem. Eng. Prog. Symp. Ser.*, **55** (No. 25), 47 (1959).
>
> L. Lapidus, "On the Dynamics of Chemical Reactors," *Chem. Eng. Prog. Symp. Ser.*, **57** (No. 36), 34 (1961).
>
> P. Fangel, "A General Method of Dimensioning the Temperature Control System of a Continuous-flow Stirred-Tank Reactor," *Chem. Eng. Sci.*, **21**, 49 (1966).

For a most interesting discussion of phase planes and limit cycles see the remarks of Hoffman and Kylstra following the latter's paper in: *Chemical Reaction Engineering* (Proc. 3rd European Symposium), p. 283. London: Pergamon Press, Inc., 1964. The exploitation of oscillatory behavior is discussed in:

> J. M. Douglas and D. W. T. Rippin, "Unsteady State Process Operation," *Chem. Eng. Sci.*, **21**, 305 (1966).

J. M. Douglas and N. Y. Gaitonde, "Analytical Estimates of the Performance of Chemical Oscillators," *Ind. Eng. Chem. Fundamentals*, **6**, 265 (1967).

7.7. K. G. Denbigh, "Velocity and Yield in Continuous Reaction Systems," *Trans. Faraday Soc.*, **40**, 352 (1944).

P. J. Trambouze and E. L. Piret, "Continuous Stirred Tank Reactors. Designs for Maximum Conversions of Raw Material to Desired Product," *A. I. Ch. E. Journal*, **5**, 384 (1959).

Sequences of reactors are discussed by Levenspiel in his text. See also:

K. Schoenemann, *Der chemische Umsatz bei kontinuierlich durchgeführten Reaktionen, Dechema Monographien*, **21**. Weinheim: Verlag Chemie, 1952.

7.8. References to optimal design up to 1960 are given in:

R. Aris, *The Optimal Design of Chemical Reactors*. New York: Academic Press Inc., 1961.

Since then much work has been done and is well summarized up to 1962 in:

D. W. van Krevelen and P. J. Hoftyzer, "Process Optimatization," *Trans. Inst. Chem. Engrs.* (London), **40**, 37 (1962).

After this time so much has been done that there is no one place where even the work on reactors can be found in survey. The notion of "attainable and nonattainable regions in chemical reaction technique" has been elegantly treated in a paper of that title by F. Horn in the 1964 *Chemical Reaction Engineering Symposium* volume (p. 293) referred to above.

For an elementary introduction to dynamic programming which makes use of the example of this section, see:

R. Aris, *Discrete Dynamic Programming*. New York: Blaisdell Publishing Co., 1964.

The optimization of the stirred tank sequence was treated by Horn in 1961 using a method that anticipated the now popular discrete maximum principle:

F. Horn, "Über das problem der optimalen Rührkesselkaskade für chemische reactionen," *Chem. Eng. Sci.*, **15**, 176 (1961).

7.9. Here again the literature is vast and we shall only mention one general review and the papers that have been referred to in the text. The former is:

O. Levenspiel and K. Bischoff, "Patterns of Flow in Chemical Process Vessels," in *Advances in Chemical Engineering*, Vol. 4. New York: Academic Press Inc., 1964.

Some later references are to be found in:

K. B. Bischoff, "Mixing and Contacting in Chemical Reactors," *Ind. Eng. Chem.*, **58**, 18 (Nov. 1966).

The recirculation model is used by Carberry in the paper referred to above (7.3). See also:

J. M. Douglas, "The Effect of Mixing on Reactor Design," *Chem. Eng. Prog., Symp. Ser.*, **60**, 1 (1964).

A. Cholette, J. Blanchet, and L. Cloutier, "Performance of Flow Reactors at Various Levels of Mixing," *Can. J. Chem. Eng.*, **38**, 1 (1960). Also an earlier paper by these authors in **37**, 105 (1959) of the same journal.

J. G. van de Vusse, "A New Model for the Stirred Tank Reactors," *Chem. Eng. Sci.*, **17**, 507 (1962).

W. Resnick and D. Wolf, "Residence Time Distribution in Real Systems," *Ind. Eng. Chem. Fundamentals*, **2**, 287 (1963).

O. A. Asbjørnsen, "Incomplete Mixing Simulated by Fluid-Flow Networks," *A.I.Ch.E.–I. Chem. E. Symp. Ser.*, **10**, 40 (1965).

P. V. Danckwerts, "The Effect of Incomplete Mixing on Homogeneous Reactions," in *Chemical Reaction Engineering* (1st European Symp.), p. 93. London: Pergamon Press, Inc., 1957. Also *Chem. Eng. Sci.*, **2**, 1 (1953) and **8**, 93 (1958).

T. N. Zwietering, "The Degree of Mixing in Continuous Flow Systems," *Chem. Eng. Sci.*, **11**, 1 (1959).

D. Y. C. Ng and D. W. T. Rippin, "The Effect of Incomplete Mixing on Conversion," in *Homogeneous Reactions in Chemical Reaction Engineering* (3rd European Symp.), p. 161. London: Pergamon Press, Inc., 1965.

D. W. T. Rippin, "Segregation in a Two Environment Model of a Partially Mixed Chemical Reactor," *Chem. Eng. Sci.*, **22**, 247 (1967).

R. J. Adler, W. M. Long, J. Rooze, and H. Weinstein, "The Use of Finite State Models for Analysis and Design of Continuous Chemical Reactors," *Proc. 6th World Pet. Conf.* (Frankfurt), 1963.

A good introduction to the notions of residence time distributions can be found in an article by:

J. C. R. Turner, "Residence Time Measurements in Chemical Plants," *Brit. Chem. Eng.*, **9**, 12 (1964).

NOTATION

(See end of this list for frequently used suffixes)

A	area of heat interchange surface
A_j	jth chemical species
a	concentration of A
b	concentration of B
C_P, C_{Pc}	heat capacity per unit volume of reactants, coolant
c, c_j	concentration of A_j
c_j^o	initial concentration of A_j in reactor
c_{Pj}	heat capacity per mole of A_j
E	activation energy
$f_N(\xi_{N+1})$	minimum total holding time, Θ_N, required when feed extent is ξ_{N+1}
ΔH	heat of reaction
h	heat transfer coefficient
h_j	partial molar enthalpy of A_j
J	$-\Delta H/C_p$
k, k', k_1, k_2	Arrhenius rate constants
L, M, N	stability parameters: $1 - \theta(\partial r/\partial \xi)_s$, $1 + (dQ/dT)_s$, $J\theta(\partial r/\partial T)_s$
M_c, M_0	controlled and uncontrolled values of M
$p(t)$	residence time distribution
Q^*	rate of heat removal from reactor
Q	Q^*/qC_P
q	flow rate
R	gas constant
$r(\xi, T)$	reaction rate
$r_m(\xi)$	maximum reaction rate at extent ξ
T	temperature
T_*, T^*	bounds on temperature
T_c^*	effective coolant temperature, $(T_f + \kappa T_{cf})/(1 + \kappa)$
$T_m(\xi)$	temperature for maximum reaction rate at extent ξ
t	time (in Sec. 7.7 it is the time of a batch reaction)
U	rate of heat generation or rejection
V	volume of reactor
X_A	conversion of A
x	$\xi - \xi_s$ (ephemerally for ξ/c_0)
Y_B	yield of B
y	$T - T_s$
α_j, α_{ij}	stoichiometric coefficients of A_j
ζ	incompatibility variable in Sec. 7.1; also dimensionless extent
Θ	total holding time of a sequence of reactors
θ	holding time

κ	heat transfer constant $Q/(T - T_{cf})$; ($2kc_0$ in Sec. 7.9 only)
μ	control parameter proportional to μ^*
μ^*	$(q_c - q_{cs})/(T - T_s)$
μ	mean residence time (Sec. 7.9)
ξ	extent of reaction
$\bar{\xi}$	mean extent
ξ_p	product extent
$\xi_s(T)$	steady state extent at temperature T
σ^2	variance of residence times
τ	t/θ or t/Θ, dimensionless time

SUFFIXES

c	coolant
e	equilibrium
f	feed
i	ith reaction
j	jth species
m	maximum
n	nth reactor in sequence
o	reference state or optimal value
s	steady state

Endnote (p. 201) Throughout this page it is well to remember that the form $r(\xi, T) = k(1 - \xi)$ implies that ξ has been normalized by dividing it by the initial concentration. As a dimensionless quantity it would have been better denoted by another symbol such as the ζ used in Chap. 4.

Adiabatic Reactors 8

8.1 General Principles

As we observed earlier, the adiabatic reactor is not so much a type, more a way of operation. We shall therefore refer to both stirred tank reactors and to tubular ones, and this chapter forms a suitable bridge between the two. We shall introduce the simplest model of the tubular reactor, but this is so elementary that the anticipation of the following chapter will cause no difficulty.

We know that an increase in temperature will decrease the equilibrium extent of an exothermic reaction. Yet to perform the exothermic reaction adiabatically is to induce a temperature increase. Similarly, an endothermic reaction has poorer equilibrium conversion at a lower temperature, and the temperature falls if it is allowed to proceed adiabatically. Thus at first blush there is something rather self-defeating about adiabatic reaction. However, adiabatic operation, involving no heat transfer equipment within the reactor, is so attractive for its simplicity that it is worth more careful examination.

As usual the endothermic reaction does not have the interesting features of the exothermic one. For clearly if the extent is to increase and the temperature to decrease, the rate of reaction will fall off markedly on both counts. The exothermic reaction rate behavior is more hopeful, for an increase of both temperature and extent leads into the region of maximum reaction rate lying below the equilibrium curve. We can perform both kinds of reaction in a number of stages with heating or cooling in between. In this way it is possible to overcome the difficulty of low equilibrium conversion.

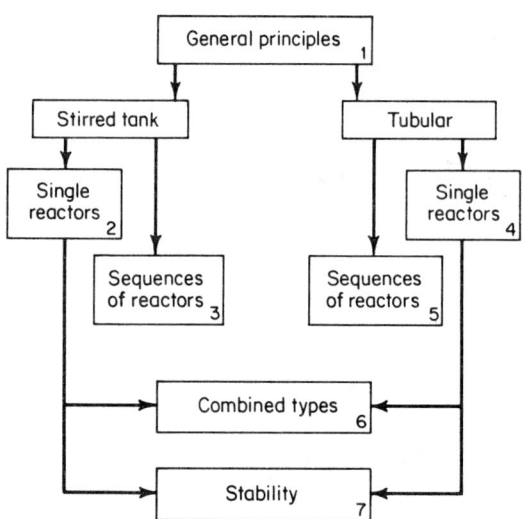

Fig. 8.1 The structure of the chapter.

The structure of this chapter (see Fig. 8.1) is based on the parallel features of the two types. We shall finally say something about the advantages of combining the two types, and make a few general observations on stability.

8.2 The Adiabatic Stirred Tank

We have already obtained the basic design equations for the stirred tank [Eqs. (7.3.1) and (7.3.2)] and these will represent an adiabatic operation if we set $Q = 0$. Thus in the steady state

$$0 = -\xi + \theta r(\xi, T), \tag{8.2.1}$$

$$0 = T_f - T + J\theta r(\xi, T), \quad J = \frac{-\Delta H}{C_p}, \tag{8.2.2}$$

and by multiplying the first equation by J and subtracting it from the second we see immediately that

$$T = T_f + J\xi. \tag{8.2.3}$$

We shall assume that J does not vary appreciably—a reasonable assumption which allows the general picture of adiabatic operation to emerge without too much clutter. This means that the product state (ξ, T) lies on a line of slope J^{-1} through the feed state $(0, T_f)$, as shown in Fig. 8.2. The point F represents the feed states and the locus of product states is the line FG whose equation is Eq. (8.2.3). The equilibrium state for adiabatic operation is at point G and there the reaction rate is zero. However, it is clearly possible

Sec. 8.2 The Adiabatic Stirred Tank

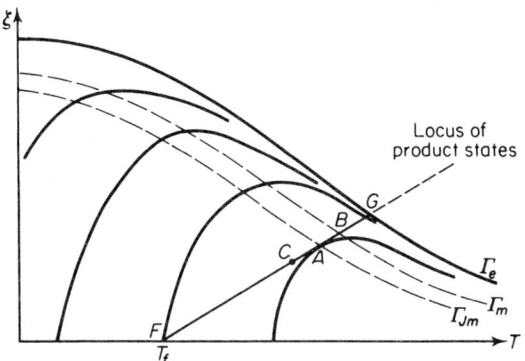

Fig. 8.2 Variation of an exothermic reaction rate in adiabatic reaction.

for the reaction rate to increase along line FG before it falls to zero at equilibrium. In fact, it will be greatest at point A where it just touches one of the loci of constant reaction rate. It will be recalled that there is the locus Γ_m on which $\partial r/\partial T$ vanishes and the reaction rate is greatest for any fixed composition. This is the locus of points of contact of horizontal tangents to the curves of constant rate, whereas point A lies on Γ_{Jm}, the locus of points of contact of tangents of slope J^{-1}. It is clear that Γ_{Jm} lies below Γ_m but may be quite close to it, if J is large. Since $r = r(\xi, T_f + J\xi)$ along the adiabatic path through $(0, T_f)$ it will be greatest when

$$\frac{d_a r}{d\xi} = \frac{\partial r}{\partial \xi} + J \frac{\partial r}{\partial T} = 0. \qquad (8.2.4)$$

This provides an equation for calculating the locus Γ_{Jm}, though unfortunately not as manageable an equation as that for Γ_m. The derivative with respect to ξ along an adiabatic path is $d_a/d\xi$. If we were to plot the reciprocal of the reaction rate along a particular adiabatic path, we would have a curve as in Fig. 8.3; the corresponding points are marked.

We can now ask a number of useful design questions.

A. Production rate specified. The production rate of any product of the reaction will be proportional to $q\xi$, and if this is specified, either ξ or q can be chosen and the other will be fixed. If $q\xi = P$, the volume required of the reactor is

$$V = q\theta = \frac{q\xi}{r(\xi, T)} = \frac{P}{r(\xi, T_f + J\xi)}. \qquad (8.2.5)$$

If T_f is given, V will be least when the product state lies on Γ_{Jm} for then the value of $r(\xi, T_f + J\xi)$ will be greatest. This result can also be obtained by differentiating Eq. (8.2.5) with respect to ξ; then, clearly, $dV/d\xi = 0$ when $d_a r/d\xi = 0$. If T_f is not given, then for any choice of ξ the temperature should be $T_m(\xi)$ making $T_f = T_m(\xi) - J\xi$, and $V = P/r_m(\xi)$. Since $r_m(\xi)$ increases

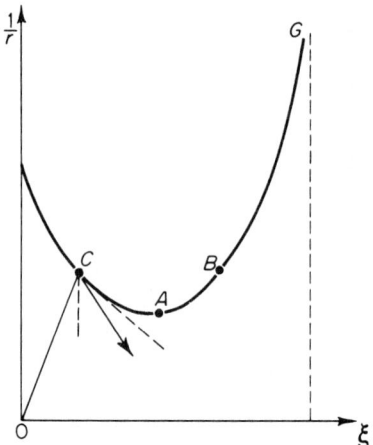

Fig. 8.3 Reciprocal reaction rate along an adiabatic path.

as ξ decreases, ξ should be made as small as possible without taking the resulting T_f (or T_m) beyond safe limits.

B. Product extent specified. If the feed temperature T_f is also specified, the required holding time is immediately given by

$$\theta = \frac{\xi}{r(\xi, T_f + J\xi)}, \qquad (8.2.6)$$

provided of course that the point $(\xi, T_f + J\xi)$ lies beneath the equilibrium curve. If the feed temperature is not specified then we should choose it so that (ξ, T) lies on the locus Γ_m, for then the reaction rate will be as large as possible. Another way of seeing this is to differentiate Eq. (8.2.6) with respect to T_f; then $d\theta/dT_f = 0$ when $\partial r/\partial T = 0$, that is, when the product state is on Γ_m.

C. Feed conditions specified. If q and T_f are given, then the volume is

$$V = \frac{q\xi}{r(\xi, T_f + J\xi)}. \qquad (8.2.7)$$

This will be minimum when ξ is such that

$$\frac{dV}{d\xi} = \frac{q}{r^2}\left(r - \xi\frac{\partial r}{\partial \xi} - \xi J\frac{\partial r}{\partial T}\right) = 0$$

or

$$\frac{d_a r}{d\xi} = \frac{r}{\xi}.$$

Since $r > 0$ this will be at a point such as C, to the left of A in Fig. 8.2. By writing the equation

$$-\frac{d_a}{d\xi}\frac{1}{r} = \frac{1/r}{\xi}$$

Sec. 8.2 The Adiabatic Stirred Tank 233

we see that point C in Fig. 8.3 is such that the tangent there and the line joining to the origin are equally inclined to the vertical. We shall see in Sec. 8.7 that this optimal solution brings the reactor to the margin of stability and this would make it an undesirable design.

D. *Elementary economic analysis.* If the feed stock is at a temperature T_0, it will cost something to heat or cool it to an assigned feed temperature. Let this cost be proportional to the amount of heat required, say $C_H q C_P |T_f - T_0|$. If the cost of the reactor is proportional to its volume V, say $C_v V$, and the value of the production $\bar{p} q \xi$ (where $\bar{p} = \sum \alpha_j p_j$—see Sec. 7.8), then the net profit is

$$P^* = \bar{p}q\xi - C_v q \theta - C_H q C_P |T_f - T_0|,$$
$$= \bar{p}q(\xi - \lambda\theta - \mu|T_f - T_0|).$$

If q is given, then ξ and T_f can be chosen so as to maximize P^*.

$$\frac{\partial}{\partial \xi}\left(\frac{P^*}{\bar{p}q}\right) = 1 - \frac{\lambda}{r^2}\left(r - \xi \frac{d_a r}{d\xi}\right)$$

$$\frac{\partial}{\partial T_f}\left(\frac{P^*}{\bar{p}q}\right) = \frac{\lambda}{r^2} \xi \frac{\partial r}{\partial T} \pm \mu.$$

The last sign is negative if the feed stock has to be preheated and positive if it has to be cooled. Taking the former to be the case, the derivative with respect to temperature vanishes when

$$\frac{\partial r}{\partial T} = \frac{\mu}{\lambda} \frac{r^2}{\xi},$$

and this is on a locus lying below and to the left of Γ_m and quite close to it for small (μ/λ), r, and large ξ. Similarly, the derivative with respect to ξ vanishes when

$$\frac{d_a r}{d\xi} = \frac{r}{\xi}\left(1 - \frac{r}{\lambda}\right).$$

Now $r > \lambda$ for if it were not $\lambda\theta = \lambda\xi/r$ would be greater than ξ and the production would be worth less than the cost of the reactor alone. It follows that this equation can only be satisfied in the region to the right of Γ_{Jm} where $d_a r/d\xi$ is negative. It also follows that if a solution to these equations is to exist, it must lie between the two curves Γ_m and Γ_{Jm}. We have not of course solved the equations, which are very messy, and there will undoubtedly be ranges of the cost variables λ and μ for which there are no solutions. But this kind of analysis and feel for where the solution must lie is an important preliminary to the numerical calculation of the maximum. If the chemical engineer avoids this kind of analysis he sows the wind of careless thinking and is liable to reap a whirlwind of wrong results.

Exercise 8.2.1. A feed of A and B in equal concentrations of 6.67 mole/l goes to a 1 liter stirred vessel and is reacted irreversibly at 120°C to form a product C. The reaction is second order and goes at a rate of kab mole/l min with $k = 3.3 \times 10^9 \exp -(20{,}000/RT)(k = 0.03$ at $120°C)$. The heat of reaction is -20 kcal/gm-mole and the heat capacity 0.65 cal/ml °C. What should be the feed flow rate and temperature if the reactor is to be adiabatic and yield a 50% conversion?

Exercise 8.2.2. With the data of the previous problem and a feed temperature of 17°C, find the flow rate and conversion that maximize the production.

Exercise 8.2.3. Discuss the stability of the steady state in Ex. (8.2.1).

Exercise 8.2.4. Suppose that the required production $P = q\xi$ is fixed and that the only two items in the cost are:

   (i) An amortized construction and maintenance cost for the reactor, $C_v V$, proportional to its volume;
   (ii) A pumping cost, $C_p q$, proportional to q.

The cost of attaining the best feed temperature is negligible. Show that the extent which minimizes the cost for a given production P is such that

$$\xi^2 \frac{-r'_m(\xi)}{r_m^2(\xi)} = \frac{C_p}{C_v}.$$

Sketch the form of the left-hand side and discuss the reasonableness of this result.

8.3 Sequences of Adiabatic Stirred Tanks

Since the maximum reaction rates on the curves Γ_m or Γ_{Jm} decrease as extent increases, it might well be possible to gain some economy by performing the reaction in two stages instead of one. For example, the first reactor might take the extent from 0 to ξ with a corresponding temperature $T_f + J\xi$, and then the process stream might be cooled down to a temperature $\bar{T}_2$ where further reaction would take place. The operation of the two reactors would be represented by two line segments such as AB and CD in Fig. 8.4, where the ordinates of A, B, C, and D are 0, ξ_1, ξ_1, and ξ_2, respectively. This might have the advantage of a greater average reaction rate and would certainly be able to reach a higher extent of reaction than the equilibrium extent for only one stage.

We shall not discuss such problems in full detail, but an elementary one is worth considering. Suppose that the extent ξ_2 is fixed and so is the feed temperature T_f, which we will now denote by T_{f1} since it is the feed to the first reactor. We ask how much reaction in the first stage and how much cooling between stages should take place if the total holding time is to be least. The temperature at the exit of the first reactor is $T_{f1} + J\xi_1$, and the feed temperature of the second reactor will be chosen to be $T_{f2} + J\xi_1$, so that

Sec. 8.3 Sequences of Adiabatic Stirred Tanks

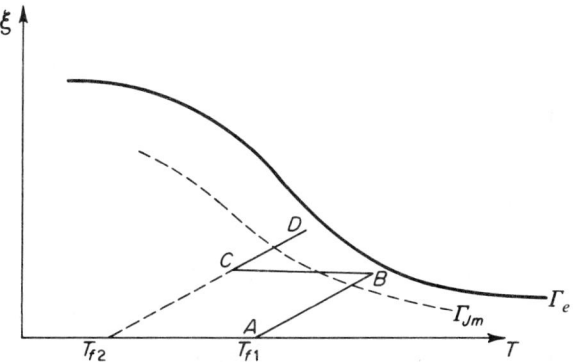

Fig. 8.4 Optimal sequence of two stirred tank reactors.

cooling brings about a temperature drop of $(T_{f1} - T_{f2})$. For simplicity, we shall assign no cost to this cooling. Then

$$\theta_1 + \theta_2 = \frac{\xi_1}{r(\xi_1, T_{f1} + J\xi_1)} + \frac{(\xi_2 - \xi_1)}{r(\xi_2, T_{f2} + J\xi_2)}. \quad (8.3.1)$$

The two choices to be made are of ξ_1 and T_{f2}. Now the choice of T_{f2} only affects the second term on the right of Eq. (8.3.1), and we know from problem B of Sec. 8.2 that, in this case, the state (ξ_2, T_2) should be on Γ_m. Since ξ_2 has been specified, this fixes T_{f2} as $T_m(\xi_2) - J\xi_2$. To find ξ_1 we notice that

$$\frac{\partial}{\partial \xi_1}(\theta_1 + \theta_2) = \frac{1}{r(\xi_1, T_1)} - \frac{\xi_1}{r^2(\xi_1, T_1)} \frac{d_a r}{d\xi_1} - \frac{1}{r(\xi_2, T_2)}, \quad (8.3.2)$$

where $T_n = T_{fn} + J\xi_n$ and $n = 1, 2$. This equation must be solved numerically, but it is easy to see what sort of solution will be obtained. Since we know the values of both T_{f1} and T_{f2}, we can draw the curves for the reciprocal rates $1/r(\xi, T_{fn} + J\xi)$, $n = 1, 2$. They are as shown in Fig. 8.5, because the reaction rate along the path through T_{f1} is at first greater than that along T_{f2}, but it has a small equilibrium extent and so the curves cross later. The final state (ξ_2, T_2) corresponds to a point D on the second curve a little beyond the minimum. If B is to represent point (ξ_1, T_1) then Eq. (8.3.2) can be interpreted geometrically to mean that intersection R of the tangent at B with the axis $\xi = 0$ should be at the same distance below B as D is above B. Then the areas of the rectangles $ABPO$ and $PCDQ$ are the two holding times, and clearly the total holding time is less than that of any single stage reactor. Similar multistage designs could be set up for other constraints and objectives.

Another way to achieve cooling between stages is to mix the product of the first stage with cold unreacted feed before passing to the next stage. This has the merit of simplicity and may avoid excessive interchanger equipment. It is shown for two stages in Fig. 8.6. A fraction λq of the feed is heated

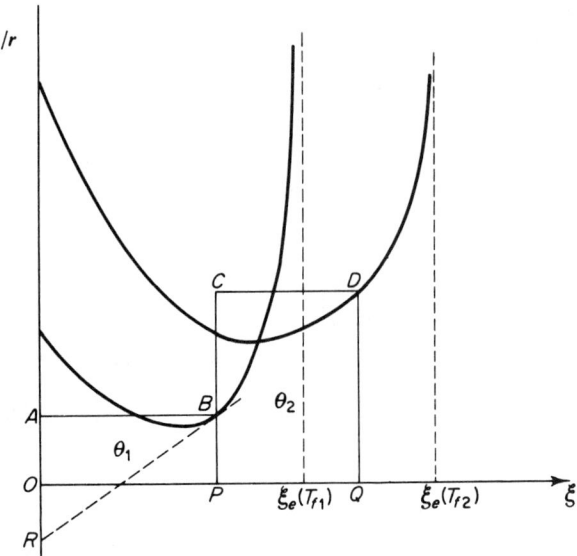

Fig. 8.5 Optimal sequence in relation to $1/r$ versus ξ curves.

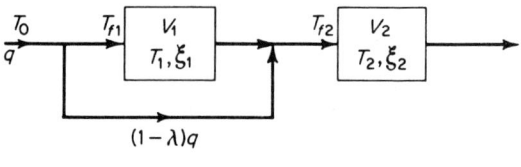

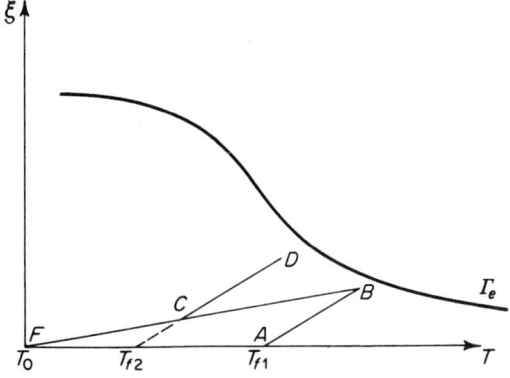

Fig. 8.6 Optimal sequence with bypass of cold feed.

Sec. 8.3   Sequences of Adiabatic Stirred Tanks   237

from T_0 to T_{f1} and fed to the first reactor of volume V_1. The reaction proceeds to the state ξ_1, $T_1 = T_{f1} + J\xi_1$ (represented by segment AB in the lower part of the figure), and this stream is mixed with the bypass stream of zero extent and temperature T_0. Since the flow rates of the two streams are λq and $(1 - \lambda)q$, respectively, heat and mass balances at the mixing point give for the feed conditions to the second tank

$$\bar{\xi}_2 = \lambda \xi_1$$
$$\bar{T}_2 = \lambda T_1 + (1 - \lambda)T_0. \tag{8.3.3}$$

This is represented by point C on line FB, where by similar triangles $FC = \lambda FB$. The adiabatic path for the second reactor would go through the point $(0, T_{f2})$ where

$$T_{f2} = \bar{T}_2 - J\bar{\xi}_2 = \lambda(T_1 - J\xi_1) + (1 - \lambda)T_0 = \lambda T_{f1} + (1 - \lambda)T_0. \tag{8.3.4}$$

In the second reactor the change in state is represented by segment CD in the lower part of Fig. 8.6. The full equations are thus

$$0 = -\xi_1 + \frac{\theta_1}{\lambda} r(\xi_1, T_{f1} + J\xi_1)$$
$$0 = \bar{\xi}_2 - \xi_2 + \theta_2 r(\xi_2, T_{f2} + J\xi_2), \tag{8.3.5}$$

where $\theta_n = V_n/q$, $n = 1, 2$.

We might now ask what the feed temperature T_{f1} and the bypass ratio λ should be if the greatest extent ξ_2 is to be attained from two given reactors of holding time θ_1 and θ_2. Using Eqs. (8.3.3) and adding Eqs. (8.3.4) we have

$$\lambda \xi_1 = \theta_1 r(\xi_1, T_{f1} + J\xi_1)$$
$$\xi_2 = \theta_1 r(\xi_1, T_{f1} + J\xi_1) + \theta_2 r[\xi_2, \lambda T_{f1} + (1 - \lambda)T_0 + J\xi_2]. \tag{8.3.6}$$

These lead to a complicated set of equations if the partial derivatives $\partial \xi_2/\partial \lambda$ and $\partial \xi_2/\partial T_{f1}$ are set equal to zero. A search on the variables T_{f1} and ξ_1 might well be the best way of proceeding, for the first equation would allow the direct calculation of λ and the second equation that of ξ_2.

More complicated problems for sequences of stirred tanks can be devised, but they follow the pattern of multibed adiabatic tubular reactors to which we now turn.

Exercise 8.3.1. Show how contours of constant $\xi/r(\xi, T)$ in the ξ, T plane might be used to calculate the holding time required of an adiabatic stirred tank. Sketch them carefully.

Exercise 8.3.2. Show that points (ξ_2, T_2) and (ξ_1, T_1) corresponding to a maximization of ξ_2 in Eqs. (8.3.5) lie above Γ_m and below Γ'_{Jm}, respectively, where Γ'_{Jm} is the locus of $dr/d\xi = r/\xi$.

8.4 The Adiabatic Tubular or Batch Reactor

For a batch reactor the very definition of reaction rate gives the equation

$$\frac{d\xi}{dt} = r(\xi, T), \qquad \xi(0) = 0. \tag{8.4.1}$$

From an enthalpy balance for the adiabatic operation

$$\frac{dT}{dt} = Jr(\xi, T), \qquad T(0) = T_0. \tag{8.4.2}$$

Hence we again have the adiabatic path

$$T(t) = T_0 + J\xi(t). \tag{8.4.3}$$

The reaction time can now be calculated by substituting Eq. (8.4.3) into Eq. (8.4.1) and integrating to give

$$t = \int_0^{\xi(t)} \frac{d\xi'}{r(\xi', T_0 + J\xi')}, \tag{8.4.4}$$

where ξ' has been written for the dummy integration variable. The batch reaction time is thus the area under the curve in Fig. 8.3. In contrast to the isothermal case, the holding time of a stirred tank may now be less than the batch reaction time. We shall exploit this later in Sec. 8.6.

The simplest model of the tubular reactor is that the stream flows uniformly through the reactor with velocity v with no mixing. In the steady state the temperature, concentrations, and extent are functions only of the distance z from the inlet of the reactor (see Fig. 8.7). Entering the section we have $vc_j(z)$ moles of A_j per unit area per unit time, $vc_j(z + dz)$ moles leave the section in the same time, and $\alpha_j\, dz\, r(c_1, \ldots, c_s, T)$ moles are formed by reaction. Hence in the limit as $dz \longrightarrow 0$

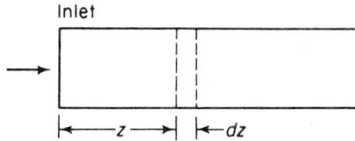

Fig. 8.7 The adiabatic tubular reactor.

$$v\frac{dc_j}{dz} = \alpha_j r(c_1, \ldots, c_s, T) \tag{8.4.5}$$

and $c_j(0) = c_{j0}$. Setting

$$c_j = c_{j0} + \alpha_j \xi, \tag{8.4.6}$$

we have

$$v\frac{d\xi}{dz} = r(\xi, T), \qquad \xi(0) = 0. \tag{8.4.7}$$

Sec. 8.4 The Adiabatic Tubular or Batch Reactor   239

This is the same as Eq. (8.4.1) if we set

$$t = \frac{z}{v}, \qquad (8.4.8)$$

the residence time of the reactants up to position z. If the length of the reactor is L, we may set

$$\theta = \frac{L}{v} \qquad (8.4.9)$$

and call it the holding time of the reactor. The balance of enthalpy again gives

$$\frac{dT}{dt} = v\frac{dT}{dz} = Jr(\xi, T), \qquad (8.4.10)$$

and there is complete formal similarity between the batch reactor and the simplest model of the tubular or plug flow reactor.

For gas reactions when the density may vary throughout the bed, the mass fraction is sometimes to be preferred above the molar concentration in representing the composition. Then, if G is the mass flow rate per unit cross-sectional area of reactor and $g_j(z)$ the mass fraction of A_j at z, a mass balance for unit time and unit area gives

mass of A_j entering element $= Gg_j(z)$,

mass of A_j leaving element $= Gg_j(z + dz)$,

mass of A_j formed by reaction $= \alpha_j m_j r\, dz$.

Thus in the limit

$$G\frac{dg_j}{dz} = \alpha_j m_j r, \qquad g_j(0) = g_{j0}. \qquad (8.4.11)$$

Letting

$$g_j = g_{j0} + \alpha_j m_j \xi'', \qquad (8.4.12)$$

we have

$$G\frac{d\xi''}{dz} = r(\xi'', T, P), \qquad \xi''(0) = 0. \qquad (8.4.13)$$

We have made the reaction rate a function of pressure as well as of temperature and extent per unit mass, for this representation is most useful with gas reactions for which pressure may be an important factor. If density is constant we recover Eq. (8.4.7) by setting $G = \rho v$ and $\xi'' = \xi/\rho$.

For ideal gases or when enthalpy is not very dependent on pressure, the total enthalpy per unit mass is $\sum g_j h_j / m_j$. If this is constant for adiabatic operation

$$\frac{d}{dz}\sum \frac{g_j h_j}{m_j} = \sum \frac{h_j}{m_j}\frac{dg_j}{dz} + \left(\sum \frac{g_j}{m_j}\frac{\partial h_j}{\partial T}\right)\frac{dT}{dz} = 0,$$

where the familiar identity (3.3A) has been invoked again (see Sec. 7.2). Using Eqs. (8.4.12) and (8.4.13) we have

$$\left(\sum \alpha_j h_j\right)\frac{d\xi''}{dz} + \left(\sum \frac{c_j}{\rho} C_{P_j}\right)\frac{dT}{dz} = 0 \tag{8.4.14}$$

or

$$G\frac{dT}{dz} = J''r(\xi'', T, P), \tag{8.4.15}$$

where

$$J'' = \frac{-\Delta H}{\hat{C}_P} = \rho J \tag{8.4.16}$$

and $\hat{C}_P$ is the heat capacity per unit mass of reaction mixture. As a function of temperature J'' is more nearly constant than J for gas mixtures. In terms of ξ'', the extent per unit mass, we have the equations

$$T = T_0 + J''\xi'' \tag{8.4.17}$$

and

$$G\frac{d\xi''}{dz} = r(\xi'', T_0 + J''\xi''), \qquad \xi''(0) = 0. \tag{8.4.18}$$

The total length of the reactor needed to achieve an extent of reaction ξ'' is

$$L = G\int_0^{\xi''} \frac{d\xi'}{r(\xi', T_0 + J''\xi')}. \tag{8.4.19}$$

Here again, ξ' has been used for the dummy integration variable; it is not the dimensionless extent.

This suggests a useful graphical presentation of the design of an adiabatic tubular reactor. In Fig. 8.8 the equilibrium line Γ_e and the curve of maximum reaction rate in an adiabatic bed, Γ_{Jm}, are shown. The broken lines are adiabatic paths. Let ξ_{Jm} and T_{Jm} be the extent and temperature at a point on Γ_{Jm}. Then the adiabatic path through such a point is

$$T = T_{Jm} + J(\xi - \xi_{Jm}) = T_0 + J\xi, \tag{8.4.20}$$

where T_0 is the intersection of this path with the T axis. Now for any path let us imagine the integral

$$\Theta(\xi; T_0) = \int_{\xi_{Jm}}^{\xi} \frac{d\xi'}{r(\xi', T_0 + J\xi')}, \tag{8.4.21}$$

with

$$T_0 = T_{Jm} - J\xi_{Jm} \tag{8.4.22}$$

to be calculated. Clearly, Γ_{Jm} itself will be the contour $\Theta = 0$, Γ_e will be the contour $\Theta = \infty$, and contours of positive and negative Θ will lie above and below Γ_{Jm}, respectively. Then if we want to calculate the holding time necessary to get from any feed condition ξ_f, T_f to some product extent ξ_p we have

Sec. 8.4 The Adiabatic Tubular or Batch Reactor 241

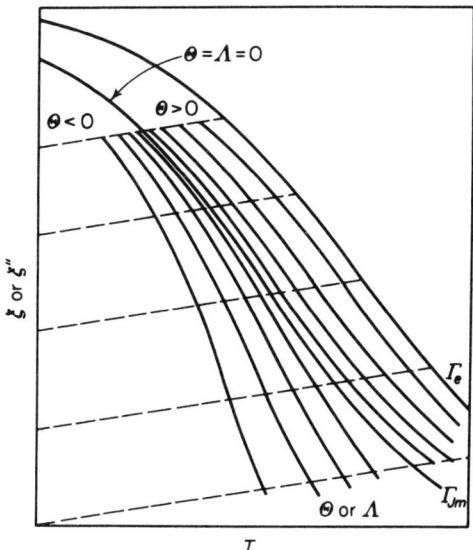

Fig. 8.8 Contours of adiabatic reactor holding time in the ξ, T or ξ'', T plane.

$$\theta = \Theta(\xi_p; T_f - J\xi_f) - \Theta(\xi_f; T_f - J\xi_f), \quad (8.4.23)$$

that is, the difference between the numbers Θ associated with the contours on which the feed and product states lie. In particular, if $\xi_f = 0$, $T_f = T_0$, and $\xi_p = \xi$, we have

$$\theta(\xi; T_0) = \Theta(\xi; T_0) - \Theta(0; T_0). \quad (8.4.24)$$

A similar set of curves could be drawn for the integral

$$\Lambda(\xi''; T_0) = \int_{\xi''_{Jm}}^{\xi''} \frac{d\xi'}{r(\xi', T_0 + J''\xi')}, \quad (8.4.25)$$

where ξ''_{Jm}, T_{Jm} is on the curve Γ_{Jm} and

$$T_0 = T_{Jm} - J'' \xi''_{Jm}. \quad (8.4.26)$$

Then Eq. (8.4.19) would give

$$L = G\Lambda(\xi''; T_0) - G\Lambda(0; T_0), \quad (8.4.27)$$

and the length of reactor required to affect any given conversion could be calculated.

For irreversible reactions which have only one rate constant, Eagleton and Douglas showed that it is possible to obtain these integrals in terms of tabulated functions. Let us consider the first order reaction with rate

$$r(\xi, T) = A \, e^{-(E/RT)} (c_0 - \xi). \quad (8.4.28)$$

Then leaving aside the limits for the moment, the integral we have to evaluate is

$$A^{-1} \int \frac{e^{E/R(T_0+J\xi')}}{c_0 - \xi'} d\xi'. \tag{8.4.29}$$

If we set

$$\frac{E}{RT} = \frac{E}{R(T_0 + J\xi')} = y, \tag{8.4.30}$$

we have

$$\xi' = \frac{E}{RJ}\frac{1}{y} - \frac{T_0}{J}, \tag{8.4.31}$$

$$d\xi' = -\frac{E}{RJ}\frac{dy}{y^2},$$

and

$$c_0 - \xi' = \frac{T_0 + Jc_0}{J} - \frac{E}{RJ}\frac{1}{y},$$
$$= \frac{T_a}{J}\frac{y - (E/RT_a)}{y}, \tag{8.4.32}$$

where

$$T_a = T_0 + Jc_0 \tag{8.4.33}$$

is the temperature that would be reached by complete adiabatic conversion. Putting all these into the integral and noting that the limits $\xi' = 0$ and $\xi' = \xi$ would give limits of E/RT_0 and E/RT for y, we obtain

$$\theta(\xi; T_0) = \frac{E}{RT_a}\frac{1}{A}\int_{E/RT}^{E/RT_0} \frac{e^y \, dy}{y[y - (E/RT_a)]}. \tag{8.4.34}$$

Splitting this integrand into partial fractions gives

$$A\theta(\xi; T_0) = \int_{E/RT}^{E/RT_0} e^y \left[\frac{1}{y - (E/RT_a)} - \frac{1}{y}\right] dy.$$

These integrals are discussed further in the Appendix; for our present purposes Eq. (A.13) gives immediately

$$A\theta = e^{E/RT_a}\left[Ei\left(\frac{E}{RT_0} - \frac{E}{RT_a}\right) - Ei\left(\frac{E}{RT} - \frac{E}{RT_a}\right)\right]$$
$$- \left[Ei\left(\frac{E}{RT_0}\right) - Ei\left(\frac{E}{RT}\right)\right]. \tag{8.4.35}$$

Other cases can be similarly treated by factorizing the concentration-dependent part of the reaction rate expression into a product of linear expressions and by using partial fractions.

Sec. 8.4   The Adiabatic Tubular or Batch Reactor

Exercise 8.4.1. If $x > 15$, $Ei(x) \sim e^x/x$, show that if $E > 30\ TT_a/(T_a - T)$ then Eq. (8.4.35) becomes approximately:

$$A\theta = \frac{R}{E}\left[\frac{T_0^2}{T_a - T_0} e^{E/RT_0} - \frac{T^2}{T_a - T} e^{E/RT}\right].$$

Exercise 8.4.2. Obtain an expression for the holding time for a zeroth order reaction.

Exercise 8.4.3. For the second order reaction $-A_1 - A_2 + A_3 + A_4 = 0$ with $c_{10} = c_{20} = c_0$ and $r = A(\exp - E/RT)(c_0 - \xi)^2$ show that

$$Ac_0\theta = \frac{(T_a - T_0)T}{T_a(T_a - T)} e^{E/RT} - \frac{T_0}{T_a} e^{E/RT_0}$$

$$- \frac{E(T_a - T_0)}{RT_a^2} e^{E/RT_a}\left[Ei\left(\frac{E}{RT_0} - \frac{E}{RT_a}\right) - Ei\left(\frac{E}{RT} - \frac{E}{RT_a}\right)\right].$$

Exercise 8.4.4. The decomposition of phosphine proceeds to completion as a first order endothermic reaction with the stoichiometric equation $4PH_3(g) \longrightarrow P_4(g) + 6H_2(g)$. If this reaction is carried out adiabatically, the temperature corresponding to a given degree of conversion can be calculated by means of an energy balance giving the following results:

Degree of conversion: 0   0.1   0.2   0.25   0.3

$T°K$:   945   885   826   797   768

The velocity constants $k(hr^{-1})$ appropriate to these temperatures are

37.2,   1.42, 3.65×10^{-2}, 3×10^{-3}, 5.83×10^{-4}.

If 50 lb per hr of phosphine at atmospheric pressure and 945°K is fed into an adiabatic reactor, determine the size of the reactor for 30% conversion to phosphorus:

(i) If the temperature changes without longitudinal mixing;
(ii) If longitudinal mixing within the reactor keeps the temperature and composition uniform.

Compare the reactor volumes in (i) and (ii) with the size of a perfectly mixed reactor maintained at 945°K.   (C.U.)

Exercise 8.4.5. The first order reaction $A \rightleftharpoons B$ has the reaction rate

$$k_1(1 - \xi) - k_2\xi, \qquad k_i = A_i \exp -\frac{E_i}{RT}, \qquad E_2 > E_1.$$

Show that the maximum reaction rate on the adiabatic path $T = T_0 + J\xi$ occurs when

$$T - T_0 = \frac{JE_1K - RT^2(1 + K)}{E_1K + E_2}$$

$$K = \frac{k_1}{k_2}.$$

How would you use this to calculate the locus $\Gamma_{Jm'}$?

Exercise 8.4.6. The irreversible reaction $A \longrightarrow B$ is heterogeneously catalyzed and has the rate $r = ka/(1 + k'a)$. Setting $a = c_0 - \xi$ and using the usual notation for the constants in k and k', show how to express $\theta(\xi; T_0)$ for this reaction rate in terms of Eq. (8.4.35) and the solution of Exercise 8.4.2.

8.5 Multistage Adiabatic Reactors

A diagram such as Fig. 8.8 could be used for a graphical design of multibed reactors; it shows the kind of design which might be optimal because it would certainly seem appropriate to work in the neighborhood of the locus of maximum reaction rate. Again, two common methods are available for the interstage cooling: (1) mixing with a cold shot of unreacted feed, and (2) interchanger cooling. With the former, the extent is decreased in passing from one stage to the next. For the latter, we assume that no reaction takes place in the heat exchanger; this is a good assumption for heterogeneously catalyzed reactions for which this method is sometimes used, as, for example, in the syntheses of methanol, ammonia, and the oxidation of sulfur dioxide.

Figure 8.9 shows the general scheme of things. If in (a) the state of the feed stock is represented by the point O, part of it will be preheated to a temperature corresponding to the point A. Reaction in the first bed takes the state to B, and mixing with cold, unreacted feed will bring the state back to a point such as C on the line OB. The reaction in the second bed takes the state to D and so on. Clearly, if the reaction segments AB, CD, and EF can be suitably chosen the reaction can be kept in the neighborhood of the

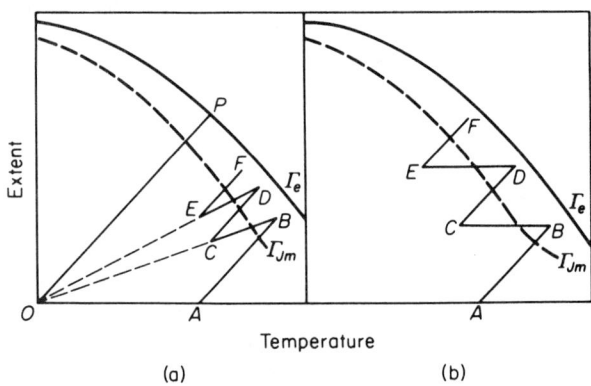

Fig. 8.9 Design of adiabatic reactors: (a) with cold shot cooling; (b) with interchanger cooling between beds.

region of maximum reaction rate. The limitation of the cold shot cooling is that it can never bring the reaction path above line OP, and hence the adiabatic equilibrium conversion for the original feed stock still represents an upper limit on conversion.

There are $2N$ choices to be made in the design of a reactor having N stages. The fraction of the cold stream to be withheld for bypass and the preheat temperature must first be decided. Then, the size of the first stage or its exit temperature or extent can be fixed (but not more than one of these, since they are related by the adiabatic condition). Next, the fraction of the bypass stream to be used will fix point C. With one choice for the size of the second stage, five decisions have been made for two stages. If the third stage is the last, there is really only one further decision which can be made, for the position of E is fixed by using the remaining amount of bypass and then a single decision fixes F. Thus six decisions are needed for three stages and evidently $2N$ would be needed for N stages. We shall not pursue the way in which these decisions could be made optimally—suffice it to say that as usual it is convenient to work backwards from the end to the beginning.

The multistage reactor with interchanger cooling is illustrated in Fig. 8.9(b). Evidently it does not suffer from the defect of being limited to the adiabatic equilibrium conversion for the cold feed-stock. Again there are $2N$ decisions to be made in the design of an N stage reactor. If the temperature of the feed to the first bed is chosen (A), one further choice will fix the outlet conditions (B). Then the amount of interchanger cooling fixes the inlet condition to the second bed (C) and a further choice, the outlet conditions at D, and so on.

Taking our cue from the optimal stirred tank sequence, we shall number the stages from the end to the beginning as in Fig. 8.10. The inlet conditions to reactor n are denoted by $\bar{\xi}_n$, $\bar{T}_n$ and the exit by ξ_n, T_n. Then in any one stage $T - J\xi$ is constant; for stage n

$$T - J\xi = \bar{T}_n - J\bar{\xi}_n = T_n - J\xi_n = S_n \qquad (8.5.1)$$

is a constant fixed by the inlet conditions. For the reaction rate in stage n

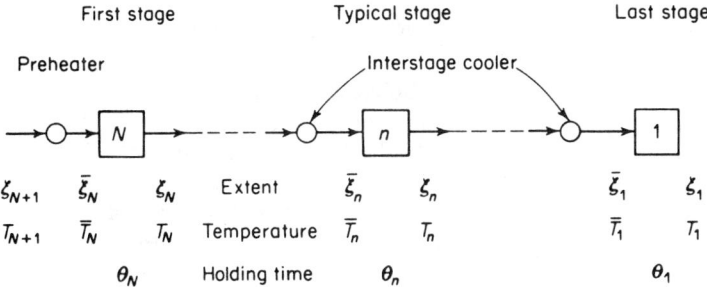

Fig. 8.10 Multibed adiabatic reactor with interstage cooling.

we may write
$$r_n(\xi) = r(\xi, S_n + J\xi), \tag{8.5.2}$$
and the holding time becomes
$$\theta_n = \int_{\xi_n}^{\bar{\xi}_n} \frac{d\xi}{r_n(\xi)}. \tag{8.5.3}$$
Between beds there is negligible reaction, so
$$\bar{\xi}_{n-1} = \xi_n, \tag{8.5.4}$$
and hence
$$S_{n-1} = S_n - (T_n - \bar{T}_{n-1}). \tag{8.5.5}$$

For N stages a suitable sequence of decisions would be as follows. Given ξ_{N+1}, Y_{N+1}:

Choose	then calculate	by Eq. 8.5....	this determines the	conditions for stage
$\bar{T}_N$				
	$\bar{\xi}_N = \xi_{N+1}$	4		
	$S_N = \bar{T}_N - J\bar{\xi}_N$	1	inlet	N
ξ_N				
	θ_N	3		
	$T_N = \bar{T}_N + J(\xi_N - \bar{\xi}_N)$	1	exit	N
$\bar{T}_{N-1}$				
	$\bar{\xi}_{N-1} = \xi_N$	4		
	$S_{N-1} = \bar{T}_{N-1} - J\bar{\xi}_{N-1}$	1	inlet	$N-1$
ξ_{N-1}				
... and so on to ...				
ξ_1				
	θ_1	3		
	T_1	1	exit	1

Actually in working out an optimal design it pays to make the decisions in precisely the reverse order. This is in keeping with our experience with stirred tanks where it was desirable to work from the end to the beginning.

Without going into complete detail, we can sketch a typical optimization of the following kind. The total value of the process stream will be increased by reaction in proportion to the increase in extent, $\xi_1 - \xi_{N+1}$. We shall take the only cost to be proportional to the total volume, and hence holding time of the reactor, and let v be the cost of this relative to the gain in extent. Thus we might seek to maximize

$$\xi_1 - \xi_{N+1} - v \sum \theta_n.$$

Sec. 8.5 Multistage Adiabatic Reactors

By using Eqs. (8.5.3) and (8.5.4) we see that this can be written as

$$P = \sum_{1}^{N}\left(\xi_n - \bar{\xi}_n - \nu \int_{\xi_n}^{\xi_n} \frac{d\xi}{r_n(\xi)}\right) = \sum_{1}^{N} p_n, \quad (8.5.6)$$

where p_n is the profit added by the stage n. If ξ_{N+1}, T_{N+1} is given it is clear from the scheme outlined above that a set of $2N$ decisions (namely the choice of $\bar{T}_n, \xi_n$ where $n = N, N-1, \ldots, 1$) would allow the profit to be calculated. If these decisions are made optimally so that the maximum value of P is attained, this maximum is a function only of ξ_{N+1} and T_{N+1}. However, it is not really dependent on T_{N+1}, for the best value of $\bar{T}_N$ will depend only on ξ_{N+1} since there is no cost associated with heating or cooling. We therefore write

$$g_N(\xi_{N+1}) = \text{Max} \sum_{1}^{N} p_n. \quad (8.5.7)$$

Had we included the cost of the interchanger then the maximum would indeed have been a function of both ξ_{N+1} and T_{N+1}. This more complicated case can be treated along the same lines as the simpler one considered here.

The technique of dynamic programming suggests that we should start with the last reactor and, given ξ_2, should chose $\bar{T}_1$ and ξ_1 so as to maximize

$$p_1 = \xi_1 - \bar{\xi}_1 - \nu\theta_1 = \int_{\xi_1}^{\xi_1}\left(1 - \frac{\nu}{r_1(\xi)}\right)d\xi. \quad (8.5.8)$$

The two decisions we have to make affect the integral in different ways, for ξ_1 appears only as the upper limit and $\bar{T}_1$ affects only $r_1(\xi) = r(\xi, \bar{T}_1 - J\xi_2 + J\xi)$. In particular, we have for the maximum of p_1

$$\frac{\partial p_1}{\partial \xi_1} = 1 - \frac{\nu}{r_1(\xi_1)} = 0, \quad (8.5.9)$$

and

$$\frac{\partial p_1}{\partial \bar{T}_1} = \nu \int_{\xi_1}^{\xi_1} \frac{r_{T_1}(\xi)\,d\xi}{r_1^2(\xi)} = 0, \quad (8.5.10)$$

where r_{T_1} denotes the derivative $\partial r/\partial T$ evaluated along the adiabatic path. The first of these conditions is easy to interpret because it says that the reaction should be stopped when the reaction rate is ν. This is sensible, for to go on beyond this point would be to react so slowly that the value of the increment in extent would be less than the cost of the additional holding time. This gives a locus, Γ_1, for optimal end points which is just the contour of constant $r(\xi, T) = \nu$. By taking any point on this contour and integrating back along an adiabatic path until the value of the integral in Eq. (8.5.10) is zero, we can find another locus, $\bar{\Gamma}_1$, on which the inlet state for stage 1 should lie. We can thus calculate

$$g_1(\xi_2) = \text{Max} \int_{\xi_2}^{\xi_1}\left(1 - \frac{\nu}{r_1(\xi)}\right)d\xi. \quad (8.5.11)$$

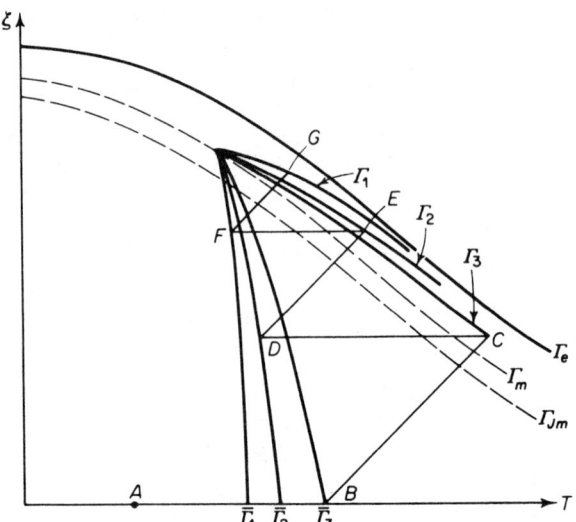

Fig. 8.11 Graphical representation of the optimal design with three beds.

The curves $\bar{\Gamma}_1$ and Γ'_1 can be used rather easily to get the optimal conditions for a single bed. If point E in Fig. 8.11 represents (ξ_2, T_2)—ignore the other curves Γ'_2, $\bar{\Gamma}_2$, etc. for the moment—then the best inlet conditions must lie at F on $\bar{\Gamma}_1$ where EF is horizontal since $\bar{\xi}_1 = \xi_2$. From F we draw an adiabatic line to meet Γ'_1 at G and this gives the optimal exit condition.

For two beds

$$g_2(\xi_3) = \text{Max}\,(p_2 + p_1) = \text{Max}\,[p_2 + g_1(\xi_2)]. \quad (8.5.12)$$

In the first maximization, we imply that all four quantities $\bar{T}_2$, ξ_2, $\bar{T}_1$, and ξ_1 must be chosen. But, given ξ_2, we already know how to choose $\bar{T}_1$ and ξ_1 so as to let p_1 have a maximum value $g_1(\xi_2)$. Hence the second maximum in Eq. (8.5.12) requires only the choice of $\bar{T}_2$ and ξ_2, and the remaining quantities will be the optimal decisions for ξ_2. We will not discuss how these choices are made (see Exercise 8.5.1) but simply note that when the optimal ξ_2 and $\bar{T}_2$ have been found, two more curves can be drawn in Fig. 8.11. These two curves are the loci of optimal inlet and exit conditions for stage 2. For three beds

$$g_3(\xi_4) = \text{Max}\,[p_3 + g_2(\xi_3)]$$

by the same principle of optimality, and optimal choice of ξ_3 and $\bar{T}_3$ must be made. This process can be continued as long as it is needed, and the results can be either tabulated or presented graphically.

Figure 8.11 shows a three-bed design. If A represents the cold feed at

Sec. 8.6 Combined Types of Adiabatic Reactor 249

zero extent, it should be preheated to B, the intersection of the horizontal through A with $\bar{\Gamma}_3$ (the locus of optimal inlet conditions for stage 3). A line of adiabatic slope is drawn to C which, lying on Γ_3, must give the optimal exit conditions from stage 3, ξ_3 and T_3. Since this product from the first stage is the feed to the remaining two stages and since they must use optimal decisions—or else the profit from them would be less than it should be—we proceed from C horizontally to D on $\bar{\Gamma}_2$, for $\bar{\xi}_2 = \xi_3$. Then by going adiabatically to E on Γ_2, horizontally to F on $\bar{\Gamma}_1$, and adiabatically to G on Γ_1, we have a complete optimal design. For accuracy the results should be tabulated, but the graphical method is useful for showing the general form of the optimal design.

A similar presentation of the optimal design can be made when the cost of interchanger cooling is included or when cold shot cooling is used. The references given at the end of this chapter cover this phase thoroughly.

Exercise 8.5.1. Show that

$$\frac{dg_1}{d\xi_2} = -\left(1 - \frac{\nu}{r(\bar{\xi}_1, \bar{T}_1)}\right),$$

and deduce that in an optimal design the reaction rate at the entrance of stage 1 should be the same as that at the exit of stage 2.

Exercise 8.5.2. Consider how you would present and use the results of an optimal design in the form of tables of $g_N(\xi_{N+1})$ and other functions.

8.6 Combined Types of Adiabatic Reactor

The shape of the reciprocal reaction rate curve in Fig. 8.3 suggests that a combination of tubular and stirred tank type reactors might have some advantages over either one of them used by itself. If we consider the feed extent to be zero and inlet temperature T_0, then for the stirred tank

$$\theta_s(\xi; T_0) = \frac{\xi}{r(\xi, T_0 + J\xi)} \tag{8.6.1}$$

is the area of a rectangle with base $(0, \xi)$ and height $1/r$ for the product condition ξ, $T = T_0 + J\xi$. On the other hand, the holding time of the tubular reactor

$$\theta_t(\xi; T_0) = \int_0^\xi \frac{d\xi'}{r(\xi', T_0 + J\xi')} \tag{8.6.2}$$

is the area under the reciprocal reaction rate curve. If the required extent corresponds to a point such as A in Fig. 8.12, then the stirred tank holding time θ_s, given by the area of the rectangle $OABC$, is obviously less than the tubular holding time θ_t, the area $OABD$. In fact, this will be true for all

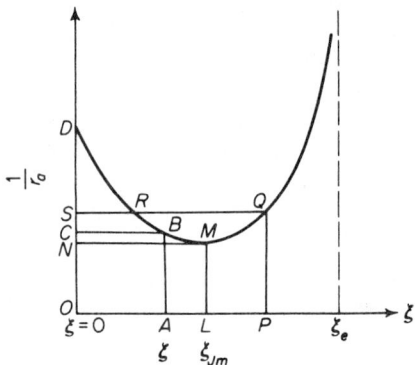

Fig. 8.12 Illustrating the calculation of reaction time in combined types of reactor.

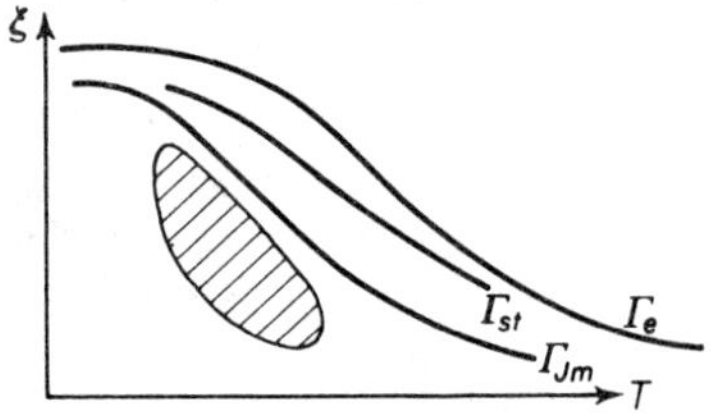

Fig. 8.13 Dependence of optimal type on product state.

extents to the left of a point P, for which the area $OPQS$ is the same as the area $OPQMBRDO$ under the curve. By calculating the function

$$\frac{\theta_t}{\theta_s} = \frac{r(\xi, T_0 + J\xi)}{\xi} \int_0^\xi \frac{d\xi'}{r(\xi', T_0 + J\xi')} \qquad (8.6.3)$$

along series of adiabatic paths until it has the value 1 we could determine a locus Γ_{st} in the ξ, T plane as shown in Fig. 8.13. For any final state to the left of this line the stirred tank would require less holding time; for a final state to the right, the tubular reactor would be preferable.

However, for any value ξ lying beyond Γ_{Jm}, that is to the right of L in Fig. 8.12, the absolute minimum reaction time would be obtained by having first a stirred tank of holding time $\theta_s(\xi_{Jm}, T_0)$ followed by a tubular reactor of holding time $\theta_t(\xi, T_0) - \theta_t(\xi_{Jm}, T_0)$. Now we have seen that combinations of stirred tank and tubular reactors can be used to represent the imperfections of mixing in a real stirred tank (Sec. 7.9). In particular, if the reaction takes place in a dispersed phase which enters a highly turbulent region near the stirrer so that there is active coalescence and breakup of drops at first, but then passes out of the reactor from a placid region where the drops maintain their identity as little batch reactors, then we have an approximation to the combination which we have found to be optimal. It seems reasonable therefore to suggest that in adiabatic operation with an exothermic reaction it is an advantage to have good mixing as early as possible in the reactor. For an endothermic reaction, where the reaction rate always decreases with increasing extent, the tubular reactor always achieves a better conversion in a given reaction time.

We could also calculate the extreme performance of a completely segregated reactor by calculating the extent of reaction, $\xi_a(t)$, under adiabatic conditions and taking the expected value

$$\bar{\xi}_a = \int_0^\infty p(t)\xi_a(t)\,dt. \tag{8.6.4}$$

The function $\xi_a(t)$ is defined implicitly by $\theta(\xi_a(t), T_0) = t$, and depends on T_0, the inlet temperature. This model is not particularly realistic because it would correspond to noncoalescing droplets that were also thermally isolated. However it does represent an extreme of behavior. For the residence time distribution of the tank, $\theta p(t) = e^{-(t/\theta)}$, the conversion $\bar{\xi}_a$ for complete segregation is less than it is for maximum mixedness when the reaction is exothermic. However, when the reaction is endothermic the performance of the reactor is rather better with complete segregation. This is due to the fact that a large fraction of the droplets have very short residence times and the mean reaction rate is weighted in favor of these. Figure 8.14 shows the

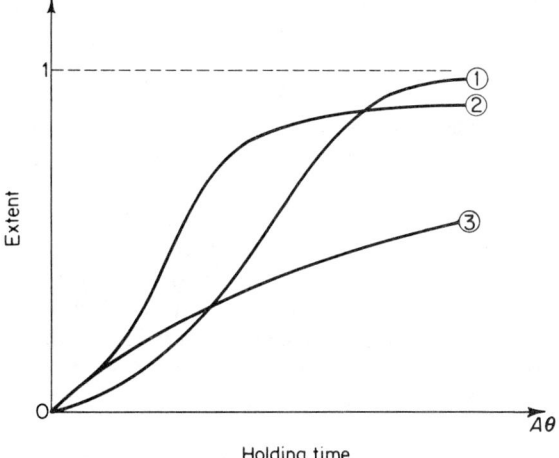

Fig. 8.14 Conversion versus residence time for an exothermic reaction in (1) a tubular reactor, (2) a completely mixed stirred tank, and (3) a completely segregated stirred tank.

extent as a function of a dimensionless holding time for a first order irreversible reaction for the tubular reactor and the two extremes of mixing in the stirred tank. The crossing of curves 1 and 2 corresponds to a point on the locus Γ_{st} of Fig. 8.13.

Carberry has investigated the effect of mixing using his tubular reactor with recycle model and has been particularly interested in yields with simultaneous reactions. There are again situations for which a degree of mixing somewhere between the two extremes of the plug flow reactor and the stirred tank is optimal.

Exercise 8.6.1. Show analytically that the value of ξ' making $\theta_s(\xi', T_0) +$

$[\theta_t(\xi, T_0) - \theta_t(\xi', T_0)]$ a minimum gives a point on Γ_{Jm}, provided that $\xi, T = T_0 + J\xi$ lies beyond it.

8.7 Stability of Adiabatic Reactors

We have already considered the stability of the adiabatic stirred tank (Sec. 7.5) and have observed that since $M = 1$, the slope condition $L > N$ suffices to ensure stability. Now recalling the definitions of L and N this is

$$1 - \theta \left(\frac{\partial r}{\partial \xi}\right)_s > J\theta \left(\frac{\partial r}{\partial T}\right)_s$$

or, by rearranging the terms and dividing by θ,

$$\frac{\partial r}{\partial \xi} + J \frac{\partial r}{\partial T} = \frac{d_a r}{d\xi} < \frac{1}{\theta} = \frac{r}{\xi}. \tag{8.7.1}$$

Figure 8.15 shows a possible form of the reaction rate along an adiabatic path. Points B and C are such that the slope $d_a r/d\xi$ of the curve exactly equals r/ξ, and between these points the stability condition is violated. Since C undoubtedly lies to the left of E where $d_a r/d\xi$ vanishes, the region of unstable states will lie below the locus Γ_{Jm}. It is shown schematically as the crosshatched region in Fig. 8.13. We notice that the design that minimizes θ for given feed conditions (problem C, Sec. 8.2) corresponds to point C in Fig. 8.15. Such a reactor would be only marginally stable and the design would not be a good one. We also see from this figure that the problem was probably not particularly well specified. For θ is the reciprocal of the slope of the line from the origin to a point on the curve. It therefore increases to a local maximum at B, decreases to a local minimum at C, and thereafter

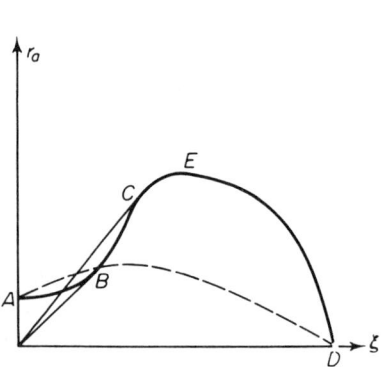

Fig. 8.15 Illustrating the stability of the adiabatic stirred tank.

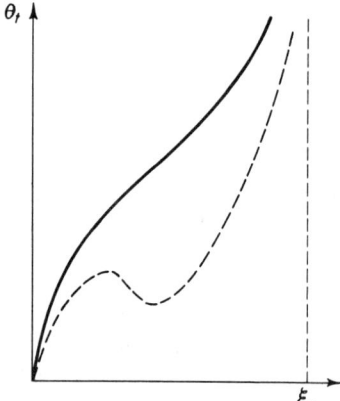

Fig. 8.16 Holding time as a function of extent on an adiabatic path.

Sec. 8.7 Stability of Adiabatic Reactors   253

increases. In practice, more would be demanded in the specification of the problem. For some adiabatic paths there may be no unstable states as, for example, a path giving a curve like the broken one in Fig. 8.15. This way of looking at stability suggests how we might discuss the stability of a tubular reactor. If by proceeding along an adiabatic path there is more than one value of the final state that gives the same value of θ, then we might suspect that not all states can be stable. In fact, if there were to be repetitions of a particular value for θ along the adiabatic path, the curve of $\theta_t(\xi, T_0)$ versus ξ would have to look like the broken curve in Fig. 8.16 rather than be monotonic like the full one. But

$$\theta_t(\xi, T_0) = \int_0^\xi \frac{d\xi'}{r(\xi', T_0 + J\xi')},$$

so that

$$\frac{d\theta_t}{d\xi} = \frac{1}{r(\xi, T)} > 0 \tag{8.7.2}$$

and the curve is monotonic. It follows that the adiabatic tubular reactor with no mixing cannot be unstable. Since the stirred tank can be unstable, we would conclude that some degree of mixing is necessary before an adiabatic reactor is unstable.

In an adiabatic packed bed of catalyst particles in which there is mass and heat transfer between the flowing reactants and the particles, the situation is rather interesting. We shall simplify the picture by assuming that the diffusion of heat and mass within the particle is not important. In this case each catalyst particle is like a little stirred tank and we may imagine that the concentrations and temperature are uniform throughout it. At a particular point z in the bed, let the concentration of a reactant A be $a(z)$ in the flowing stream and $\hat{a}(z)$ within the particle. Similarly let $T(z)$ and $\hat{T}(z)$ be the temperatures outside and within the particle. Then if A is disappearing irreversibly by a first order reaction, it is doing so at a rate $V_p \rho_b S_g \hat{k}(\hat{T})\hat{a}$ within a single particle, where V_p is the particle volume, $\rho_b S_g$ the catalytic area per unit volume, and $\hat{k}$ the rate constant per unit area. (The notation is that of Chap. 6.) The rate of transfer of A to the particle will be $k_c S_x(a - \hat{a})$, where S_x is the external area and k_c the mass transfer coefficient. Hence in the steady state

$$k_c S_x(a - \hat{a}) = V_p \rho_b S_g \hat{k}(\hat{T})\hat{a}$$

or

$$a - \hat{a} - \theta \hat{k}(\hat{T})\hat{a} = 0, \tag{8.7.3}$$

where

$$\theta = \frac{V_p \rho_b S_g}{k_c S_x}. \tag{8.7.4}$$

This is the equation of a stirred tank with feed concentration a, temperature $\hat{T}$, and composition $\hat{a}$. The quantity θ is a "holding time" for the particle.

If the heat of reaction is ΔH and is negative, then heat will be generated at a rate $(-\Delta H)V_p\rho_b S_g \hat{k}(\hat{T})\hat{a}$. In the steady state this balances the rate of heat transfer $US_x(\hat{T} - T)$, where U is the heat transfer coefficient (notation of Sec. 6.6). Thus,

$$T - \hat{T} + \lambda \theta \hat{k}(\hat{T})\hat{a} = 0 \tag{8.7.5}$$

where $\lambda = (-\Delta H)k_c/U$. This again is a heat balance equation of exactly the same form as in a stirred tank. Eliminating $\hat{a}$ in Eq. (8.7.5) we have

$$\hat{T} - T = a\lambda \frac{\theta \hat{k}(\hat{T})}{1 + \theta \hat{k}(\hat{T})} \tag{8.7.6}$$

as the equation which determines the value of $\hat{T}$ when T and a are given. Now we know from our study of the stirred tank that there are certainly values of θ, λ, T, and a which could give this equation more than one solution. Moreover, we saw that a slow change in feed conditions (here a and T) might produce a sudden jump in the internal temperature of a stirred tank (here $\hat{T}$), so that we may expect sudden jumps in the catalyst particle temperature.

The pseudo-homogeneous reaction rate for the packed reactor is

$$r = (1 - \epsilon)\rho_b S_g \hat{k}(\hat{T})\hat{a}, \tag{8.7.7}$$

where ϵ is the fractional free space. Now if we solve Eqs. (8.7.5) and (8.7.3) for $\hat{T}$ and $\hat{a}$ in terms of a and T, we have the effective rate of reaction in terms of the concentration and temperature of the flowing stream $r(a, T)$. If the feed concentration of A is a_0, the adiabatic bed equations

$$v\frac{da}{dz} = -r(a, T) \qquad a(0) = a_0$$
$$v\frac{dT}{dz} = Jr(a, T) \qquad T(0) = T_0 \tag{8.7.8}$$

still hold when the pseudo-homogenous reaction rate is used. It follows that

$$T = T_0 + J(a_0 - a),$$
$$= T_m - Ja, \tag{8.7.9}$$

where $T_m = T_0 + Ja_0$ is the adiabatic temperature that would be reached by complete conversion of the feed. Substituting from Eq. (8.7.9) into Eq. (8.7.7) gives

$$\frac{\hat{T} - T}{T_m - T} = \frac{\lambda}{J} \frac{\theta \hat{k}(\hat{T})}{1 + \theta \hat{k}(\hat{T})}. \tag{8.7.10}$$

The right-hand side is the familiar sigmoidal function of $\hat{T}$ shown in Fig. 8.17. The left-hand side is a family of lines pivoting about the point of ordinate 1 and abscissa T_m. Since T is the intersection of one such line with the horizontal axis, the lines corresponding to $T(z)$ will sweep toward the vertical

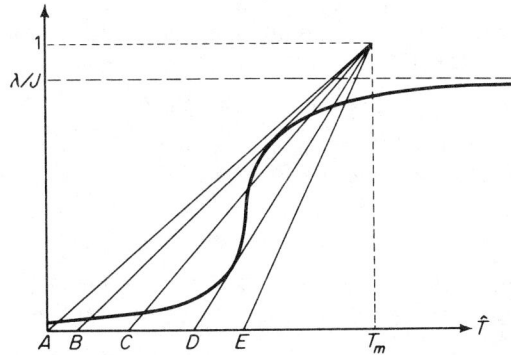

Fig. 8.17 Possible steady states for a catalyst particle in the adiabatic bed.

and might quite easily pass across the thigh of the sigmoidal curve. In this case there would be the possibility (when T corresponded to point C, say) of more than one steady state. In fact, there is a whole spectrum of possible steady states, depending on where the transition is made from the calf to the back of the curve.

Exercise 8.7.1. Show that for sufficiently small catalyst particles no problem with multiple steady states can occur. What would be the effect of increasing the reactant flow rate on the probability of multiple steady states? Put teeth into your answer by using Exercise 7.3.3.

REFERENCES

8.4. For further examples of analytical solutions see:

J. M. Douglas and L. C. Eagleton, "Analytical Solutions for some Adiabatic Reactor Problems," *Ind. Eng. Chem. Fundamentals*, **1**, 116 (1962).

8.5. The earliest results on the optimal sequence of adiabatic beds are given in a dissertation by:

F. Horn, *Optimalprobleme bei kontinuierlichen chemischen Prozessen*, 1958.

They are also contained in:

F. Horn and L. Küchler, "Probleme bei reaktionstechnischen Berechnungen," *Chem. Ingr. Tech.*, **32**, 382 (1959).

A detailed treatment of the problem of cold shot cooling is:

K. Y. Lee and R. Aris, "Optimal Adiabatic Bed Reactors for Sulphur

Dioxide with Cold Shot Cooling," *Ind. Eng. Chem. Process Design and Develop.*, **2,** 300 (1963).

This treatment needs modification in the light of:

J. P. Malenge and J. Villermaux, "Optimal Adiabatic Bed Reactor with Cold Shot Cooling," *Ind. Eng. Chem. Process Design and Develop.*, **6,** 535 (1967).

See also:

H. Bakemeier and R. Krabetz, "Ein Beitrag zur Optimalisierung von Abschnittsreaktoren," *Chem. Ingr. Tech.*, **34,** 1 (1962).

R. Aris, *The Optimal Design of Chemical Reactors.* New York: Academic Press Inc., 1961.

8.6. A. Cholette and J. Blanchet, "Optimum Performance of Combined Flow Reactors under Adiabatic Conditions," *Can. J. Chem. Eng.*, **39,** 192 (1961).

R. Aris, "On Optimal Adiabatic Reactors of Combined Types," *ibid.*, **40,** 87 (1962).

J. M. Douglas, "The Effect of Mixing on Reactor Design," *Chem. Eng. Prog. Symp. Ser.*, **60** (No. 48), 1 (1964).

B. Gillespie and J. J. Carberry, "Influence of Mixing on Isothermal Reactor Yield and Adiabatic Reactor Conversion," *Ind. Eng. Chem. Fundamentals*, **5,** 164 (1966).

J. J. Carberry, "Yield in Chemical Reactor Engineering," *Ind. Eng. Chem.*, **58,** 41 (Oct. 1966).

R. P. King, "Calculation of Optimal Conditions for Chemical Reactors of the Combined Type," *Chem. Eng. Sci.*, **20,** 537 (1965).

8.7. N. R. Amundson and S. L. Liu, "Stability of Adiabatic Packed Bed Reactors. An Elementary Treatment," *Ind. Eng. Chem. Fundamentals*, **1,** 200 (1962).

N. R. Amundson and L. Raymond, "Some Observations on Tubular Reactor Stability," *Can. J. Chem. Eng.*, **42,** 173 (1964).

C. Van Heerden, "The Character of the Stationary State of Exothermic Processes," *in Chemical Reaction Engineering* (1st Europ. Symp.), p. 133. London: Pergamon Press, Inc., 1957.

NOTATION

A Arrhenius pre-exponential factor
A_j jth species

Sec. 8.7 Notation

a	concentration of reactant A (Sec. 8.7)
$\hat{a}$	concentration of A within catalyst particle (Sec. 8.7)
c_j, c_{j0}	concentration, initial concentration of A_j
E	activation energy
G	mass flow rate per unit area of reactants
g_j, g_{j0}	mas fraction, initial mass fraction of A_j
$g_N(\xi_{N+1})$	maximum profit from N stage reactor (Sec. 8.5)
J	$-\Delta H/C_P$
J''	$(-\Delta H)\rho/C_P$
$k, \hat{k}$	rate constant, rate constant for catalyst
k_c	mass transfer coefficient to catalyst (Sec. 8.7)
L	length of reactor
m_j	molar weight of A_j
P	profit (or, incidentally, pressure)
p_n	$\xi_n - \bar{\xi}_n - \nu\theta_n$ (Sec. 8.5)
p_j	price of one mole of A_j
$\bar{p}$	$\sum \alpha_j p_j$
q	volume flow rate of reactants
R	gas constant
$r(\xi, T)$	reaction rate
$r_a(\xi; T_0)$	reaction rate on adiabatic path, $r(\xi, J\xi + T_0)$
$r_m(\xi)$	maximum reaction rate for a given extent
$r_n(\xi)$	adiabatic reaction rate in nth bed, $r(\xi, J\xi + S_n)$
S_g	catalyst surface area per gram
S_n	$T_n - J\xi_n = \bar{T}_n - J\bar{\xi}_n$
S_x	external surface area of catalyst particle
T	temperature
$\hat{T}$	temperature within catalyst particle
T_f	feed temperature
T_m	maximum temperature $T_0 + Ja_0$ (Sec. 8.7)
$T_n, \bar{T}_n$	exit and inlet temperatures to bed n
T_0	temperature on adiabatic path at which $\xi = 0$
T_a	temperature after complete adiabatic reaction
t	batch time of $\rho z/G$ for tubular reactor
U	heat transfer coefficient to catalyst particle
V	volume
V_p	volume of catalyst particle
v	velocity of reactants in tubular reactor
y	integration variable, $E/R(T_0 + J\zeta)$
z	distance from inlet of tubular reactor
α_j	stoichiometric coefficient of A_j
Γ_e	locus of equilibrium states
Γ_m	locus of $\partial r/\partial T = 0$

Γ_{Jm}	locus of maximum reaction rate in adiabatic reactor
Γ_n	locus of optimal exit states for stage n
$\bar{\Gamma}_n$	locus of optimal inlet states for stage n
$\Theta(\xi; T_0)$	$\int_{\xi_{Jm}}^{\xi} d\xi'/r(\xi', T_0 + J\xi')$
$\theta, \theta_s, \theta_t$	holding time in general, of stirred tank, of tubular reactor
θ_n	holding time of stage n
$\hat{\theta}$	$V_p \rho_b S_g / k_c S_x$ (Sec. 8.7)
$\Lambda(\xi''; T_0)$	$\int_{\xi_{Jm}''}^{\xi''} d\xi'/r(\xi', T_0 + J''\xi')$
λ	$(-\Delta H)k_c/U$ (Sec. 8.7)
λ_n	fraction of total flow through stage n
ν	cost of holding time
ξ	extent, moles per unit volume
ξ''	extent, moles per unit mass
ξ_{Jm}	extent of a point on Γ_{Jm}
ξ_m	extent of a point on Γ_m
$\xi_n, \bar{\xi}_n$	exit and inlet extents of stage n
$\bar{\xi}_a$	mean extent with complete segregation
ρ	density
ρ_b	bulk density of catalyst
$d_a/d\xi$	$\partial/\partial\xi + J\,\partial/\partial T$, derivative along adiabatic path

Endnote (p. 235) In Fig. 8.4 the curve labelled Γ should be labelled Γ_e. The locus Γ_m (if shown) would lie between Γ_{Jm} and Γ_e and pass through D.

The Tubular Reactor 9

9.1 Types of Tubular Reactor

In complete contrast to their behavior in a stirred tank reactor, reactants pass through a tubular reactor with as little mixing as possible. In the ideal case, each element of the reaction mixture would have a reaction time precisely equal to the residence time of the reactor. The tubular reactor is therefore a convenient way of making the batch reactor continuous, for the flow through the reactor will correspond to the course of time in a batch reactor. Naturally, there is more to the tubular reactor than this, but the analogy is an extremely important one and the idealization, though gross, is useful. This type of reactor is sometimes known as the plug flow reactor (P.F.R.) in contrast to the continuous stirred tank reactor (C.S.T.R.). We can soften the idealization somewhat by saying that in the stirred tank reactor the contents are as well mixed as possible, whereas in the tubular reactor the intention is that they should be as little mixed as possible. Basically, then, the tubular reactor is a conduit of some sort through which the reactants flow and in which no attempt is made to increase the natural degree of mixing.

But when this is said, there still remains a wide variety of different types of tubular reactor. In some the reactants pass slowly through a mile-long pipe, in others they flow at near sonic speeds through a short pipe in a high temperature furnace. A reactor packed with catalyst particles may vary in size from a high pressure ammonia converter 50 ft high to the so-called differential reactor of the research worker, perhaps only a centimeter in length. The classification illustrated by Fig. 9.1 is therefore a very rough one.

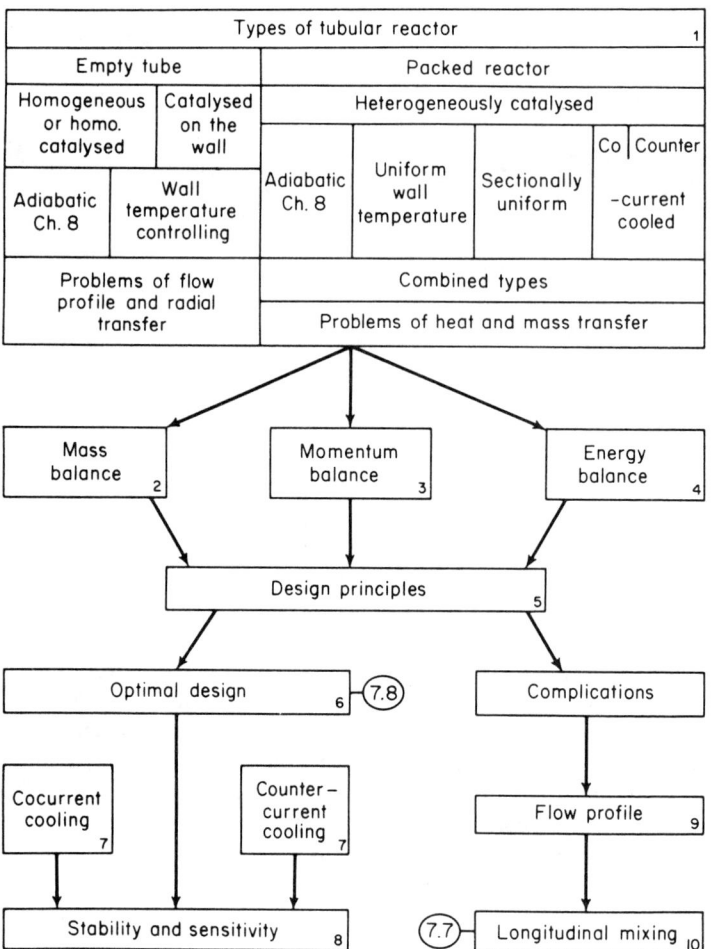

Fig. 9.1 The structure of the chapter.

We may first divide tubular reactors into those designed for homogeneous reactions, and therefore basically just an empty tube, and those designed for a heterogeneously catalyzed reaction, and hence to be packed with a catalyst. Both types can of course be operated adiabatically, and it was the simplest model of these that we discussed in the last chapter. If the temperature of the reactor is to be controlled this is through the wall, and the associated problems of heat transfer now arise. These include transfer at the wall and subsequent radial diffusion across the flowing reactants. In the empty tubular reactor there may be considerable variations in flow rate across the tube. For example, in the slow laminar flow the fluid

Sec. 9.1 Types of Tubular Reactor

at the center is moving at twice the average speed of flow and though in turbulent flow the mean speed is much more uniform across the tube it is important to know how this departure from ideal conditions will affect the performance of the reactor. To some extent the radial mixing will counteract the influence of nonuniformities of flow profile, but in the empty tube good radial diffusion will imply that there is also diffusion in the direction of flow. In the packed bed the flow profile and the radial and longitudinal diffusion are all governed by the packing. These departures from the ideal plug-flow assumption are considered in Secs. 9.9 and 9.10. An ingenious use of secondary flow to decrease the longitudinal dispersion has been described by Adler, but this and the problems which arise when the reaction is catalyzed on the wall of the empty tube are beyond the scope of this text (see references at the end of this chapter).

In a packed bed reactor there are of course the problems of heat and mass transfer to and from the catalyst particles. These we have faced separately in Chap. 6, where we saw that a pseudo-homogeneous reaction rate expression could be devised to accommodate such effects. There remain the important effects of radial and longitudinal mixing, mentioned above, and of heat transfer from the packed bed to the reactor wall. This wall is not necessarily the outer, curved wall of a packed cylinder (cooling tubes may pass through the bed as illustrated later in Fig. 9.12). Obviously the approximations that have to be made are considerable, and the available correlations are never as complete as the designer would wish. Still, there is a lot of material that can be used with a good degree of confidence.

There are also different ways of bringing about temperature control through the wall. The simplest in theory—though not necessarily in practice—would be to hold the wall of the reactor at a uniform temperature. Variation of temperature control along the reactor could be obtained by doing this in a number of different sections. For an endothermic reaction where heating is required, this is often accomplished by putting the tube in a furnace or by allowing it to pass from one furnace to another at a different temperature. For an exothermic reaction we think in terms of cooling jackets. The coolant stream can also flow along the reactor wall either in the same direction as the reactants or in the opposite direction. Here the coolant temperature will vary with position and will require a further equation to determine it. This is discussed in Sec. 9.7.

In all these cases, the correct design must grow from the equations of mass, energy, and momentum balance to which we now turn in the next few sections. From these we proceed to the design problem (Sec. 9.5) and hence to elementary considerations of optimal design (Sec. 9.6). The stability and sensitivity of a tubular reactor is a vast and fascinating subject. Since the steady state equations are ordinary differential equations, the equations describing the transient behavior are partial differential equations. This means that a full discussion of stability is very much more difficult than it was for the stirred tank reactor. All we can hope to do in Sec. 9.8 is to indicate the kind of problem which arises and how it can be related to our understanding of the nature of the reaction rate expression.

9.2 The Mass Balance

If there is no change of density during the reaction, the concentration can be suitably measured in moles per unit volume and the linear velocity will be denoted by v. If the reactor is packed, the velocity v is taken to be the volume flow rate divided by the total cross-sectional area of the reactor. (The mean velocity through the interstices of the packed bed is v/ϵ, where ϵ is the void fraction.) Also, let $c_j(z)$ be the concentration of A_j at a distance z from the inlet and $r(c_j, T)$ be the pseudo-homogeneous reaction rate in moles per unit reactor volume per unit time. If A_r is the cross-sectional area of the reactor then in a unit of time

$vA_r c_j(z)$ moles of A_j will flow into the section between z and $z + dz$,

$vA_r c_j(z + dz)$ moles of A_j will flow out of it, and

$(A_r\, dz)\alpha_j r(c, T)$ moles of A_j will be formed within it.

Hence

$$vA_r[c_j(z + dz) - c_j(z)] = \alpha_j r(c, T) A_r\, dz,$$

and dividing by $A_r\, dz$ and letting $dz \longrightarrow 0$ gives

$$v\frac{dc_j}{dz} = \alpha_j r(c, T). \tag{9.2.1}$$

This is a differential equation for $c_j(z)$ and is subject to the initial conditions that specify the feed composition:

$$c_j = c_{j0} \quad \text{at} \quad z = 0. \tag{9.2.2}$$

If we put

$$c_j(z) = c_{j0} + \alpha_j \xi(z) \tag{9.2.3}$$

and substitute into Eq. (9.2.1), we have a single equation

$$v\frac{d\xi}{dz} = r(\xi, T) \tag{9.2.4}$$

subject to

$$\xi(0) = 0. \tag{9.2.5}$$

When simultaneous reactions $\sum_{j=1}^{S} \alpha_{ij} A_j = 0$, $i = 1, \ldots, R$, are taking place, the only change that has to be made in the basic balance is to recognize that the rate of formation of A_j will be due to all reactions and so will be $(A_r\, dz)\sum_{i=1}^{R}\alpha_{ij}r_i(c, T)$. Then Eq. (9.2.1) becomes

$$v\frac{dc_j}{dz} = \sum_{i=1}^{R}\alpha_{ij}r_i(c, T) \qquad (9.2.6)$$

and is subject to the same boundary conditions as Eq. (9.2.1) at $z = 0$. If now we set

$$c_j(z) = c_{j0} + \sum_{i=1}^{R}\alpha_{ij}\xi_i(z) \qquad (9.2.7)$$

and substitute into Eq. (9.2.6), we have

$$\sum_{i=1}^{R}\alpha_{ij}\left[v\frac{d\xi_i}{dz} - r_i(\xi, T)\right] = 0.$$

But since the reactions are independent, the only set of multipliers λ_i that make $\sum \alpha_{ij}\lambda_i = 0$ for all j is the trivial set $\lambda_i = 0$. Hence the S equations [Eq. (9.2.6)] can be replaced by R equations

$$v\frac{d\xi_i}{dz} = r_i(\xi, T) \qquad (9.2.8)$$

subject to

$$\xi_i(0) = 0. \qquad (9.2.9)$$

By writing

$$t = \frac{z}{v} \qquad (9.2.10)$$

for the time the reactants have taken to reach position z, we see that the equations are precisely those for a batch reactor with reaction time t. The term "space velocity" is widely used for a quantity having the dimensions of the reciprocal of time. In some cases it is just the reciprocal of the residence time v/L, or of a residence time based on the inlet velocity $v_0/L = G/\rho_0 L$. In other cases it may be defined quite specially to the particular reaction being considered. For example, in a cracking reaction it might be in lb of oil fed per hr/lb of catalyst in the reactor, i.e., $Gg_0/L(1 - \epsilon)\rho_b$, where g_0 would be the inlet mass fraction of oil and ρ_b the bulk density of catalyst. In all cases, however, it should be proportional to G/L, the constant of proportionality having the dimensions of the reciprocal of density.

If the density is not constant, the linear velocity will not be constant, for

$$G = \rho(z)v(z) \qquad (9.2.11)$$

is the mass flow rate per unit area; this must be independent of z or there would be an accumulation of mass at some point. In the basic balance we must now say that $v(z)c_j(z)$ is the flux of A_j across a section at z, and

$v(z + dz)c_j(z + dz)$ that at $z + dz$. Thus making the same balance as at first and letting $dz \longrightarrow 0$, we have

$$\frac{d}{dz}[v(z)c_j(z)] = \alpha_j r(c, T). \tag{9.2.12}$$

This equation can be put into a number of different forms. First, notice that it is not particularly useful to substitute from Eq. (9.2.3) for then

$$\frac{d}{dz}(v\xi) = r - \frac{c_{j0}}{\alpha_j}\frac{dv}{dz},$$

and this is only independent of j if v is constant. The equations show, however, that the change in vc_j is proportional to α_j so that we can put

$$v(z)c_j(z) = v_0 c_{j0} + \alpha_j f(z), \tag{9.2.13}$$

where $f(z)$ is a flux of extent. Then

$$\frac{df}{dz} = r(c, T), \tag{9.2.14}$$

but the awkward relation

$$c_j = \frac{v_0 c_{j0} + \alpha_j f(z)}{v(z)} \tag{9.2.15}$$

has to be substituted in the rate expression. It is probably better to work in the mass fraction as concentration unit and write

$$c_j = \frac{\rho g_j}{m_j}. \tag{9.2.16}$$

Then

$$\frac{d}{dz}(v\rho g_j) = \alpha_j m_j r \tag{9.2.17}$$

and

$$v\rho = G = \text{the total mass flux per unit area}, \tag{9.2.18}$$

which is constant. Thus

$$G\frac{dg_j}{dz} = \alpha_j m_j r. \tag{9.2.19}$$

We can now substitute

$$g_j(z) = g_{j0} + \alpha_j m_j \xi''(z) \tag{9.2.20}$$

to give

$$G\frac{d\xi''}{dz} = r(\xi'', T), \tag{9.2.21}$$

with

$$\xi''(0) = 0. \tag{9.2.22}$$

For simultaneous reactions

$$G \frac{dg_j}{dz} = \sum_{i=1}^{R} \alpha_{ij} m_j r_i(g, T), \qquad (9.2.23)$$

and the substitution of

$$g_j(z) = g_{j0} + \sum_{i=1}^{R} \alpha_{ij} m_j \xi_i''(z) \qquad (9.2\ 24)$$

gives

$$\sum_{i=1}^{R} \alpha_{ij} m_j \left[G \frac{d\xi_i''}{dz} - r_i(\xi'', T) \right] = 0.$$

This is true for each j so that we may divide through by m_j and invoke the independence of the reactions to obtain the R equations,

$$G \frac{d\xi_i''}{dz} = r_i(\xi_1'', \xi_2'', \ldots, \xi_R'', T). \qquad (9.2.25)$$

In all these equations we have made the reaction rate a function of composition and temperature only. If it is also a function of pressure we should include this as a further variable. To connect pressure, temperature, and density we have an equation of state. For an ideal gas

$$P_j = c_j RT = \frac{\rho_j RT}{m_j}, \qquad (9.2.26)$$

and the total pressure is

$$P = cRT = \frac{\rho RT}{\bar{m}}, \qquad (9.2.27)$$

where $\bar{m}$ is the mean molar weight given by

$$\bar{m} = \frac{1}{\sum (g_j/m_j)}. \qquad (9.2.28)$$

Now

$$\frac{1}{\bar{m}} = \sum \frac{g_j}{m_j} = \sum \frac{g_{j0}}{m_j} + \sum \alpha_j \xi'',$$

$$= \frac{1}{\bar{m}_0}(1 + \bar{\alpha} \bar{m}_0 \xi''),$$

where $\bar{m}_0$ is the initial mean molar weight. Thus

$$P = \rho \frac{RT}{\bar{m}_0}(1 + \bar{\alpha} \bar{m}_0 \xi'')$$

or

$$\frac{P}{P_0} = \frac{\rho}{\rho_0} \frac{T}{T_0}(1 + \bar{\alpha} \bar{m}_0 \xi''). \qquad (9.2.29)$$

In particular, if P and T are constant

$$\frac{\rho_0}{\rho} = \frac{v}{v_0} = (1 + \bar{\alpha}\bar{m}_0 \xi'') \qquad (9.2.30)$$

and

$$c_j = \frac{\rho g_j}{m_j} = \rho \left(\frac{g_{j0}}{m_j} + \alpha_j \xi''\right),$$

$$= \frac{\rho_0}{m_j} \frac{g_{j0} + \alpha_j m_j \xi''}{1 + \bar{\alpha}\bar{m}_0 \xi''} \qquad (9.2.31)$$

which allows reaction rates expressed in terms of concentrations c_j to be written as functions of ξ''.

If the gas mixture is not ideal, we must use compressibility factors as given in the literature and write

$$P = \frac{Z}{\bar{m}} \rho \mathrm{R} T \quad \text{and} \quad \frac{Z}{\bar{m}} = \sum \frac{g_j Z_j}{m_j}. \qquad (9.2.32)$$

Other equations can be developed (see Exercise 9.2.1) but we shall allow these to suffice and close the section with two examples. In the first, we consider the irreversible reaction $-A_1 + \alpha_2 A_2 + \ldots + \alpha_S A_S = 0$ with feed of pure A_1. If the reaction is at constant pressure and temperature and the rate is kc_1^n, we have from Eq. (9.2.31)

$$c_1 = \frac{\rho_0}{m_1} \frac{1 - m_1 \xi''}{1 + \bar{\alpha} m_1 \xi''}, \qquad (9.2.33)$$

since $g_{10} = 1$ and $m_0 = m_1$. Thus

$$G \frac{d\xi''}{dz} = k \left(\frac{\rho_0}{m_1}\right)^n \left(\frac{1 - m_1 \xi''}{1 + \bar{\alpha} m_1 \xi''}\right)^n. \qquad (9.2.34)$$

But

$$x = \frac{g_1}{g_{10}} = 1 - m_1 \xi''$$

is the fraction of the reactant remaining when the extent is ξ'', and substituting this we have

$$\frac{dx}{dz} = -\frac{k\rho_0}{G} \left(\frac{\rho_0}{m_1}\right)^{n-1} \left(\frac{x}{1 + \bar{\alpha} - \bar{\alpha} x}\right)^n,$$

or, if $x = X$ when $z = L$,

$$\lambda(X) = \frac{k\rho_0 L}{G} \left(\frac{\rho_0}{m_1}\right)^{n-1} = \int_X^1 \left(\frac{1 + \bar{\alpha}}{x} - \bar{\alpha}\right)^n dx, \qquad (9.2.35)$$

where λ is a dimensionless reactor length.

Consider next an SO_2 converter with a feed of the following composition in mole percent: $SO_3 = 0$, $SO_2 = 7.8$, $O_2 = 10.8$, and $N_2 = 81.4$. The reaction is $A_1 - A_2 - \frac{1}{2}A_3 = 0$, where A_1 stands for SO_3, A_2 for SO_2, A_3 for O_2, and A_4 for N_2. Thus

Sec. 9.2   The Mass Balance   267

$$\bar{m}_0 = \sum x_{j0} m_j = (0.078 \times 64) + (0.108 \times 32) + (0.814 \times 28),$$
$$= 31.24,$$

and

$$g_{10} = 0, \qquad g_{20} = 0.160, \qquad g_{30} = 0.111, \qquad g_{40} = 0.729.$$

Thus $g_1 = 80\xi''$, $g_2 = 0.160 - 64\xi''$, $g_3 = 0.111 - 16\xi''$, and $g_4 = 0.729$ are the mass fractions. But

$$x_j = \frac{g_j/m_j}{\sum(g_j/m_j)} = \frac{(g_{j0}/m_j) + \alpha_j \xi''}{\sum(g_{j0}/m_j) + \bar{\alpha}\xi''},$$

so that

$$x_1 = \frac{\xi''}{0.032 - \tfrac{1}{2}\xi''},$$

$$x_2 = \frac{0.0025 - \xi''}{0.032 - \tfrac{1}{2}\xi''},$$

$$x_3 = \frac{0.0035 - \tfrac{1}{2}\xi''}{0.032 - \tfrac{1}{2}\xi''},$$

$$x_4 = \frac{0.026}{0.032 - \tfrac{1}{2}\xi''}.$$

Suppose now that, following Calderbank [*Chem. Eng. Prog.*, **49**, 585 (1953)], we take the rate to be

$$R = k_1(P_{SO_2})^{1/2}(P_{O_2}) - k_2(P_{SO_3})(P_{SO_2})^{-1/2}(P_{O_2})^{1/2},$$

in kg-mole of SO_3 formed per kg of catalyst per second, where

$$\ln k_1 = 12.07 - \frac{31{,}000}{RT}, \qquad \ln k_2 = 22.75 - \frac{53{,}600}{RT}.$$

If G is the mass flow rate per unit area of the reactants and ρ_b the bulk density of the catalyst, $r = \rho_b R$, and (assuming ideal behavior)

$$\frac{G}{\rho_b}\frac{d\xi''}{dz} = k_1 P^{3/2} x_2^{1/2} x_3 - k_2 P x_1 x_2^{-1/2} x_3^{1/2},$$

$$= k_1 P^{3/2} \frac{(2.5 - 1000\xi'')^{1/2}(3.5 - 500\xi'')}{(32 - 500\xi'')^{3/2}}$$

$$\times \left\{ 1 - \frac{10^{3/2}}{K_P P^{1/2}} \frac{\xi''(32 - 500\xi'')^{1/2}}{(2.5 - 1000\xi'')(3.5 - 500\xi'')^{1/2}} \right\}.$$

This is clearly the sort of equation that will have to be integrated numerically—most certainly will this be necessary if T and P vary with position.

Exercise 9.2.1. Show that the mole fraction x_j is governed by the equation

$$vc\frac{dx_j}{dz} = (\alpha_j - \bar{\alpha}x_j)r.$$

Deduce that x_j can be written in terms of a dimensionless extent as

$$x_j = \frac{x_{j0} + \alpha_j \xi'}{1 + \bar{\alpha}\xi'}$$

and that

$$v_0 c_0 \frac{d\xi'}{dz} = r.$$

Exercise 9.2.2. Show that for $n = 1$, the dimensionless length given in Eq. (9.2.35) is

$$\lambda(X) = (1 + \bar{\alpha}) \ln \frac{1}{X} - \bar{\alpha}(1 - X).$$

Deduce that if $\bar{\alpha} > 0$, a longer reactor is needed to attain the same degree of conversion than is the case if $\bar{\alpha} = 0$.

Exercise 9.2.3. Find $\lambda(X)$ when $n = 2$.

Exercise 9.2.4. Show that $\lambda(X)$ increases with $\bar{\alpha}$ for a fixed X, whatever the order of reaction may be.

Exercise 9.2.5. Consider the reaction $-A_1 + A_2 + A_3 = 0$ taking place in an isothermal tubular reactor at constant pressure. Let $r = kc_1 - k'c_2 c_3$ and take a feed of pure A_1. Derive an expression for the length of the reactor required to attain a fraction f of the equilibrium conversion. *Hint:* Let $\lambda = k\rho_0 L/G$, $\kappa^2 = km_1/(km_1 + k'\rho_0)$.

Exercise 9.2.6. A gross over-simplification of a cracking reaction lumps together a range of reactants R and supposes that they crack into a range of products P. If the ratio of the mean molecular weight of reactants, m_R, to that of the products into which they crack, m_P, is a constant $\mu = m_R/m_P$, and g is the mass fraction of R remaining, show that at constant temperature and pressure the vapor density ρ is related to its initial value ρ_0 by

$$\rho = \frac{\rho_0}{g + \mu(1 - g)}.$$

9.3 The Pressure Drop Through the Reactor

The primary purpose of the momentum balance is to obtain an equation for the pressure variation through the reactor. A great deal of data has been taken on the pressure drop over finite beds and this has been correlated with the flow rate and other variables. It is usually possible to assume that the pressure gradient is related to the local variables in the same manner and to write

Sec. 9.4   The Energy Balance   269

$$-\frac{dP}{dz} = f\frac{\rho v^2}{g_c d_p},\quad (9.3.1)$$

where f = friction factor,
g_c = gravitational constant,
$d_p = 2V_p/3S_x$ = diameter of sphere having the same specific area.

Ergun has correlated much avaiable data on packed beds by the equation

$$f = \frac{1-\epsilon}{\epsilon^3}\left(1.75 + 150\frac{1-\epsilon}{\text{Re}}\right),\quad (9.3.2)$$

where the particle Reynolds number is

$$\text{Re}_p = \frac{Gd_p}{\mu} = \frac{vd_p}{\nu},\quad (9.3.3)$$

with

μ = viscosity of flowing reactants
$\nu = \mu/\rho$ = kinematic viscosity.

Thus the pressure gradient is proportional to the velocity at low Reynolds numbers and to its square in the turbulent regime where the Reynolds number is large. If a mixture of particle sizes is involved, say equivalent diameters $d_{p1}, \ldots, d_{pn}$ in volume fractions $v_1, \ldots, v_n$, then d_p is given by

$$\frac{1}{d_p^3} = \sum_{i=1}^{n}\frac{v_i}{d_{pi}^3}.\quad (9.3.4)$$

If the particle diameter, d_p, is an appreciable fraction of the radius, d, of the tube in which the bed is packed, the pressure drop is decreased. There is some evidence in favor of dividing the pressure drop calculated above by a factor of $(1 + 3.5\, d_p/d)$.

It must be remembered that these correlations are only approximate, though they are useful for estimating the importance of pressure variations. In the empty tube the same kind of correlation is used, with d_p taken to be the pipe diameter and f the familiar Fanning friction factor.

9.4 The Energy Balance

In making the energy balance, we shall neglect the potential and kinetic energy of the flowing reactants and the work done by resistance and pressure and simply make an enthalpy balance over the element shown in Fig. 9.2. We have:

Flux of enthalpy in at $z = \sum v(z)c_j(z)h_j[T(z)]$;
Flux of enthalpy out at $z + dz = \sum v(z+dz)c_j(z+dz)h_j[T(z+dz)]$;
Rate of heat removed from section = $Q^*(z)\,dz$.

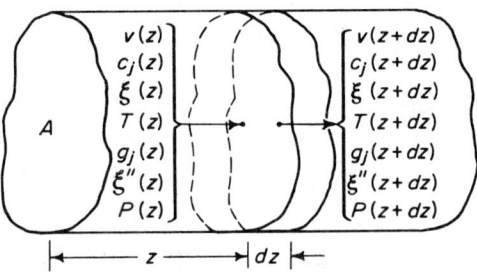

Fig. 9.2 The simplest model of the tubular reactor.

Here we have assumed ideal behavior so that the partial molal enthalpies h_j are only functions of temperature and we are thinking in terms of cooling the reactor by making Q^* positive when heat is removed. Thus

$$(v \sum c_j h_j)_{z+dz} - (v \sum c_j h_j)_z = -Q^* \, dz,$$

and dividing by dz and taking the limit $dz \longrightarrow 0$

$$\frac{d}{dz}\left\{v(z) \sum c_j(z) h_j[T(z)]\right\} = -Q^*(z). \tag{9.4.1}$$

But the left-hand side can be written as

$$\sum \left(\frac{d}{dz} v c_j\right) h_j + \sum v c_j \frac{\partial h_j}{\partial T}\frac{dT}{dz} = \left(\sum \alpha_j h_j\right) r + v \sum c_j c_{Pj} \frac{dT}{dz}$$

by Eq. (9.2.12) and the definition of c_{Pj}. If we put

$$J = -\frac{\Delta H}{C_P}, \qquad C_P = \sum c_j c_{Pj}, \qquad Q = \frac{Q^*}{C_P}, \tag{9.4.2}$$

we have

$$v \frac{dT}{dz} = Jr - Q, \tag{9.4.3}$$

an equation for temperature.

Alternatively we can write the total enthalpy per unit mass as $\sum g_j \hat{h}_j = \sum g_j h_j / m_j$, and using Eq. (9.2.19) in a similar way we have

$$G \frac{dT}{dz} = J''r - Q'', \tag{9.4.4}$$

where $J'' = \rho J = -\Delta H / \hat{C}_P$,
$Q'' = \rho Q = Q^*/\hat{C}_P$,
$\hat{C}_P$ = specific heat of reactants.

In the case of simultaneous reactions we have

$$v \frac{dT}{dz} = \sum_{i=1}^{R} J_i r_i - Q \tag{9.4.5}$$

or

$$G\frac{dT}{dz} = \sum_{i=1}^{R} J_i'' r_i - Q'', \qquad (9.4.6)$$

where

$$J_i = \frac{-\Delta H_i}{C_P} = \frac{J_i''}{\rho}.$$

9.5 The Basic Design Problems

Now that we have a basic set of equations, we may ask what can be done with them. Let us take the case in which pressure is a known function of position and varies so little that it may be replaced by an average value. Then for a single reaction we have the pairs of equations,

$$G\frac{d\xi''}{dz} = r(\xi'', T) \qquad (9.5.1)$$

$$G\frac{dT}{dz} = J'' r(\xi'', T) - Q'', \qquad (9.5.2)$$

or, if the density is constant,

$$v\frac{d\xi}{dz} = r(\xi, T) \qquad (9.5.3)$$

$$v\frac{dT}{dz} = J r(\xi, T) - Q. \qquad (9.5.4)$$

At $z = 0$

$$\xi = \xi'' = 0, \qquad T = T_0. \qquad (9.5.5)$$

We note that the reaction rate can be eliminated between the two expressions to give

$$\begin{aligned} T &= T_0 + J'' \xi'' - \int_0^z \frac{Q''}{G} dz, \\ &= T_0 + J\xi - \int_0^z \frac{Q}{v} dz, \end{aligned} \qquad (9.5.6)$$

and these express a total heat balance over the reactor from the inlet to the section at z.

Three basic types of design may now arise, some of which we have met already. They are:

(i) Temperature specified at each point;
(ii) Heat removal rate specified at each point;
(iii) Heat removal rate subject to a further equation.

Under the first heading we have:

(ia) The isothermal reactor. In this case $T = T_0$ is fixed, and the mass balance Eq. (9.5.1) or (9.5.3) can be integrated by itself using the methods of Chap. 5;

(ib) The temperature programmed reactor. If $T(z)$ is specified $(0 \leq z \leq L)$, the mass balance equation can again be integrated by itself, though now a numerical technique will probably be necessary, since r is a function of both z and ξ. It is not too practical to imagine that we can specify $T(z)$ but it is worthwhile to ask what the optimal temperature profile should be, for even if it is unattainable it gives an absolute standard against which other designs can be measured. If $T(z)$ is given, the heat removal program required to attain it is given by

$$Q(z) = J r(\xi, T) - vT'(z). \qquad (9.5.7)$$

Under the second heading we have:

(iia) Adiabatic reactor. By setting $Q^* = 0$ we have the adiabatic relation in Eq. (9.5.6). We need not add to the treatment already given in Sec. 8.4;

(iib) The reactor with specified heat removal. If $Q^*(z)$ is a specified function of position then Eqs. (9.5.1) and (9.5.2) or Eqs. (9.5.3) and (9.5.4) can be integrated simultaneously.

Finally, we may specify the way in which the heat is removed and derive another equation to express this. If the coolant temperature at point z is $T_c(z)$ it is often possible to take

$$Q^*(z) = \frac{U a_c}{A_r}(T - T_c) \qquad (9.5.8)$$

where U = heat transfer coefficient,
a_c = perimeter of heat exchange surface in cross section,
A_r = cross-sectional area of reactor.

We then have two further cases:

(iiia) $T_c(z)$ specified. The simplest such case would be where the reactor has an isothermal cooling jacket and so $T_c(z) = T_{c0}$, a constant;

(iiib) $T_c(z)$ subject to a further equation expressing a heat balance on the coolant.

In the next section we shall explore the problem of choosing $T(z)$ optimally in an effort to see what the maximum performance of a tubular reactor may be. In Sec. 9.7 we shall look at the cocurrent and countercurrent cooled reactors, and other problems of the third sort. For the moment, however, we need to go back to the basic model and enquire how realistic its assumptions may be.

In the first place, how far are we justified in assuming that conditions are uniform across the section of the reactor? Tierney and others studied the variation of the fractional free area across the packed bed. They found that with very irregular particles, such as Berl saddles, the slight increase in porosity near the wall was lost in a single particle diameter, and the average value was maintained to the center of the bed. With more regular shapes the free area drops rapidly and then approaches the mean by two or three damped oscillations. Thus for cylinders in a bed 14 times the diameter of the particle, the fractional free area at 0.5, 1.0, 1.5, 2.0, 2.5, and 3.0 particle diameters from the wall might be 0.15, 0.31, 0.20, 0.27, 0.22, and 0.25, the average porosity being 0.25. Evidently, irregularities are not important for irregular particles or when the bed diameter is significantly greater than the particle size. The drag of the wall offsets the greater free area there, so that in the flow profile the greatest velocity may be found around a particle diameter or so into the bed. It is extremely difficult to say anything at all definite about this question in view of the complexity of the shape of the cross section of many industrial reactors and the conditions under which the catalyst beds may be packed. At least it can be said that the nonuniformity is so speculative that it would not be worthwhile to incorporate it in a more detailed model.

The conditions will only be maintained uniformly across the section of the bed if there is good lateral dispersion. This is often represented by a radial eddy diffusion coefficient E_r. In a packed bed division and recombination of streams around the particles promote this radial dispersion, and this has been analyzed as a random walk problem to predict a radial Peclet number $Pe_r = d_p v / E_r$ of about 8. For Reynolds numbers above 10, a value for the Peclet number of the order of 8 to 15 has been found in a variety of nonreacting systems, both for heat and mass transfer. This confirms the assumption that it is principally a hydrodynamic mixing effect. For decreasing Reynolds numbers the Peclet number appears to increase, but it must be ultimately brought down by the relationship between viscosity and molecular diffusivity which would apply if there were no flow at all. In Fig. 9.3 the areas have been marked vaguely so as to indicate the variability and uncertainty of the data. (The axial dispersion shown also in this figure is discussed below in Sec. 9.10.) Some work has also been done on the effect of the ratio of particle diameter to tube diameter, and it has been shown that the value of

$$Pe_r \left[1 + 19.4 \left(\frac{d_p}{d} \right)^2 \right]^{-1}$$

is again of the order of 10.

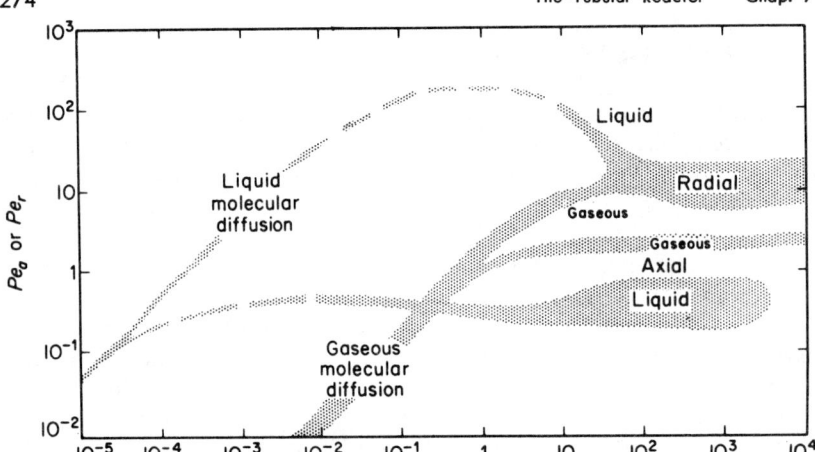

Fig. 9.3 The radial and axial Peclet numbers in the packed bed as functions of the Reynolds number. [Adapted from Wilhelm's Figs. 4 and 5 in *Pure and Applied Chemistry*, **5**, 403 (1962), by permission of the International Union of Pure and Applied Chemistry and Butterworths Scientific Publications.]

If we discard the assumption that conditions across the reactor are uniform, we have to go to a two-dimensional model with a second space variable in the radial direction. Even this model, which would have to assume radial symmetry, is beyond our scope here: some references are given at the end of this chapter.

Exercise 9.5.1. The first order reaction $-A_1 + A_2 = 0$ is to produce P mole/hr of A_2 with a feed rate of G lb/hr. If the temperature is held constant, the rate of reaction is

$$r = k(g_{10} - m_1 \xi'') - k'(g_{20} + m_2 \xi'').$$

If V is the volume of the reactor show that

$$(kg_{10} - k'g_{20})V = G\xi_e'' \ln\left(1 - \frac{P}{G\xi_e''}\right)^{-1},$$

where ξ_e'' is the equilibrium extent of the reaction.

Exercise 9.5.2. Show that for a fixed temperature the volume is smallest when G is as large as possible, but that this volume can never be less than

$$\frac{P}{kg_{10} - k'g_{20}}.$$

Exercise 9.5.3. In the case of constant density show that the limiting performance of a sequence of N stirred tanks (each of holding time L/Nv) as $N \longrightarrow \infty$ is the same as that of a tubular reactor of length L.

Sec. 9.6   Optimal Designs for Tubular Reactors   275

Exercise 9.5.4. Laboratory experiments on the dehydration of ethyl alcohol indicate that the reaction, $C_2H_5OH \longrightarrow C_2H_4 + H_2O$, is second order with respect to the alcohol concentration. The rate constant is 0.52 l gm-mole^{-1} sec^{-1} at 150°C. It is proposed to construct a small scale tubular reactor which will operate at 2 atm and 150°C to give 35% conversion of the alcohol when the feed rate is 9.9 kg/hr. If the reactor has a diameter of 10 cm, what length will be required? Ideal gas behavior and piston flow through the reactor may be assumed and the heat of reaction may be neglected.   (C.U.)

Exercise 9.5.5. Calculate the residence time required to bring about a 35% decomposition of acetaldehyde in a tubular reactor at 520°C and 1 atm according to the following reaction: $CH_3CHO \longrightarrow CH_4 + CO$. The reaction is second order with a rate constant of 0.43 l/mole sec at 520°C and may be assumed to be irreversible.   (C.U.)

9.6 Optimal Designs for Tubular Reactors

We shall consider only the most elementary formulations of optimal design and avoid the heavy mathematics that is often associated with the more sophisticated problems. For a single reaction, the basic problem is to get as large a conversion in as short a reactor as possible. For simultaneous reactions, the problem is to get as good a yield of the main product as possible.

Consider first an isothermal tubular reactor with constant density of reactants for which the equation is

$$v\frac{d\xi}{dz} = r(\xi, T). \qquad (9.6.1)$$

The extent $\xi(L)$, achieved by a reactor of length L, is given by

$$\frac{L}{v} = \int_0^{\xi(L)} \frac{d\xi'}{r(\xi', T)} \qquad (9.6.2)$$

so that L/v is the area under the curve of $1/r(\xi, T)$ between $\xi = 0$ and $\xi = \xi(L)$. Figure 9.4 shows a plot of $1/r$ versus ξ for a reversible exothermic reaction at four temperatures $T_1 > T_2 > T_3 > T_4$. The vertical asymptotes of the curves are at the equilibrium extents $\xi_e(T_i)$ of the several T_i. Suppose the required extent $\xi(L)$ corresponds to point P. Then if the temperature were T_3 or T_2, the holding time would be given by the area $OBB'P$ or $OCC'P$. The required extent could not be achieved at temperature T_1 since it lies beyond $\xi_e(T_1)$. Now the area $OBB'P$ is clearly less than $OAA'P$ so that a smaller holding time is required if the temperature is T_3 rather than T_4. On the other hand, P lies so close to the equilibrium at T_2 that the area $OCC'P$

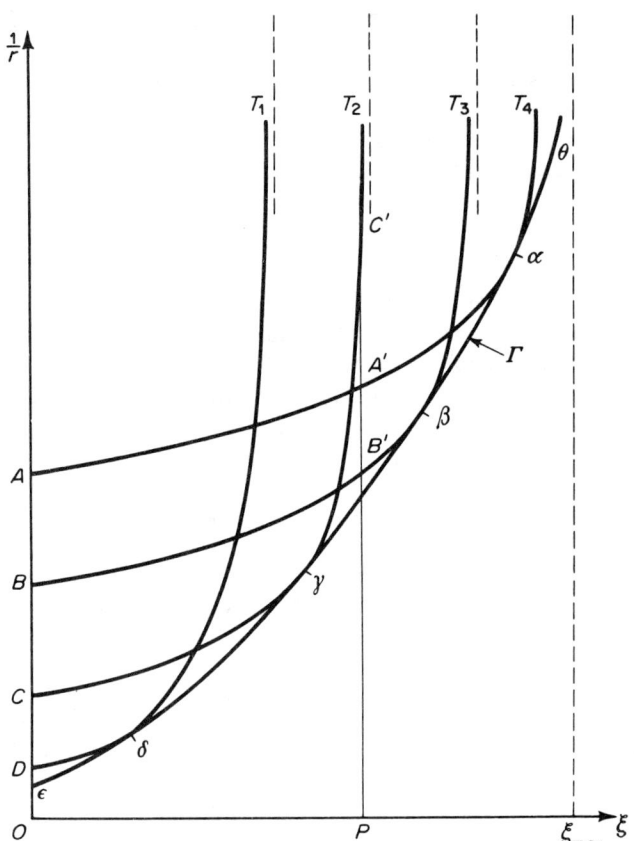

Fig. 9.4 The curves of reciprocal reaction rate versus extent for a reversible exothermic reaction at several temperatures.

is also greater than $OBB'P$. It is therefore clear that there will be some temperature between T_2 and T_4 for which L/v is least. In general, we can say that for any reversible exothermic reaction there will be an optimal temperature somewhat below the equilibrium temperature for $\xi(L)$ for which L/v is least. For an endothermic reaction, the temperature should be made as high as possible (see Exercise 9.6.1).

If the integral in Eq. (9.6.2) can be worked out explicitly, then it may be possible to obtain an equation for the optimal temperature. For example, consider the first order reversible reaction $A \rightleftharpoons B$, with rate law $r = ka - k'b$, taking place in an isothermal reactor whose feed is pure A. If the required fractional conversion is X, then the feed concentration can be written a_0, the current concentrations are $a = a_0 - \xi$ and $b = \xi$, and

Sec. 9.6 Optimal Designs for Tubular Reactors

$\xi(L) = a_0 X$. Thus

$$\frac{L}{v} = \int_0^{a_0 X} \frac{d\xi}{ka_0 - (k+k')\xi} = \frac{1}{(k+k')} \ln \frac{k}{k - (k+k')X}.$$

Now in looking for the minimum of the right-hand side it is convenient to remember that k' is proportional to a power of k, for

$$k' = A' e^{-E'/RT} = A'(A^{-1}k)^{E'/E} = \kappa k^\sigma.$$

Hence

$$\frac{L}{v} = \frac{1}{k + \kappa k^\sigma} \ln [1 - X(1 + \kappa k^{\sigma-1})]^{-1}, \quad (9.6.3)$$

and in seeking the minimum with respect to temperature we can equally well find the minimum with respect to k. If we differentiate with respect to k and set this derivative equal to zero we have an equation

$$-(1 + \sigma \kappa k^{\sigma-1}) \ln [1 - X(1 + \kappa k^{\sigma-1})]^{-1} + \frac{(k + \kappa k^\sigma) X(\sigma - 1) \kappa k^{\sigma-2}}{1 - X(1 + \kappa k^{\sigma-1})} = 0.$$

Let us rearrange this equation so that all the terms involving X are on one side and note that

$$X_e = (1 + \kappa k^{\sigma-1})^{-1} = \frac{k}{k + k'} = \frac{K}{1 + K} \quad (9.6.4)$$

is the equilibrium extent at the temperature for which k and k' are evaluated. Thus

$$\kappa k^{\sigma-1} = \frac{1 - X_e}{X_e}$$

and

$$\frac{X_e - X}{X} \ln \frac{X_e}{X_e - X} = \frac{(\sigma - 1)(1 - X_e)}{X_e + \sigma(1 - X_e)} = \frac{1 - X_e}{pX_e + (1+p)(1 - X_e)} \quad (9.6.5)$$

if we use the same parameter,

$$p = \frac{E}{-\Delta H} = \frac{1}{\sigma - 1}, \quad (9.6.6)$$

as was used in Sec. 4.3 in discussing the maximum reaction rate. Figure 9.5 shows the curves of the two sides of the equation plotted as functions of X_e. The curves for the left-hand side are functions of the parameter X, the required conversion. Those for the right-hand side depend only on the parameter σ, the ratio of the activation energies, or equivalently on p. Given this ratio σ and the required conversion X, we have only to look for the intersection of the corresponding curves to see what the equilibrium extent should be. From the known variation of X_e with temperature [Eq. (9.6.4)] we can then find the optimal temperature. The corresponding minimum length is

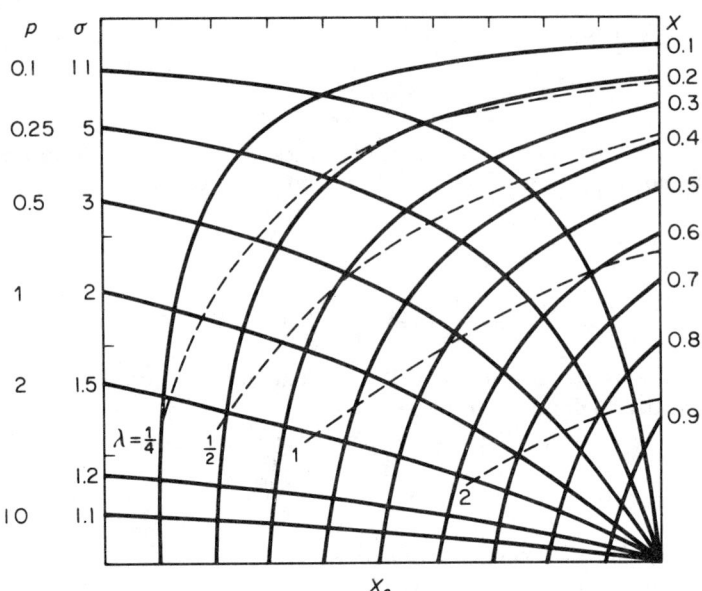

Fig. 9.5 Curves to determine the minimum length and optimal temperature for a first order reactor.

given by Eq. (9.6.3). The broken lines in Fig. 9.5 are the contours of

$$\lambda = \frac{L}{\kappa^p v} = \frac{LA^{1+p}}{vA'^p} = \frac{X_e^{1+p}}{(1-X_e)^p} \ln\left(1 - \frac{X}{X_e}\right)^{-1}, \quad (9.6.7)$$

which is a dimensionless length independent of temperature. For given kinetics these numbers can be translated into dimensional quantities.

So much for the isothermal tubular reactor, which we could treat similarly for any prescribed kinetics. We could also divide the reactor up into two or more isothermal sections and choose the several temperatures and the intermediate extents of reaction optimally. We may also use the argument that the optimal is the boundary between the possible and the impossible to show that the solution to the problem of minimizing the length of the reactor for a fixed $\xi(L)$ also maximizes the extent of reaction for a fixed L.

Returning to Fig. 9.4 we observe that the curves of $1/r$ versus ξ, for constant temperature, are all tangent to the curve Γ; $\alpha, \beta, \gamma,$ and δ are the points of contact with the curve Γ, which goes from ϵ on $\xi = 0$ to θ approaching the asymptote $\xi = \xi_{\max}$. This is of course the curve of $1/r_m(\xi)$ versus ξ and the points of contact of the curves $1/r(\xi, T_i)$ are at the values of ξ for which $T_i = T_m(\xi)$, where $i = 1, 2, 3,$ and 4. Now in the problem of minimizing the total holding time of a sequence of stirred tanks we observed that the temperature in each should be $T_m(\xi_n)$. In fact as the total number of tanks

Sec. 9.6 Optimal Designs for Tubular Reactors   279

is increased, the curves Γ_n of Fig. 7.28 close down on this very curve Γ. This suggests that at every point the temperature should be chosen so as to maximize the reaction rate. There are a number of ways of proving this elementary fact. It is a trivial example in the calculus of variations and has served as an introductory case in the application of dynamic programming and of Pontryagin's maximum principle. It is dangerous to treat it as intuitively obvious however, for, as we shall remark later, in the comparably simple case for simultaneous reactions the rate of production of the valuable product is not maximized at every point as it is here.

Before proving that the temperature condition which minimizes L/v for a prescribed $\xi(L)$ is that for which everywhere maximizes the rate, we should recall that, since $T_m(\xi)$ can become very large for a small ξ, the bounds on the permitted temperature may be important. Let us suppose that T_1 and T_4 in Fig. 9.4 correspond to the upper and lower bounds on temperature, T^* and T_*. Then part $\epsilon\delta$ of curve Γ is unattainable since it corresponds to temperatures above $T^* = T_1$; similarly part $\alpha\theta$ corresponds to forbidden temperatures below $T_* = T_4$. To the left of point δ, the curves for all temperatures lower than $T_1 = T^*$ will lie above $D\delta$, so that the permitted temperature which maximizes $r(\xi, T)$ in this region is $T = T^*$. Similarly, to the right of α the curves for all temperatures higher than $T_4 = T_*$ will lie to the left of αT_4. It follows that the curve $D\delta\gamma\beta\alpha T_4$ represents the reciprocal of the maximum reaction rate subject to $T_* \leq T \leq T^*$. In writing $r^\circ(\xi)$ for the maximum reaction rate and $T^\circ(\xi)$ for the optimal temperature we shall assume that these restrictions have been observed, that is if $T_m(\xi_*) = T^*$ and $T_m(\xi^*) = T_*$,

Let
$$T^\circ(\xi) = \begin{cases} T^* \\ T_m(\xi), \\ T_* \end{cases} \quad r^\circ(\xi) = \begin{cases} r(\xi, T^*) \\ r_m(\xi) \\ r(\xi, T_*) \end{cases} \text{if} \begin{cases} \xi \leq \xi_* \\ \xi_* \leq \xi \leq \xi^* \\ \xi^* \leq \xi. \end{cases} \quad (9.6.8)$$

$$\frac{L^\circ}{v} = \int_0^\xi \frac{d\xi'}{r^\circ(\xi')} \quad (9.6.9)$$

be the allegedly minimum holding time, namely the area under the curve $D\delta\gamma\beta\alpha\theta$ up to any abscissa ξ. Let $T(\xi)$ be any other policy and L/v the consequent holding time; then

$$\frac{L}{v} = \int_0^\xi \frac{d\xi'}{r[\xi', T(\xi')]}. \quad (9.6.10)$$

But
$$\frac{L - L^\circ}{v} = \int_0^\xi \left\{ \frac{1}{r[\xi', T(\xi')]} - \frac{1}{r^\circ(\xi')} \right\} d\xi'$$

is certainly positive since $r[\xi', T(\xi')] < r^\circ(\xi')$ for all $T(\xi)$ other than $T^\circ(\xi')$.

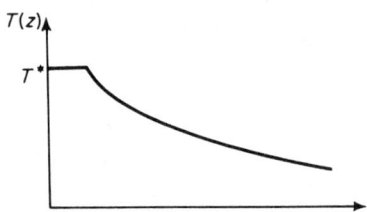

Fig. 9.6 The optimal temperature profile for a single exothermic reaction.

Hence $L°$ is the optimal length. The optimal temperature profile is as shown in Fig. 9.6.

Let us illustrate this by considering again the first order reversible reaction. Using the same notation as before and the results of Sec. 4.3 we know that for this reaction

$$r_m(\xi) = \frac{p^p}{(1+p)^{1+p}} \frac{A^{1+p}}{A'^p} a_0 \left(1 - \frac{\xi}{a_0}\right)^{1+p} \left(\frac{\xi}{a_0}\right)^{-p}. \quad (9.6.11)$$

We shall avoid having to invoke the bounds on temperature by supposing that the feed is not pure A but corresponds to an extent of reaction $\xi_f \geq \xi_*$. Then $r°(\xi) = r_m(\xi)$ and Eq. (9.6.9) gives

$$\lambda = \frac{L° A^{1+p}}{vA'^p} = \frac{(1+p)^{1+p}}{p^p} \int_{x_f}^{x} x^p (1-x)^{-(1+p)} dx. \quad (9.6.12)$$

However this integral is convergent even when $x_f \longrightarrow 0$, so that we may calculate the integral with zero as the lower bound and obtain integrals with x_f or x as lower bound as the diffrence of two integrals. Figure 9.7 shows the way in which this integral varies with the required conversion for several values of p.

The profile for a single reaction as shown in Fig. 9.6 can be cut off at any point $z = L$ to give the optimal temperature control for a reactor of that length. This means that the control policy depends only on the local state and not on the length of the remainder of the reactor. With more than one reaction, this feature is generally lost. For example, if we have a reaction $A \longrightarrow B \longrightarrow C$ and the objective is to obtain the greatest yield of B and minimize the waste product C, then a falling temperature profile is required if the activation energy of the second reaction is greater than that of the first. Again, if the feed is pure A an infinite temperature is required at first, so that the upper bound T^* is important. In this case, the lower temperature bound is also important, for, since an increase of temperature promotes the reaction with the greater activation energy ($B \longrightarrow C$) at the expense of the other ($A \longrightarrow B$), an almost complete conversion of A to B could be obtained by having an exceedingly long reactor at an exceedingly low temperature. With $T_* > 0$, there is an optimum length beyond which the yield of B decreases. Some profiles are shown in Fig. 9.8. It is clear that as the length increases, a shorter and shorter initial length is held at T^*, and the drop to T_* is increasingly steep. The curves are drawn with $z = L$ at a common abscissa so that the details are less confused, the points $A, B, \ldots, E$ representing the inlets to beds of five different lengths. E is the inlet of the longest bed which it is profitable to use at the given minimum temperature T_*.

Sec. 9.6   Optimal Designs for Tubular Reactors   281

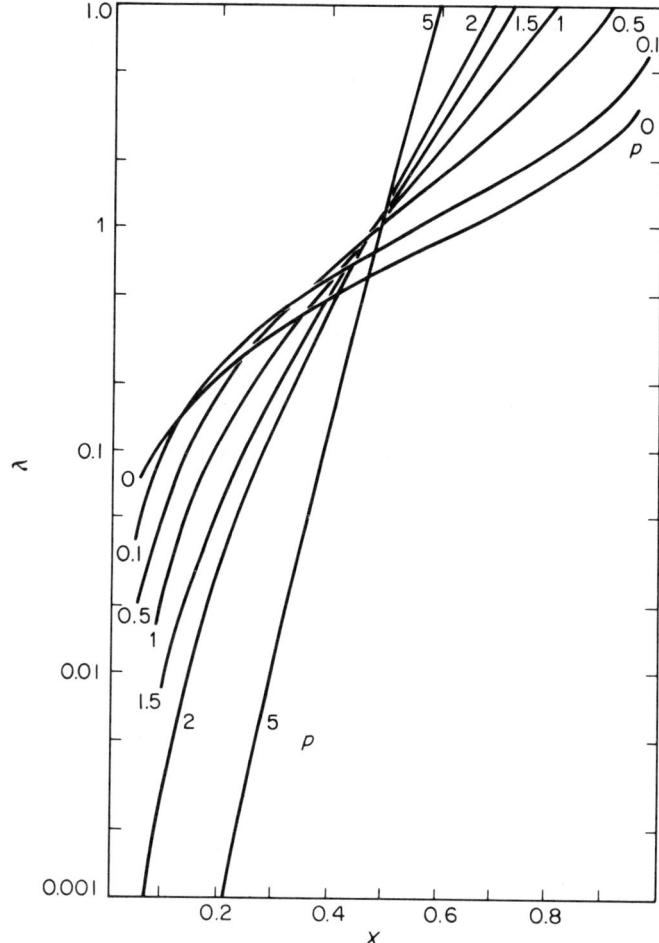

Fig. 9.7 The dimensionless optimal length for a first order reaction as a function of the degree of conversion.

A vast range of interesting optimization problems can be propounded for tubular reactors, many of them showing features of considerable complexity. Jackson and Coward have shown some interesting features of the optimal policy even for the simple example of consecutive reactions given above. Although the optimal profile can rarely be attained in an actual reactor, it is still important to be able to get some idea of what the optimal performance and control of a reactor is like. Without any such idea the designer can only compare one design with another. The optimal does at least give an absolute standard of comparison even though it cannot be attained.

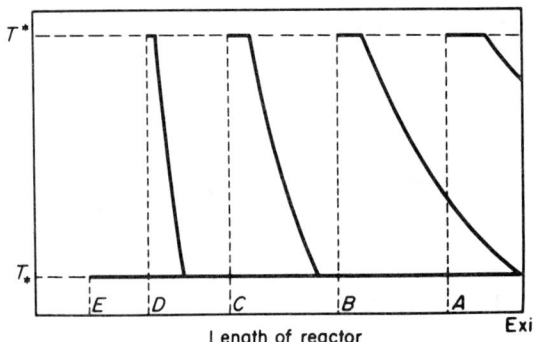

Fig. 9.8 Optimal temperature profiles for consecutive reactions and various lengths of reactor.

Exercise 9.6.1. Draw a diagram similar to Fig. 9.4 for the reversible endothermic reaction. Hence show that the isothermal reactor giving the highest conversion is at the highest permitted temperature.

Exercise 9.6.2. Establish Eq. (9.6.7).

Exercise 9.6.3. Horn [*Chem. Ing. Tech.*, **33**, 413 (1961)] has written Eq. (9.6.12) in the form

$$\frac{L^\circ A^{1+p}}{vA'^p} = \left(\frac{1+p}{p}\right)^{1+p}\left(\frac{X}{1-X}\right)^p I_1\left(\frac{1+p}{p}, \frac{1-X}{X}\right),$$

where

$$I_1(\alpha, \beta) = \int_0^1 \frac{t^{\alpha-1}\,dt}{t^{\alpha-1} + \beta}$$

is one of several integrals that he has tabulated. Obtain this form of the equation by letting $x_f = 0$ and substituting $t = [x(1-X)/(1-x)X]^p$.

Exercise 9.6.4. Show that if the inlet extent of reaction for a first order reversible reaction is not zero, then Eq. (9.6.5) for the optimal temperature must be modified to read

$$\frac{X_e - X_i}{X_e}\frac{X_e - X}{X - X_i}\ln\frac{X_e - X_i}{X_e - X} = \frac{1 - X_e}{1 + p - X_e}.$$

X_i, X, and X_e are, respectively, the values of $\xi/(a_0 + b_0)$ at the inlet, outlet of the reactor, and at equilibrium.

Exercise 9.6.5. Let $T^\circ(\xi_1, \xi_2)$ denote the optimal isothermal reactor temperature when the feed extent is ξ_1 and the required product extent ξ_2. The reactor is to be divided into two isothermal parts and ξ_1, ξ_2, and ξ_3 will denote, respectively, the feed extent to the first, the intermediate extent attained in the first and fed to the second, and the final specified extent from the second. By plotting the curves of $1/r[\xi, T°(\xi_1, \xi)]$ and $1/r[\xi, T°(\xi, \xi_3)]$ against ξ, show that the optimal value of ξ_2 is where they intersect. Can you develop this into a graphical method for N stages?

Exercise 9.6.6. The temperature in a tubular reactor must lie between 550°K and 750°K, and a first order reaction $A - B = 0$ is to take place in it. The feed concentrations are $a_f = 0.1$ and $b_f = 0.9$ mole/kg, and the required product is to have $a = b = 0.5$ mole/kg. If the mass flow rate is $G = 10$ kg/m² min and the rate of reaction $r = k_1 b - k_2 a$ mole/m³ min, what is the least length of reactor required in the following three cases?

(i) $k_1 = \exp(19 - 12{,}000/T)$, $\quad k_2 = 0$,
(ii) $k_1 = \exp(19 - 12{,}000/T)$, $\quad k_2 = \exp(37 - 24{,}000/T)$,
(iii) $k_1 = \exp(40 - 24{,}000/T)$, $\quad k_2 = \exp(19 - 12{,}000/T)$.

9.7 Tubular Reactors Cooled or Heated from the Wall

In this section we shall be concerned with more realistic models of tubular reactors. The isothermal reactor is obviously the simplest type, but it implies that either there are no large heat effects or that they can be completely dominated by temperature control. The reactor with an optimal temperature profile is clearly the most desirable, but this means that the rate of heat exchange can be regulated precisely at each point. Between these two extremes there is a range of designs about which something should be said. We shall not always solve the equations in detail but we shall try to show the important features of the behavior of the reactor by means of examples.

It will be convenient to divide this section more formally than the others, according to the following scheme. Under Sec. 9.7.1 we discuss the general form of the equations when the wall temperature is constant and illustrate this by considering an endothermic cracking reaction. Under Sec. 9.7.2 we shall consider cocurrent and countercurrent cooled reactors and use the ammonia synthesis reactor as an illustration. These correspond to the two subcases of the third type of design problem mentioned in Sec. 9.5.

9.7.1 Reactors with constant wall temperature

We have chosen the signs in the basic equations of Sec. 9.5 so that the terms are positive when the reaction is exothermic and the reactor is being cooled. Even though the reaction may be endothermic (i.e., $J < 0$) and the temperature T_c much higher than that of the reactants, T, we shall retain the suffix c to indicate the temperature of the external cooling or heating fluid. If it is independent of z, as we are assuming in this section, then combining Eq. (9.5.8) with Eqs. (9.5.1) and (9.5.2) or Eqs. (9.5.3) and (9.5.4) gives

$$G\frac{d\xi''}{dz} = r(\xi'', T) \tag{9.7.1}$$

$$G\frac{dT}{dz} = J''r(\xi'', T) - \frac{Ua_c}{\hat{C}_p A_r}(T - T_c), \tag{9.7.2}$$

or

$$v\frac{d\xi}{dz} = r(\xi, T) \tag{9.7.3}$$

$$v\frac{dT}{dz} = Jr(\xi, T) - \frac{Ua_c}{C_p A_r}(T - T_c). \tag{9.7.4}$$

In these equations, A_r is the cross-sectional area of the reactor and a_c the perimeter of the heat interchange area in any section. The ratio A_r/a_c has the dimensions of length and is known as the *hydraulic radius;* it will be denoted by p in this section. The dimensionless quantity

$$\frac{U}{vC_p} = \frac{U}{G\hat{C}_p} = S \tag{9.7.5}$$

is the ratio of the capacity of the wall to transmit heat to the capacity of the reactant stream to convect it; this is a so-called *Stanton number.* The two heat balance equations can therefore be written as

$$G\frac{dT}{dz} = J''r(\xi'', T) - G\frac{S}{p}(T - T_c) \tag{9.7.6}$$

and

$$v\frac{dT}{dz} = Jr(\xi, T) - v\frac{S}{p}(T - T_c). \tag{9.7.7}$$

For the empty tube the heat transfer coefficient can be estimated by the traditional correlations. If d is the diameter of the pipe, the Nusselt number Ud/k is commonly related to the Prandtl and Reynolds numbers by the *Dittus-Boelter relation*

$$\frac{Ud}{k} = 0.023\left(\frac{dv}{\nu}\right)^{0.8}\left(\frac{C_p\nu}{k}\right)^q, \tag{9.7.8}$$

where $k =$ thermal conductivity of the fluid,
$\nu =$ kinematic viscosity,
$q = 0.4$ when the tube is being cooled,
$q = 0.3$ when it is being heated.

Our Stanton number

$$\begin{aligned}S = \frac{U}{vC_p} &= \left(\frac{Ud}{k}\right)\left(\frac{\nu}{vd}\right)\left(\frac{k}{\nu C_p}\right), \\ &= 0.023\left(\frac{dv}{\nu}\right)^{-0.2}\left(\frac{C_p\nu}{k}\right)^{q-1}.\end{aligned} \tag{9.7.9}$$

The first of these is the Reynolds number based on the tube diameter and the second is the Prandtl number. Sometimes q is given an average value of 0.333 so that

$$S = 0.023(\text{Re})^{-1/5}(\text{Pr})^{-2/3}, \tag{9.7.10}$$

a form of the *Colburn equation*.

For a packed bed it is found that heat transfer to the wall is enhanced by the packing. Beek (in his review article mentioned in the References) has suggested a correlation of the form

$$S = A(\text{Re}_p)^{-2/3}(\text{Pr})^{-2/3} + B(\text{Re}_p)^{-1/5}(\text{Pr})^{-3/5}. \tag{9.7.11}$$

The coefficients A and B depend on the packing and he gives the values:

$$\begin{aligned} A &= 2.58, & B &= 0.094 & \text{for cylinders} \\ A &= 0.203, & B &= 0.220 & \text{for spheres.} \end{aligned} \tag{9.7.12}$$

In this case the Reynolds number Re_p is evaluated using the particle diameter, rather than the tube diameter, as the typical length:

$$\text{Re}_p = \frac{Gd_p}{\rho v}. \tag{9.7.13}$$

It has been assumed that the metal wall of the reactor has little heat transfer resistance but if it has appreciable thickness this must be taken into account. If k_w is the thermal conductivity of the wall and d_i and d_o its inner and outer diameters we may define a sort of Stanton number for the wall:

$$S_w = \frac{2k_w}{vC_p d_i \ln(d_o/d_i)}. \tag{9.7.14}$$

Then the overall value of the Stanton number will be S_0:

$$\frac{1}{S_0} = \frac{1}{S} + \frac{1}{S_w}, \tag{9.7.15}$$

where S is calculated from Eq. (9.7.9) or (9.7.11) and T_c is now the temperature on the outside of the wall.

Let us consider an exothermic reaction in which the density can be taken to be constant. Then the equations to be solved are

$$v\frac{d\xi}{dz} = r(\xi, T) \tag{9.7.16}$$

$$v\frac{dT}{dz} = Jr(\xi, T) - v\frac{S}{p}(T - T_c). \tag{9.7.17}$$

These will usually have to be integrated numerically, but this being done we could plot the values of ξ and T as points in the ξ, T plane. If the reaction is exothermic, J is positive. If the coolant temperature T_c is less than the inlet temperature T_0, the term $(T - T_c)$ will always be positive. In fact, if we divide Eq. (9.7.17) by Eq. (9.7.16) we have

$$\frac{dT}{d\xi} = J - S\frac{v(T - T_c)}{pr(\xi, T)}, \tag{9.7.18}$$

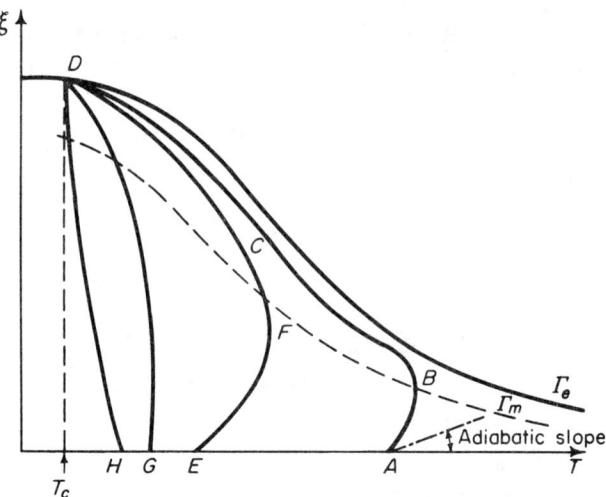

Fig. 9.9 Reaction path of a cooled tubular reactor with an exothermic reaction.

and since each factor of the second term is positive, the value of $dT/d\xi$ is always less than J. This we should expect since J is the value of the rate of change of temperature with extent under adiabatic conditions, i.e., with $S = 0$. In Fig. 9.9 the familiar ξ, T plane is shown with the equilibrium and maximum reaction rate curves, Γ_e and Γ_m. If point A represents a fairly high inlet temperature, the locus representing the solution $\xi(z)$, $T(z)$ might be a curve such as $ABCD$. During the early part of the reaction, the rate of generation of heat by the reaction exceeds the rate of cooling. But a peak temperature is reached at B, where $Jr(\xi, T) = (vS/p)(T - T_c)$, and thereafter the temperature decreases since the rate of reaction (and so also the rate of heat generation) drops off rapidly. For a finite length L, the final point $\xi(L), T(L)$ might be such as C. Only if the reactor is infinitely long would equilibrium be reached at the coolant temperature; this state is represented by point D. A slightly lower inlet temperature E would give a similar path with a maximum temperature at F, but very much lower inlet temperatures such as G or H might give profiles of falling temperature throughout the reactor. Let it be emphasized that this figure illustrates some of the possibilities. The precise form of the profile will depend on the particular values of the parameters.

However, a temperature which increases to a local maximum, the hot spot, and then falls off is not uncommon. It clearly has some advantage for the best rate of reaction will be when the path is in the neighborhood of the curve Γ_m. Thus the paths ABC and EF would achieve a given conversion in a notably shorter time than such paths as GD and HD. It is likely therefore that for any set of parameters v, J, S, p, and T_c there will be an optimal inlet temperature. The drawback to the temperature profile with a hot spot is that this maximum temperature may be excessive. Figure 9.19 (see Sec. 9.8) shows some computed curves and this kind of sensitivity will be discussed in the next section.

Certain endothermic cracking reactions often take place in reactors which approximate to the condition of constant wall temperature, for the reactor takes the form of a tube coiled into a high temperature furnace. If the feed is quite cold, the early part of the reactor may be little more than a heat exchanger since the reaction rate can be neglected, but later, when the reaction really gets going, the temperature change is governed by the competition of the heating effect of the furnace and the heat absorbed by the reaction. In this case, T_c will be greater than T and we shall write $J = -J'$, where J' is now positive, so that

$$v \frac{d\xi}{dz} = r(\xi, T) \tag{9.7.19}$$

$$v \frac{dT}{dz} = -J' r(\xi, T) + \frac{Sv}{p}(T_c - T). \tag{9.7.20}$$

In Fig 9.10 the ξ, T plane for an endothermic reaction is shown with the equilibrium curve Γ_e. There will also be a locus DBE on which the right-

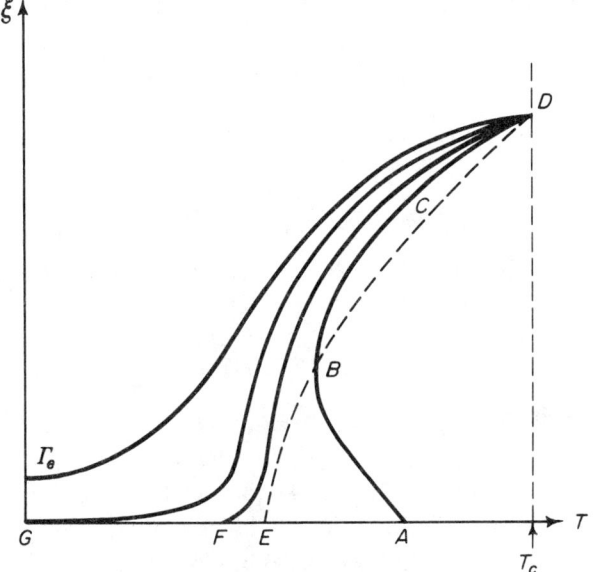

Fig. 9.10 Reaction path of a heated tubular reactor with an endothermic reaction.

hand side of Eq. (9.7.20) vanishes. It can be shown that the slope of this locus is always positive (see Exercise 9.7.2). Hence if the inlet temperature corresponds to a point A to the right of E, the temperature will begin by decreasing, reaching a minimum at B, and thereafter increasing as the heat transfer dominates the heat absorption by the endothermic reaction. On the other hand, an inlet temperature corresponding to a point F to the left of E will give a monotone increasing temperature throughout the reactor. If the inlet temperature is very low, say at G, then there will be virtually no reaction at first and the path will follow the T axis until the reaction rate becomes appreciable.

This case may be illustrated by the cracking of acetone vapor to methane and ketene which is the first stage in the production of acetic anhydride. The main cracking reaction is $(CH_3)_2CO \longrightarrow CH_4 + CH_2:CO$. It is an endothermic reaction with $\Delta H = 19$ kcal/mole at 700°C and proceeds as a first order irreversible reaction with a rate constant $k = \exp(34.34 - 68,000/RT)$. The vaporized acetone is preheated to 650°C and fed through a 4-in. diameter tube which winds back and forth in a furnace of temperature $T_c = 780$°C. Other data will be adapted from Jeffreys' very full discussion of this design problem (see References) as needed.

Since the reaction is of the first order and will be taken to be at constant pressure with a feed of pure acetone vapor, the rate is kc where c is the molar concentration of acetone vapor. But we may readily adapt Eq. (9.2.30) to the case of varying temperature and write

$$c = c_0 \frac{T_0}{T} \frac{1-x}{1+x}, \qquad (9.7.21)$$

for the stoichiometric coefficient of acetone is -1 and $\bar{\alpha} = 1$; $x = m_1\xi''$. Then

$$\frac{dx}{dt} = v_0 \frac{dx}{dz} = \frac{G}{\rho_0} m_1 \frac{d\xi''}{dz} = \frac{1}{c_0} r = k(T) \frac{T_0}{T} \frac{1-x}{1+x}, \qquad (9.7.22)$$

where v_0 is the inlet velocity and Eq. (9.7.1) has been used; $t = z/v_0$ is the residence time based on the inlet velocity. Similarly,

$$v_0 \frac{dT}{dz} = \frac{G}{\rho_0} \frac{dT}{dz} = \frac{-\Delta H}{\hat{C}_p \rho_0} c_0 k(T) \frac{T_0}{T} \frac{1-x}{1+x} - \frac{U}{\rho_0 \hat{C}_p p}(T - T_c).$$

From Jeffreys' data:

$$U = 58 \text{ Btu/hr ft}^2 \text{ °F},$$
$$\rho_0 \hat{C}_p = C_v = 0.0434 \text{ Btu/ft}^3 \text{ °F},$$
$$c_0 = 0.0129 \text{ lb-mole/ft}^3,$$
$$\Delta H = 34,200 \text{ Btu/lb-mole},$$
$$v_0 = 173.5 \text{ ft/sec},$$
$$p = \tfrac{1}{4}\pi d^2/\pi d = \tfrac{1}{12} \text{ ft}.$$

Sec. 9.7 Tubular Reactors Cooled or Heated from the Wall

Thus (remembering that T is in °K)

$$\frac{dT}{dt} = v_0 \frac{dT}{dz} = -566k(T)\frac{T_0}{T}\frac{1-x}{1+x} + 4.46(1053 - T). \quad (9.7.23)$$

Equations (9.7.22) and (9.7.23) have the boundary conditions $T = 923$ and $x_0 = 0$.
These are slightly simplified forms of the equations which Jeffreys solves by finite differences using a step length of $\Delta t = 0.0722$. The added complication that Jeffreys faces in his very full treatment of the design of this plant is that there are side reactions by which the ketene can be decomposed and the acetone can be dehydrogenated. The complete set of reactions is

$$(CH_3)_2CO \longrightarrow CH_4 + CH_2:CO,$$

$$2(CH_2:CO) \longrightarrow C_2H_4 + 2CO,$$

$$(CH_3)_2CO \longrightarrow 3H_2 + CO + 2C.$$

Pilot plant studies give the following relationship between x the conversion, y the total moles present, and w the fraction of the acetone that is cracked in the main reaction:

x	0.1	0.15	0.20	0.25
y	1.116	1.191	1.27	1.362
w	0.845	0.80	0.75	0.70

If only the main reaction had been taking place as we assumed above, the value of y would have been $(1 + x)$ and w would have been 1. The observed relationship between x and y is not quite linear and the factor of $(1 + x)$ in the denominator of the reaction rate expressions in Eqs. (9.7.22) and (9.7.23) must be replaced by the empirical $y(x)$. Over the given range, the relationship between w and x is very closely linear, being $w = 0.95 - x$. This must be used in deciding what final conversion is needed to produce the required amount of ketene. Thus the rate of production of ketene is $P = F_0 xw$, where $F_0 = v_0 c_0 A_r$ is the feed rate of acetone vapor. Now from the expression given above, $F_0 = 173.5 \times 0.0129 \times (\pi/3) = 0.0234$ lb-mole/sec; hence the production rate of ketene will be $0.0234 \times (0.95 - x)$. Figure 9.11 (based on Jeffreys' calculations) shows the following quantities as functions of x, the fractional conversion acetone: $y(x)$, the production of ketene in lb-mole/sec, the temperature in °C, and the length of the reactor in feet.

9.7.2 Cocurrent and countercurrent cooled tubular reactors

Figure 9.12 shows several arrangements of the tubular reactor in which the cooling stream flows along the reactor wall so that T_c is a function of z as well as the extent and temperature within the reactor. In arrangements (a) and (b) the coolant is an independent fluid, but in (c) and (d) the cold

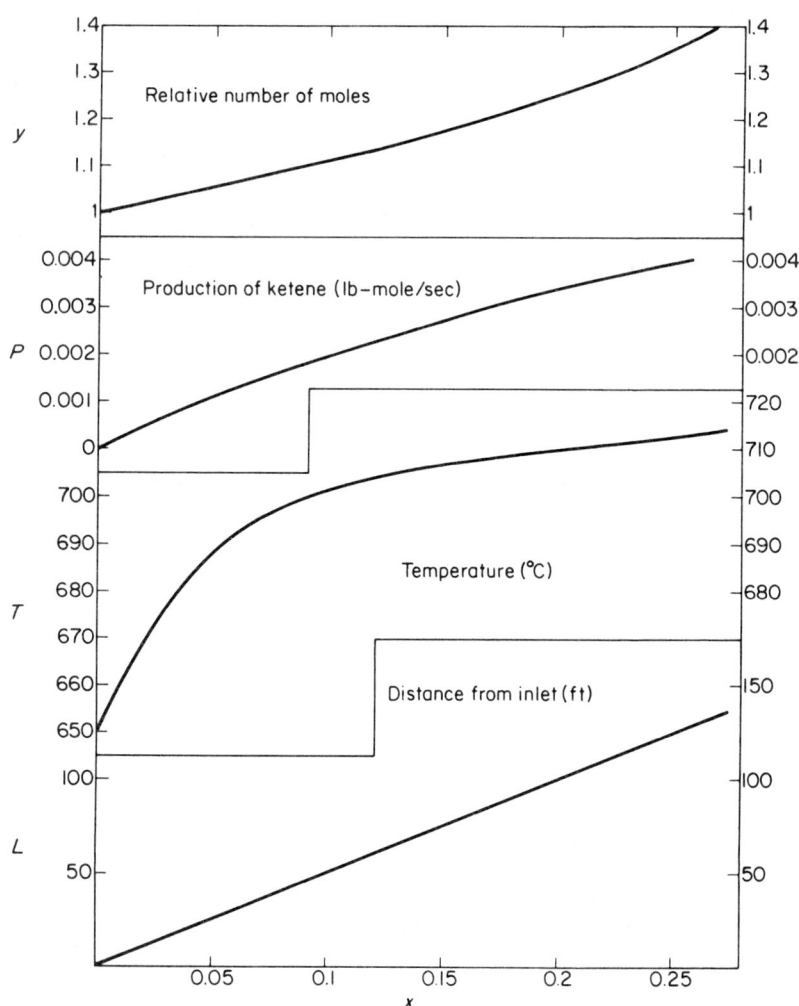

Fig. 9.11 The cracking of acetone to form ketene.

feed is itself used as coolant. Evidently arrangement (c) is a little awkward since the feed has to be brought back to the inlet of the reactor after it has served as coolant, and the countercurrent arrangement is more practical.

We can obtain an additional equation for $T_c(z)$, the coolant temperature, by making a heat balance over the reactor and cooling jacket as a whole between the inlet and the section through z. To do this, let G_c denote the mass flow rate of the coolant per unit cross-sectional area of the reactor and $\hat{C}_{pc}$ the heat capacity per unit mass of the coolant. If A_c is the cross-sectional

Sec. 9.7 Tubular Reactors Cooled or Heated from the Wall   291

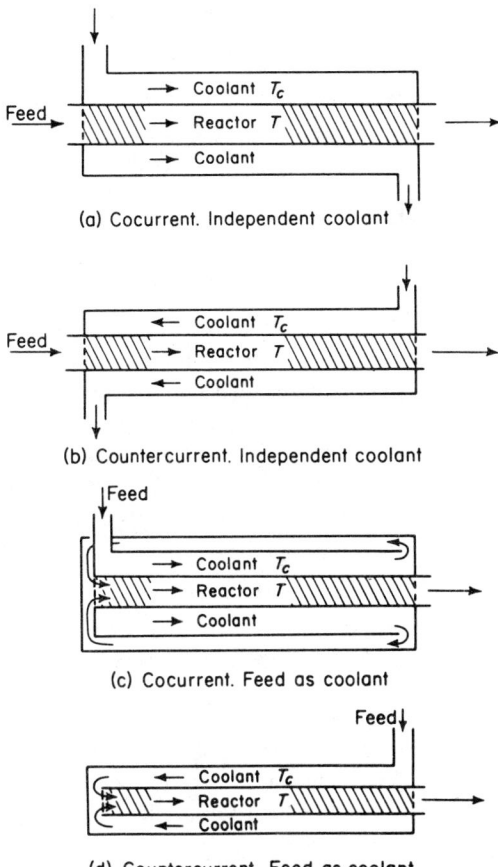

Fig. 9.12 Arrangements for cooling the tubular reactor.

area of the jacket (to be distinguished from a_c, the perimeter of the exchange area), then the linear velocity of the coolant is $v_c = A_r G_c / A_c \rho_c$ and the estimation of the Stanton number will generally require a knowledge of this. Then between the plane of the inlet to the reactor—where the extent of reaction is zero, the reactant temperature T_0, and the coolant temperature T_{c0}—and the plane z—where their values are $\xi''(z)$, $T(z)$, and $T_c(z)$, respectively—we have the following rates of heat generation in the cocurrent arrangement of Fig. 9.12(a):

(i) Rate of heat generation by reaction,
$$\int_0^z (-\Delta H) r A_r \, dz = (-\Delta H) G A_r \xi''(z);$$

(ii) Net rate of flow of the sensible heat of reactants,
$$GA_r\hat{C}_p[T(z) - T_0];$$
(iii) Net rate of flow of the sensible heat of coolant,
$$G_c A_c \hat{C}_{pc}[T_c(z) - T_{c0}].$$
Balancing these and dividing through by $G_c A_r \hat{C}_{pc}$ we have
$$\begin{aligned} T_c(z) &= T_{c0} + \gamma[J''\xi''(z) - T(z) + T_0], \\ &= T_{c0} + \gamma[J\xi(z) - T(z) + T_0], \end{aligned} \quad (9.7.24)$$
where
$$\gamma = \frac{G\hat{C}_p}{G_c \hat{C}_{pc}} \frac{A_r}{A_c} \quad (9.7.25)$$

is the ratio of the capacities of the two streams to convey heat. We note that if $\gamma = 0$, the jacket is isothermal, while if it is infinite, the adiabatic relation $T = T_0 + J\xi$ is obtained. If the countercurrent form of cooling is used, then the sign in front of γ must be changed to correspond to the fact that G and G_c are now in opposite senses. In this case

$$\begin{aligned} T_c(z) &= T_{c0} - \gamma[J''\xi''(z) - T(z) + T_0], \\ &= T_{c0} - \gamma[J\xi(z) - T(z) + T_0]. \end{aligned} \quad (9.7.26)$$

Finally, if the reactant stream itself is used as coolant, then $\gamma = 1$. In all this we have assumed the constancy of the specific heat, but it is not difficult to remove this restriction.

With these expressions for T_c we can substitute in Eq. (9.7.6) or (9.7.7) and obtain the following heat balance equations to be solved simultaneously with the mass balance. The letters a, b, c, and d below will refer to the four arrangements of Fig. 9.12. In the cocurrent, independent coolant case

$$G\frac{dT}{dz} = J''r(\xi'', T) - G\frac{S}{p}[-\gamma J''\xi'' + (1 + \gamma)T - (T_{c0} + \gamma T_0)]. \quad (9.7.27a)$$

For the countercurrent case

$$G\frac{dT}{dz} = J''r(\xi'', T) - G\frac{S}{p}[\gamma J''\xi'' + (1 - \gamma)T - (T_{c0} - \gamma T_0)], \quad (9.7.27b)$$

but T_0 cannot be specified in advance. The known temperature of the coolant is $T_c(L)$ and this can only be matched by trial and error on T_{c0}. We can of course write

$$T_{c0} = T_c(L) + \gamma[J''\xi''(L) - T(L) + T_0].$$

In the third case, we have $\gamma = 1$ and the condition that T_0 is related to $T_0(L)$. The simplest assumption is that it is isolated in its return to the inlet end of the reactor and, though this is not entirely realistic, we may write

$$G\frac{dT}{dz} = J''r(\xi'', T) - G\frac{S}{p}[-J''\xi'' + 2T - (T_{c0} + T_0)]. \quad (9.7.27c)$$

In the fourth case, $\gamma = 1$ and $T_{c0} = T_0$, since both denote the temperature at the inlet plane. Hence

$$G\frac{dT}{dz} = J''r(\xi'', T) - G\frac{S}{p}J''\xi''. \quad (9.7.27d)$$

Again, the temperature which can be specified from the physical conditions of the reactor is not T_0, but

$$T_c(L) = T(L) - J''\xi''(L). \quad (9.7.28)$$

It is therefore necessary to assume a value of T_0 and integrate Eq. (9.7.27d) simultaneously with

$$G\frac{d\xi''}{dz} = r(\xi'', T)$$

until $T_c(L)$ can be calculated from $T(L)$ and $\xi''(L)$. Repeated trials will allow the correct value of T_0 to be selected. We shall see later in this section and in the next one that the solution to this iteration may not be unique. The equations for constant density follow immediately by dropping the double primes on J'' and ξ'' and replacing G by v.

Consideration of the familiar plane of ξ and T shows the form that the solution of these equations must have. Let us take the case of constant density and the countercurrent arrangement with feed as coolant. Then

$$v\frac{d\xi}{dz} = r(\xi, T)$$

$$v\frac{dT}{dt} = Jr(\xi, T) - v\frac{S}{p}J\xi,$$

so that the solution can be represented by a curve in the ξ, T plane satisfying

$$\frac{dT}{d\xi} = J - J\frac{vS}{p}\frac{\xi}{r(\xi, T)}. \quad (9.7.29)$$

Figure 9.13 shows the ξ, T plane with the equilibrium and maximum reaction rate curves Γ_e and Γ_m. The broken curve (LBG) is the locus $r(\xi, T) = vS\xi/p$ for a particular value of the number vS/p; it is the locus of maximum temperatures since on it $dT/d\xi = 0$. Suppose that point A represents a particular value of T_0, the inlet temperature; $dT/d\xi = J$ when $\xi = 0$, so the reaction path takes off in the adiabatic direction, AE. However, as ξ increases, the second term on the right of Eq. (9.7.29) begins to grow and $dT/d\xi$ decreases. At B, $dT/d\xi = 0$ and the reaction path has a vertical tangent after which the temperature decreases. Hence the solution must give a path such as ABC lying between Γ_e and the adiabatic line AE. A slightly higher inlet temperature might give a path such as FGH and a lower value of T_0 a path such as KLM. It is evident that for a suitable value of T_0 the path can cross the locus of maximum reaction rate, so that there is probably an optimal value of T_0. Just how far one would get on a particular path in a reactor of length L would

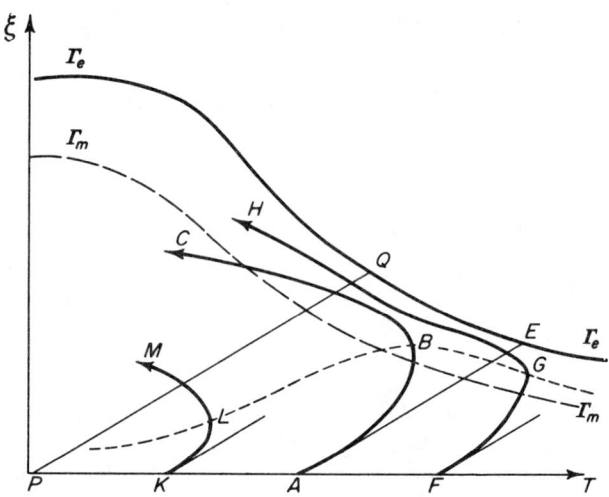

Fig. 9.13 The shape of the solutions for the countercurrent tubular reactor for various T_0.

be found by integrating the two equations. If the feed temperature is specified, say T_f, this is equivalent to specifying $T_c(L)$ since the feed is used as coolant. Then the reaction path must stop on the line

$$T - J\xi = T_f. \qquad (9.7.30)$$

This is shown in Fig. 9.13 as line PQ, which is the adiabatic line through the feed temperature. As we should expect therefore such a reactor could never achieve a greater extent than the adiabatic equilibrium extent for the given feed. If the design problem specifies T_f and calls for a given exit extent $\xi(L)$, then we can start from the exit condition $\xi(L)$, $T(L) = T_f + J\xi(L)$ and integrate the equations backwards to the axis $\xi = 0$.

The curves in Fig. 9.13 are all drawn for a given value of vS/p. Those in Fig. 9.14 are for various values of vS/p. It would be more appropriate to introduce a characteristic time, θ, such as the holding time of the reactor and use the dimensionless group $v\theta S/p$, but the shapes of the curves are unaltered. The broken curves are the loci of the maximum temperature (i.e., $r(\xi, T) = (vS/p)\xi$) for three values of (vS/p); the lower curves correspond to the larger values. For a given inlet temperature T_0 represented by point A, we know that the solution curve must lie above the adiabatic line A-E and have its vertical tangent on the appropriate broken curve. For a very large value of vS/p (i.e., good exchange of heat between feed and reaction mixture) the solution might look like the curve ABC. For a slightly smaller value of vS/p it would stay closer to the adiabatic line, as does ADF, while for a low value of vS/p it might be like AGH. In the limit $vS/p \longrightarrow 0$, the

Sec. 9.7   Tubular Reactors Cooled or Heated from the Wall   295

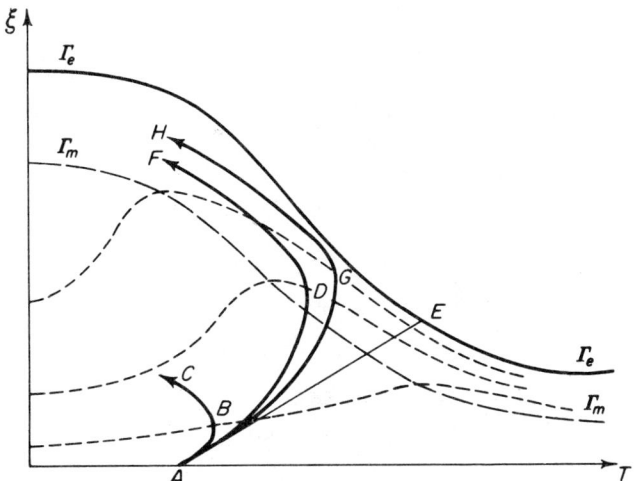

Fig. 9.14 The shape of the solutions for the countercurrent tubular reactor for various values of vS/p.

solution is the adiabatic path AE. Again we see that there is likely to be an optimum value of vS/p because the curve ADF seems to lie much more appropriately in the region of high reaction rates than do the others.

Many other design problems might be discussed, but this type of reactor will be best illustrated by mentioning an excellent study of the ammonia synthesis converter by Baddour et al. This reactor is commonly of the form shown in Fig. 9.15 in which the feed (perhaps already heated up by an interchanger section as in the T.V.A. reactor considered by Baddour) passes up through a number of tubes arranged in a catalyst bed and then downward through the bed as the reaction mixture. Since the reaction is heterogeneously catalyzed, it does not take place in the tubes but only in the catalyst bed, where the heat generated flows radially to be exchanged through the tube walls. It is easy to see that conditions will be far from uniform across the section of the reactor, yet in spite of this the simple one-dimensional model has proved quite successful.

Modifying the equations of Baddour slightly to suit our notation we have the reaction $\frac{1}{2}N_2 + \frac{3}{2}H_2 = NH_3$. If y is the mole fraction of NH_3, then the mass balance gives

$$\frac{dy}{dz} = \beta r(y, T) \qquad (9.7.31)$$

$$\frac{C_p}{C_{p_0}}\frac{dT}{dz} = J\beta r(y, T) - \frac{S}{p}(T - T_c), \qquad (9.7.32)$$

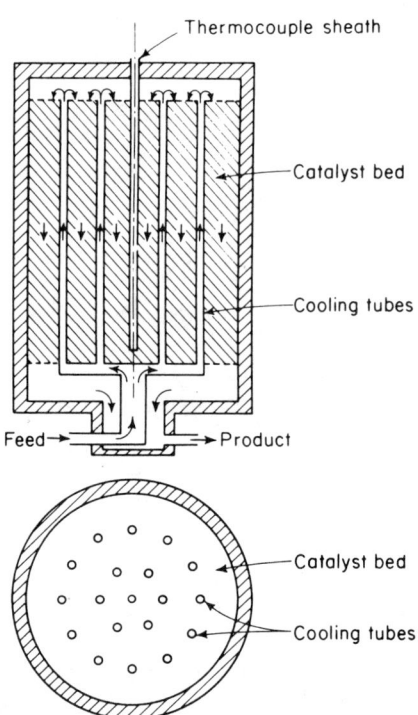

Fig. 9.15 The countercurrent cooled tubular reactor.

where

$$\beta = \frac{f(1+y)^2}{v_0 c_0 T(1+y^*)},$$

$$r(y,T) = \frac{k'(T)}{P^{1/2}} \left[\frac{(K_p P)^2 \mathscr{A}(\mathscr{B}-y)^{1.5}(\mathscr{C}-y)}{y} - \frac{\mathscr{D}y}{(\mathscr{B}-y)^{1.5}} \right],$$

$y^*, y_{\text{II}}^*, y_{\text{N}}^*$ = inlet mole fractions of NH_3, H_2 and N_2, respectively,
f = catalyst activity factor,
$k'(T) = 1.75 \times 10^{16} \exp -(20,300/T)$,
P = pressure (assumed constant),
K_p = equilibrium constant (atm^{-1}),
$\mathscr{A} = (1.5 - y_{\text{II}}^*)^{1.5} \dfrac{0.5 - y_{\text{N}}^*}{(1+y^*)^{2.5}}$,
$\mathscr{B} = \dfrac{y_{\text{II}}^* + 1.5 y^*}{1.5 - y_{\text{II}}^*}$,
$\mathscr{C} = \dfrac{y_{\text{N}}^* + 0.5 y^*}{0.5 - y_{\text{N}}^*}$,
$\mathscr{D} = \dfrac{1 + y^*}{(1.5 - y_{\text{II}}^*)^{1.5}}$.

Sec. 9.7 Tubular Reactors Cooled or Heated from the Wall

The equations take into account the variation of C_p with composition and $-\Delta H$ with temperature by means of the ratios

$$\frac{C_p}{C_{p0}} = 1 + \frac{\sum \alpha_j C_{pj}}{C_{p0}} \frac{y - y^*}{1 + y}$$

$$J = \left(\frac{-\Delta H}{C_p}\right)_0 \left[1 + \frac{\sum \alpha_j C_{pj}(T - T_b)}{-\Delta H_0}\right] \frac{1 + y^*}{(1 + y)^2},$$

where T_b is the base temperature for the calculation of ΔH_0 and C_{p0}. The Stanton number is calculated using C_{p0}. The temperature in the equations is in °R; in the figures it will be shown as °C.

In their study of these equations, Baddour and his co-workers first compared the solution with an observed temperature profile under known operating conditions. Figure 9.16, based on Fig. 3 of their paper, shows a good agreement for the conditions:

Conditions		Plant	Equations
Mole fraction of H_2 in feed	(y_H^*)	0.65	0.6375
Mole fraction of N_2 in feed	(y_N^*)	0.219	0.2125
Mole fraction of NH_3 in feed	(y^*)	0.052	0.050
Mole fraction of inerts in feed	(y_i^*)	0.079	0.080
Space velocity (v_0/L) (hr^{-1})		13,800	13,800
Pressure (P) (atm)		286	300
Catalyst volume $(A_r L)$ (ft^3)		144	144

It would be very difficult to predict the Stanton number with any accuracy so that the appropriate value of the heat transfer conductance (the product of the heat transfer coefficient and total area of transfer, Ua_cL) was adjusted to give the best fit. This was found to be

$$Ua_c L = 55,000 \text{ Btu/hr °F}.$$

The estimated Stanton number gave a value of 57,300, in reasonable agreement, and the calculated ammonia production rate of 142 tons a day was 19% higher than observed. The authors discuss the reasons for these discrepancies with care and conclude that the agreement is satisfactory.

The value of a model of this kind, in which a fair degree of confidence can be placed, is that it allows us to examine the effect of changing the operating conditions. Figure 9.17 shows the variation of production rate with the inlet temperature T_0 for a fixed feed composition. With the standard space velocity of 13,800 hr^{-1} it is clear that the optimum T_0 is about 425°C. By increasing the flow rate so that the space velocity is 18,000 hr^{-1}, a greater production can be achieved (this time with $T_0 = 430$°C) but it is somewhat more sensitive to the value of T_0. A decrease of flow rate lowers the production and makes it less sensitive to variations in T_0. (Also shown on this figure is the stability limit about which we shall have more to say shortly.)

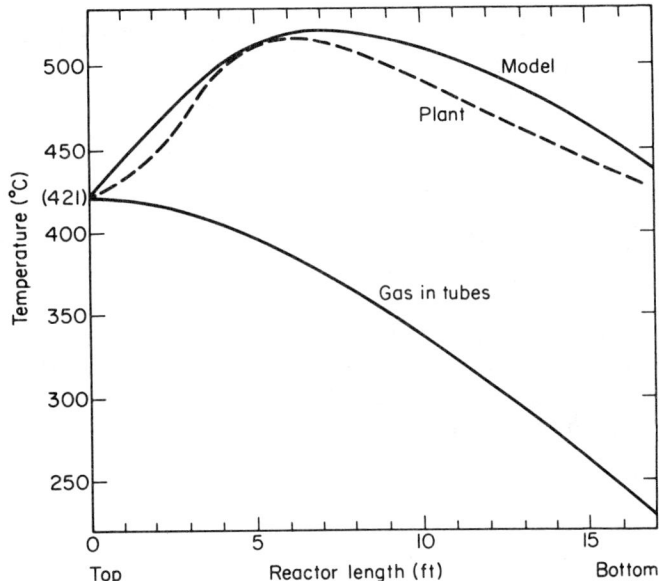

Fig. 9.16 Comparison of solution of Eqs. (9.7.30) and (9.7.31) with an actual ammonia converter temperature profile. [Based on a figure of Baddour *et al.*, *Chem. Eng. Sci.*, **20**, 281 (1966) by courtesy of the authors and Pergamon Press.]

The effect of the variables on production can be summarized by tabulating the sensitivities. If P is the production when a parameter has the value q and if a small change Δq in q produces a change ΔP in P, the sensitivity of P to q at this point may be defined as

$$\sigma = \frac{\Delta P/P}{\Delta q/q}.$$

It is clearly the ratio of the percentage change induced in P to that made in q and is an estimate of $\partial(\ln P)/\partial(\ln q)$. From the calculations of Baddour about the standard state, we find the following sensitivities of the maximum production rate:

Operating variable	Standard value	Sensitivity
Space velocity	13,800 hr^{-1}	0.7
Inlet NH$_3$ mole fraction	0.05	−0.30
Inert mole fraction	0.08	−0.15
Catalyst activity	1.00	0.35

The sensitivity of the maximum production rate to the heat transfer area is very slight.

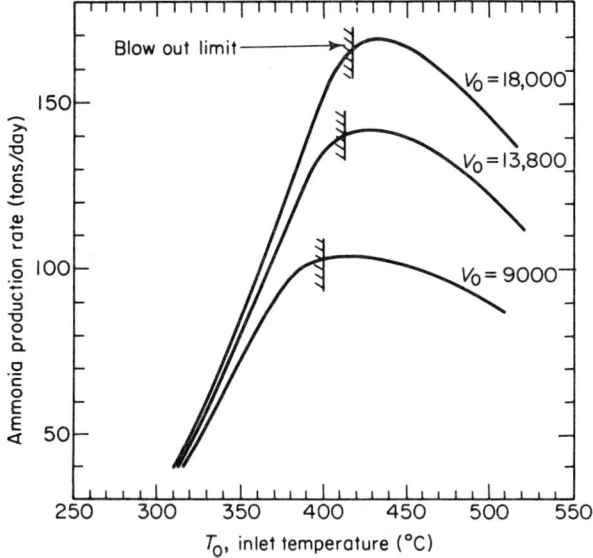

Fig. 9.17 The variation of production with T_0 for several flow rates. [Based on a figure of Baddour et al., *Chem. Eng. Sci.*, **20**, 281 (1966) by courtesy of the authors and Pergamon Press.]

This kind of detailed study is most valuable at every stage of the design of a plant and it is becoming increasingly common for a plant to be simulated on the computer while it is being built.

We shall conclude this section by reporting Baddour's results on the stability of the plant, as this will form an excellent introduction to the considerations of the next section. In solving the equations for the countercurrent converter of fixed size, a value of T_0 is chosen and the value of this feed temperature, $T_f = T_c(L)$, is calculated at the end of the integration. We then know that, for this value of T_f, the value of T_0 will be as chosen and the profiles will be those calculated. If this calculation is made for several values of T_0, all other parameters being kept constant, then T_f can be plotted against T_0, and for any prescribed T_f, the unknown T_0 can be read off. Figure 9.18 shows the results of Baddour's calculations on the standard case and it is at once evident that for T_f between 224°C and 270°C there can be two values of T_0. The smaller of these is unstable because the slope of the curve is negative here and an increase in T_f actually decreases T_0. The minimum feed temperature of 224°C is called the blow-out feed temperature for there is no intersection on the figure at all. Actually, the curve goes out of the figure to the left and quickly drops to the asymptote $T_0 = T_f$; this means that there is virtually no reaction taking place since the feed is too cold. Stable operation which is also profitable is to be found only on the part of the curve to

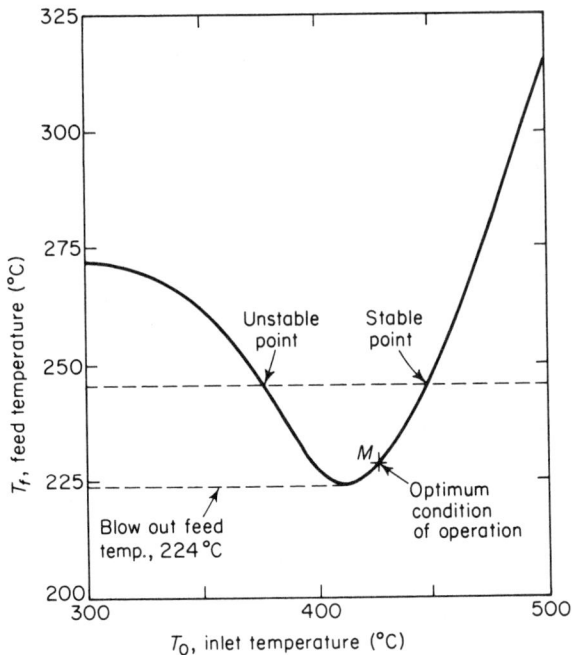

Fig. 9.18 Variation of T_f with T_0. [Based on a figure of Baddour et al., *Chem. Eng. Sci.*, **20**, 281 (1966) by courtesy of the authors and Pergamon Press.]

the right of the minimum. The maximum production rate actually corresponds to the point M. The feed temperature below which the reaction will not sustain itself is obviously of great importance: the hatched line in Fig. 9.17 shows the variation of this for the three different space velocities. The sensitivities of the blow-out feed temperature to various parameters based on Baddour's calculations are as follows:

Operating variable	Standard value	Sensitivity
Space velocity	13,800 hr^{-1}	0.1
Inlet NH$_3$ mole fraction	0.05	0.028
Inlet inert gas mole fraction	0.08	0.015
Catalyst activity	1.0	-0.09
Heat transfer conductance	55,000 Btu/hr °F	0.74

Since the blow-out feed temperature is 224°C we see that a change in space velocity from 13,800 to 14,000 hr^{-1} would increase it by $(224)(0.1)(200)/(13,800) = 3.25°C$. On the other hand, a decrease of catalyst activity to 0.9 would increase it by about 20°C.

Another operating variable of importance is the peak temperature;

in Fig. 9.16 this appears to be about 529°C. Baddour's calculations show that this is very insensitive to the space velocity and that its sensitivities to inlet NH_3, inlet of inert gases, activity, and heat transfer conductance are -0.04, -0.02, -0.04, and $+0.076$, respectively.

We now turn to a more general description of the sensitivities and stability of the tubular reactor.

Exercise 9.7.1. Consider the solution of Eqs. (9.7.3) and (9.7.7) for the extent and temperature in a wall-cooled reactor with an exothermic reaction; show that the maximum reaction rate always occurs before the maximum temperature. Show also that the point (ξ, T) at which the reaction rate is greatest must lie between Γ_{Jm} and Γ_m.

Exercise 9.7.2. Show that the slope of the locus on which the right-hand side of Eq. (9.7.20) vanishes is always positive.

Exercise 9.7.3. Obtain a differential equation for $T_c(z)$ by writing a differential balance over a section of the cooling jacket and show that this leads to Eqs. (9.7.24) and (9.7.26).

Exercise 9.7.4. Given the Temkin equation for ammonia synthesis,
$$r = k_1 P_{N_2} P_{H_2}^{1.5} P_{NH_3}^{-1} - k_2 P_{NH_3} P_{H_2}^{-1.5},$$
justify the rate expression in Eq. (9.7.31).

Exercise 9.7.5. In the arrangement of Fig. 9.12(c) let T'_c denote the temperature in the outer shell which is insulated on the outer side. Show that
$$T + T_c - T'_c - J''\xi'' = T_c(0) = T(L) - J''\xi''(L).$$
Obtain a set of differential equations and indicate how you would go about solving them.

9.8 Sensitivity and Stability

With the example of the ammonia converter in mind we can proceed to a more general discussion of sensitivity and stability. *Parametric sensitivity* exists when some operating characteristic shows an extreme change when only a slight change is made in an operating variable. It is closely related to the notion of stability. We speak of a steady state of the tubular reactor being stable, when a sufficiently small perturbation from it tends to die away. Thus in the concept of stability the operating variables are assumed to be constant when perturbations are introduced into the state. It may be necessary to change the operating variables to create such a perturbation, but once it has been introduced the operating variables are held constant and the perturbation is followed to see if it decays. A stable operating condition may still be very sensitive to a particular variable. On the other hand, an unstable condition is such that the least perturbation will lead to a finite change and such a condition may be regarded as infinitely sensitive to any operating variable. Sensitivity can be fully explored in terms of steady state solutions. A complete discussion of stability really requires the study of the transient equations. For the stirred tank this was possible since we had only to deal with ordinary differential equations; for the tubular reactor the full treatment of the partial differential equations is beyond our scope here. Nevertheless, just as much could be learned about the stability of a stirred tank from the heat generation and removal diagram, so here something may be learned about stability from features of the steady state solution.

The classic example of parametric sensitivity was given in 1956 by Amundson and Bilous and is shown in Fig. 9.19. The curves shown are the temperature profiles in a reactor with uniform wall temperature T_{co} and fixed inlet temperature T_0. The reaction chosen was a first order, irreversible, exothermic

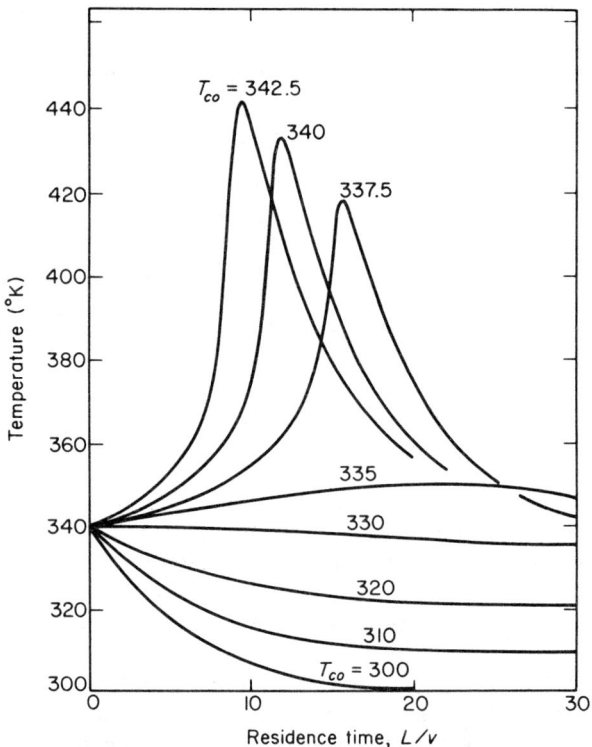

Fig. 9.19 Sensitivity of the tubular reactor to wall temperature. [From Bilous and Amundson, *A.I.Ch.E. Journal*, **2**, 116 (1956), with permission.]

Sec. 9.8. Sensitivity and Stability

one and the curves are thus the solutions of

$$v\frac{d\xi}{dz} = k(T)(1 - \xi)$$

$$v\frac{dT}{dz} = Jk(T)(1 - \xi) - v\frac{S}{p}(T - T_{co}),$$

with $\xi = 0$, $T = T_0$ at $z = 0$, and for several values of T_{co}.

When T_{co} is low, for example $T_{co} = 300°C$ scarcely any reaction is taking place and the temperature simply falls off to the jacket temperature. By the time $T_{co} = 330°C$ this fall off has become quite slow and at $T_{co} = 335°C$ sufficient reaction is taking place so that the temperature increases a few degrees before falling off. However, a change of only 2.5°C in T_{co} produces a dramatic change in profile. A great deal of reaction is taking place and the peak temperature is now nearly 80°C above inlet. It is very reasonable therefore to speak of the maximum temperature as being extremely sensitive to T_{co} in the neighborhood of 335°C. The reaction has been "ignited" by a sufficiently high temperature in the jacket. This is the sort of behavior we have come to expect of the exothermic reaction; it is a reflection of the twist in the surface which we noticed in Fig. 4.7.

A countercurrent cooled tubular reactor such as that in Fig. 9.15 is said to be operating *autothermically* when the heat of reaction is sufficient to raise the temperature of the incoming stream from T_f to T_0. If it is not autothermic it might be necessary to add heat at $z = 0$ by means of an electric heater and this in fact is often done during start-up. By solving the equations

$$v\frac{d\xi}{dz} = r(\xi, T) \tag{9.8.1}$$

$$v\frac{dT}{dz} = Jr(\xi, T) - v\frac{S}{p}J\xi \tag{9.8.2}$$

for $\xi = 0$ and $T = T_0$ at $z = 0$, we can calculate $T_f = T_c(L) = T(L) - J\xi(L)$ for each T_0. Instead of plotting T_f against T_0 as Baddour did, let us plot

$$\Delta T = T_0 - T_c(L)$$

against T_0 for fixed values of L and the other parameters. By writing Eq. (9.8.2) in the dimensionless form

$$\frac{d(T/T^*)}{d(z/L)} = \frac{L}{T^*}\frac{dT}{dz} = \frac{JLr(\xi, T)}{vT^*} - \frac{SL}{p}\frac{J\xi}{T^*}, \tag{9.8.3}$$

where T^* is some characteristic temperature, we see that the key parameter (apart from the kinetic parameters in r) is

$$\hat{N} = \frac{SL}{p}. \tag{9.8.4}$$

We might call this the number of heat transfer units in the reactor for it is $Ua_cL/GA_r\hat{C}_p$, the ratio of the total heat transfer capacitance to the capacity of the reactants to convect heat. Thus $\Delta T = T_0 - T_c(L)$ is a function of T_0, $\hat{N}$, and of other parameters which we take to be constant; we may write it $\Delta T(T_0; \hat{N})$. For any given T_f, the appropriate value of T_0 is found by solving

$$T_0 - T_f = \Delta T(T_0, \hat{N}). \qquad (9.8.5)$$

Figure 9.20 shows some typical curves. If $\hat{N}$ is very low as for $\hat{N}_1 = 0.5$ there is only one solution when $T_f = 400°C$, namely at point A. This is so close to the feed temperature T_f that it is clear that little reaction is taking place. On the other hand, with $\hat{N}_3 = 2$ there is only one intersection (E) and that is at so high a temperature (above 700°C) that evidently the reactor is generating a lot of heat. However, when $\hat{N}_2 = 1$ there are three intersections at B, C, and D, so that three steady states are possible.

The profiles of temperature and extent for the steady states D and B are shown in Fig. 9.21. In the upper part, corresponding to D, there is much heat generated and a conversion of nearly 70%; in the lower one, corresponding to B, the conversion is only a hundredth of this. We should not be surprised by now to learn that the state corresponding to the intersection at C is un-

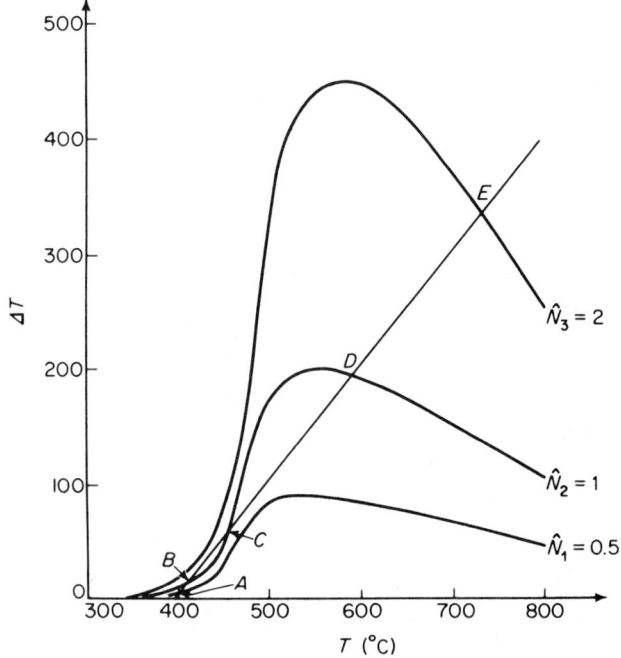

Fig. 9.20 Temperature increase of the feed as a function of inlet bed temperature.

Sec. 9.8 Sensitivity and Stability

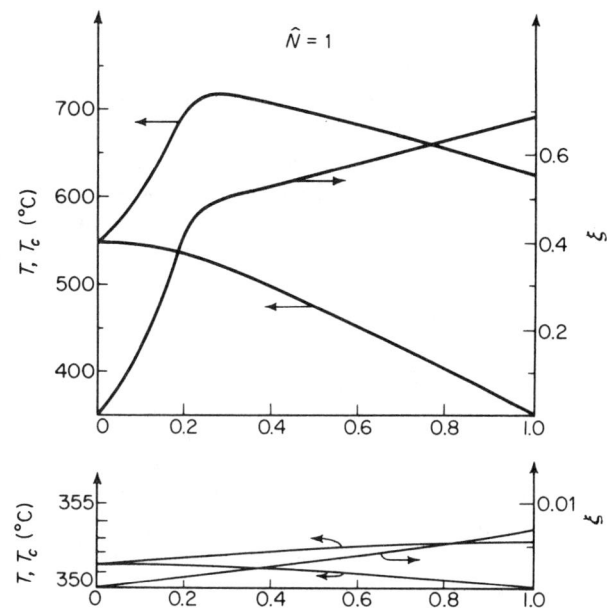

Fig. 9.21 Profiles of two possible steady states for a given feed temperature.

stable. We could use all of the old arguments showing that a slow increase and a decrease of T_f would give a hysteresis loop which avoids entirely intersections such as C. In fact, Fig. 9.20 is really a heat generation and rejection diagram and the argument used in Sec. 7.5 can be used again. As before, this suffices to show that the state corresponding to C is unstable; it does not prove the stability of the other states.

The way in which the intersections lie with respect to the variations of $\hat{N}$ is also instructive. If a design with $\hat{N} = 1$ operating at point D is acceptable and if fouling of the cooling tubes reduces the heat transfer coefficient, there will come a point before $\hat{N}$ reaches 0.5 when the intersection will slip down to the lower part of the curve. That the designer should be aware of this possibility is of the first importance. The diagram also shows what remedial action to take. If we want the intersection to stay on the upper part of the curve as $\hat{N}$ decreases, the feed temperature must be increased to ensure that there is a suitable intersection. Even so it is clear that ΔT at the point of intersection will fall and with it so will the conversion.

Exercise 9.8.1. Sketch the behavior of T_0 as T_f is slowly increased and then slowly decreased.

9.9 The Effect of Flow Profile

To take the effect of the flow profile properly into account we would have to solve partial differential equations in two or even three space variables. This is not only difficult, but the analysis of the results would be difficult also and there is some incentive to look for a simple situation in which it is possible to estimate the effect rather quickly. The ordinary laminar flow profile would seem to be a good candidate for study, for it is certainly far from uniform, with the central stream moving at twice the mean speed. If we assume that there is no diffusion, so that there is no transfer between different parts of the stream, we should be able to see the effect of velocity variations at their worst. In practice, there will be a certain amount of diffusion and reactants that would otherwise pass rapidly through the reactor will diffuse to slower streams and the distribution of residence times will be more uniform than when there is no diffusion at all.

Consider a cylindrical tubular reactor of radius a through which reactants are flowing with mean linear velocity v. If the flow profile is parabolic, the linear velocity at a radius r from the axis of the tube is $v(r) = 2v(1 - r^2/a^2)$ and the volume flow rate through an annulus between the radii r and $r + dr$ is $2\pi r\, dr\, v(r)$. If the length of the reactor is L, the mean residence time of the reactor is $L/v = \theta$ say, but the residence time of the fluid flowing between r and $r + dr$ is $L/v(r) = \theta/2(1 - r^2/a^2)$. Since

$$t = \frac{\theta}{2(1 - r^2/a^2)}, \tag{9.9.1}$$

then

$$dt = \frac{\theta r\, dr}{a^2(1 - r^2/a^2)^2}, \tag{9.9.2}$$

and the fraction of the fluid passing through in times between t and $t + dt$ is the same as the fraction of fluid passing through between radii r and $r + dr$. But this gives the residence time distribution, for

$$p(t)\, dt = \frac{2\pi r\, dr\, v(r)}{\pi a^2 v} = 4\frac{r}{a^2}\left(1 - \frac{r^2}{a^2}\right) dr,$$

$$= 4\frac{r}{a^2}\left(1 - \frac{r^2}{a^2}\right)\frac{dt}{\theta r} a^2 \left(1 - \frac{r^2}{a^2}\right)^2.$$

Thus

$$p(t) = \frac{1}{\theta} 4\left(1 - \frac{r^2}{a^2}\right)^3 = \frac{\theta^2}{2t^3}.$$

Since the central stream is moving with velocity $2v$, the least possible residence time is $L/2v = \theta/2$, hence a full specification of the residence time distribution

Sec. 9.9 The Effect of Flow Profile

is

$$p(t) = \begin{cases} 0, & 0 \le t < \theta/2 \\ \theta^2/2t^3, & \theta/2 < t. \end{cases} \quad (9.9.3)$$

Consider now a second order irreversible reaction for which

$$\frac{d\xi}{dt} = k(c_0 - \xi)^2, \qquad \xi(0) = 0. \quad (9.9.4)$$

Then the reactants with residence time t will react to extent

$$\xi(t) = c_0 \frac{kc_0 t}{1 + kc_0 t}. \quad (9.9.5)$$

In the completely segregated flow we are assuming, the mean extent of the reaction will be

$$\bar{\xi} = \int_{\theta/2}^{\infty} p(t)\xi(t)\,dt = c_0 \int_{\theta/2}^{\infty} \frac{kc_0 t}{1 + kc_0 t} \frac{\theta^3}{2t^3} \frac{dt}{\theta}. \quad (9.9.6)$$

Let

$$\kappa_2 = kc_0\theta, \qquad \tau = \frac{t}{\theta}; \quad (9.9.7)$$

then

$$\frac{\bar{\xi}}{c_0} = \int_{1/2}^{\infty} \left(1 - \frac{1}{1 + \kappa_2 \tau}\right) \frac{d\tau}{2\tau^3}$$

$$= \frac{\kappa_2^2}{2} \ln \frac{\kappa_2}{2 + \kappa_2} + \kappa_2.$$

Thus the ratio of $\bar{\xi}$ to the extent that would be attained in a plug flow tubular reactor is

$$X_2(\kappa_2) = \frac{\bar{\xi}}{\xi(\theta)} = (1 + \kappa_2)\left(1 + \frac{\kappa_2}{2} \ln \frac{\kappa_2}{2 + \kappa_2}\right). \quad (9.9.8)$$

This function is the curve $n = 2$ in Fig. 9.22 with $\kappa_2 = \kappa$, and it evidently

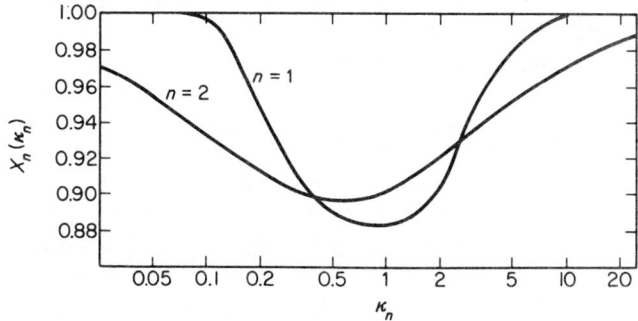

Fig. 9.22 Ratio of the extent of reaction with a laminar flow profile to that with plug flow for first and second order reactions.

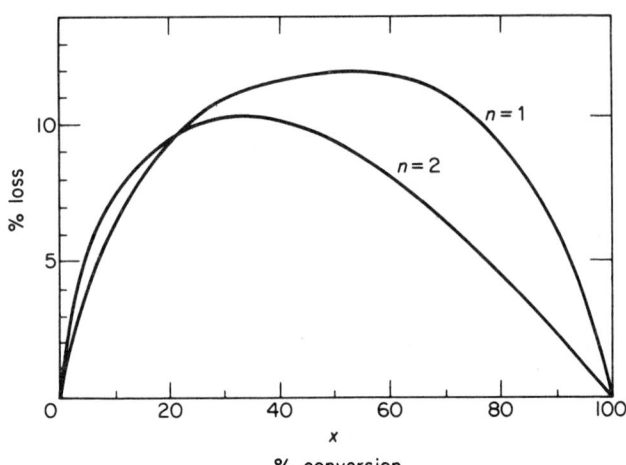

Fig. 9.23 Percentage loss from the effect of the laminar flow profile as a function of percentage conversion.

drops to about 0.9 when κ is around 0.5. We should expect X_2 to be near 1 when κ is small, for then very little is taking place and the residence time distribution is unimportant. Similarly, when κ is very large the reaction is virtually complete and the variation in residence times matters little. It is instructive to plot the percentage loss $100(1 - X_2)$ against the percentage conversion in laminar flow, $100\bar{\xi}/c_0 = 100x$ as in Fig. 9.23. In a design it is then possible to make an approximate allowance for the extra length that is needed. For example, if a conversion of 70% is required, the estimated loss as compared with plug flow is 6.5%. Hence the length of the reactor will have to be chosen so that in plug flow $\xi(\theta)/c_0 = 0.765$, i.e., $\kappa = 0.765/0.235 = 3.26$.

In Figs. 9.22 and 9.23 the results for a first order reaction are also shown as curves of $X_1(\kappa_1)$ versus $\kappa_1 = k\theta$. The only reason for having started with the second order reaction is that the integral in Eq. (9.9.6) comes out in simple terms. For the first order irreversible reaction

$$\begin{aligned}X_1(\kappa_1) &= \frac{1}{1 - e^{-k\theta}} \int_{\theta/2}^{\infty} (1 - e^{-kt}) \frac{\theta^3}{2t^3} \frac{dt}{\theta}, \\ &= \frac{1 - 2E_3(\tfrac{1}{2}\kappa_1)}{1 - \exp - \kappa_1},\end{aligned} \quad (9.9.9)$$

where E_3 is the form of exponential integral defined in the Appendix [Eq. (A.16)].

Exercise 9.9.1. If at a radius $a\rho$ the linear velocity is $vU(\rho)$, where v is the mean velocity, and the kinetics of an isothermal reaction are given by $d\xi/dt = r(\xi)$, show that the ratio of the mean extent of reaction to the extent in a plug flow reactor of the same mean residence time θ is

$$X = \int_0^1 2\rho U(\rho) \frac{\xi[\theta/U(\theta)]}{\xi(\theta)} d\rho,$$

where $\xi(t)$ is given implicitly by

$$t = \int_0^{\xi(t)} \frac{d\xi}{r(\xi)}.$$

Exercise 9.9.2. Show how to use the graph in Fig. 9.22 for a reversible first order reaction.

Exercise 9.9.3. Show that when the fractional conversion x in Fig. 9.23 is close to 1, the fractional loss $1 - X_2$ is $\frac{1}{3}x(1 - x) + 0[(1 - x)^3]$ in the case of the second order reaction.

Exercise 9.9.4. Show that the distribution function of residence times for laminar flow in a tubular reactor has the form $2t_m^2/t_p^3$, where t_p is the time of passage of any fluid annulus and t_m the minimum time of passage. Diffusion and entrance effects may be neglected. Hence show that the fractional conversion to be expected in a second order reaction with velocity constant k is $2B[1 + B \ln B/(B + 1)]$ where $B = akt_m$ and a is the initial concentration of both reactants. (C.U.)

Exercise 9.9.5. Sugar is being inverted in a long cylindrical reactor. The reaction, which takes place in aqueous solution, is irreversible and pseudo–first order, being catalyzed by excess mineral acid. There is no appreciable change in volume or temperature during reaction and diffusion may be neglected. The size of the reactor is such that the assumption of piston flow in the reactor leads to an estimated conversion of 86.5% of the sugar to inverted sugar. If the flow is, in fact, laminar, obtain:

(i) an expression for the distribution of residence times in the reactor;

and hence,

(ii) the expected degree of conversion with laminar flow. (C.U.)

9.10 Axial Dispersion in Tubular Reactors

The Peclet number of radial dispersion was found to be between 8 and 15, as we have noted above. However, the axial Peclet number is about 2, which shows that the axial eddy diffusion coefficient E_a is anywhere from four to seven times the radial coefficient E_r. A very simple model gives some indication why this should be so. The flow through a packed bed has been described as "seaweed flow" to denote the fact that it passes rapidly between regions of large free space through the narrow interstices of the packing. In a regular rhombohedral packing of spheres, for example, the fractional free area is about 9% in a plane through the sphere centers. Dividing the layer between two such planes into thirds, the free volume for the middle third is 41% as compared with 18% for the upper and lower thirds, the overall voidage being 26%. It is thus reasonable to imagine the reactants being passed rather rapidly from one relatively open region to the next and to conceive the passage of the fluid as being similar to that through a chain of stirred tanks. Indeed, we have seen in Sec. 7.9 that a sharp burst of tracer injected into the first of n tanks of total holding time Θ spreads out into a bell-shaped distribution with mean residence time Θ and a variance of residence times Θ^2/n. A similar tracer injection into a flow of mean velocity v and longitudinal diffusion coefficient E_a would give a mean residence time of L/v in a tube of length L and variance $2E_aL/v^3$. If we equate the means and variances in the two cases, we find that $Lv/nE_a = 2$. But it has been experimentally found that vd_p/E_a is approximately 2, so that $n = L/d_p$ is about equal to the number of layers of particles in the bed. It is therefore reasonable to describe the longitudinal mixing induced by the packing in terms of an axial dispersion coefficient, E_a.

The interaction of transverse diffusion and flow profile can also be represented by an effective dispersion coefficient. Consider a steady parabolic flow through a tube of circular cross section. If the fluid in a cross-sectional plane were suddenly colored and if there were no diffusion, this disk would be drawn out into a paraboloid of revolution, the rim not moving from its original position and the tip advancing with twice the mean speed of flow. But if such a paraboloid could be produced, the effect of lateral diffusion would be to smear it out. Thus the dye near the tube wall would diffuse inwards and so pass into faster moving streams and that at the center would diffuse outwards to slower streams. This reduces the spread of residence times by speeding up the slower tracer and slowing down the faster. In fact, if there were extremely rapid lateral diffusion, but no longitudinal diffusion, it would be virtually impossible for a parabolic shell to form and the tracer would move as a compact blob. A more detailed analysis shows that the mean concentration of the tracer is very much like that obtained in plug flow with longitudinal diffusion. The center of gravity moves with the mean speed of flow and the variance is such as would be given by an axial diffusion coefficient of $E_a = a^2v^2/48D$. Here a is the radius of the tube, v the mean speed, and D the molecular diffusion coefficient of the tracer. The coefficient 1/48 is characteristic of the flow profile and the shape of the tube cross section. This expression shows that as E_a is reduced to D, the diffusion coefficient across the tube increases. Of course, it is not possible to have rapid diffusion across the tube without some longitudinal diffusion as well. If

Sec. 9.10   Axial Dispersion in Tubular Reactors   311

this is allowed for we have a general form,

$$E_a = D + \frac{\lambda a^2 v^2}{D}, \qquad (9.10.1)$$

where λ is a shape factor. This makes it plausible to use an axial dispersion coefficient rather widely and we must now ask what effect this will have on the yield of the reactor.

Consider an isothermal first order irreversible reaction $A \longrightarrow B$, the rate of which is kc, c being the concentration of A. With no diffusion we have as usual

$$v \frac{dc}{dz} = -kc.$$

However, if the axial dispersion can be described by an effective diffusion coefficient E_a giving a diffusive flux of $E_a(-dc/dz)$, then a balance over the element between z and $z + dz$ will give:

total flux of A into section: $vc(z) - E_a \left(\dfrac{dc}{dz}\right)_z$;

total flux of A out of section: $vc(z + dz) - E_a \left(\dfrac{dc}{dz}\right)_{z+dz}$;

rate of disappearance of A by reaction: $kc(z)\, dz$.

Hence in the limit,

$$E_a \frac{d^2 c}{dz^2} - v \frac{dc}{dz} - kc = 0. \qquad (9.10.2)$$

If c_0 is the concentration in the feed, the flux to the plane $z = 0$ is vc_0, and from it is $vc - E_a dc/dz$, so that if there is no diffusion back into the feed stream

$$vc - E_a \frac{dc}{dz} = vc_0 \qquad (9.10.3)$$

at $z = 0$.

Finally, at the exit $z = L$

$$\frac{dc}{dz} = 0. \qquad (9.10.4)$$

These two boundary conditions have been subject to much discussion as the student may see by looking up some of the references given. We shall not attempt to justify them here, but only point out that they are consistent with the limit of $E_a \longrightarrow \infty$. In this limit the diffusion is so dominant that the composition should be uniform and we should have the stirred tank equations. If we integrate Eq. (9.10.2) from $z = 0$ to $z = L$ we have

$$\left[E_a \frac{dc}{dz} - vc \right]_0^L = k \int_0^L c(z)\, dz.$$

Using the boundary conditions Eqs. (9.10.3) and (9.10.4) gives

$$-vc(L) + vc_0 = kL\bar{c},$$

where $\bar{c}$ is the average concentration. But $L/v = V/q = \theta$, the holding time of the reactor, and if the concentration is uniform $c(L) = \bar{c} = c$, so that

$$c_0 - c - \theta k c = 0,$$

which is the stirred tank equation.

Let us make the length variable dimensionless by taking

$$x = \frac{z}{L} \tag{9.10.5}$$

and introduce two dimensionless numbers

$$P_a = \frac{vL}{E_a}, \qquad M = \frac{kL}{v}. \tag{9.10.6}$$

Then

$$\frac{1}{P_a}\frac{d^2c}{dx^2} - \frac{dc}{dx} - Mc = 0, \tag{9.10.7}$$

with conditions

$$-\frac{1}{P_a}\frac{dc}{dx} + c = c_0 \quad \text{at} \quad x = 0. \tag{9.10.8}$$

and

$$\frac{dc}{dx} = 0 \quad \text{at} \quad x = 1. \tag{9.10.9}$$

This is just one of the types of equation treated in Sec. 5.1; it has in fact been solved in Exercise 5.1.2. The solution is a linear combination of exponentials, $\exp m_i x$, $i = 1, 2$, where

$$\left.\begin{matrix} m_1 \\ m_2 \end{matrix}\right\} = \tfrac{1}{2}P_a \pm \{\tfrac{1}{4}P_a^2 + P_a M\}^{1/2}. \tag{9.10.10}$$

Of these roots, m_2 is negative and m_1 positive and the solution may be written as

$$c(x) = c_0 P_a \frac{m_1 e^{-m_2(1-x)} - m_2 e^{-m_1(1-x)}}{m_1^2 e^{-m_2} - m_2^2 e^{-m_1}}. \tag{9.10.11}$$

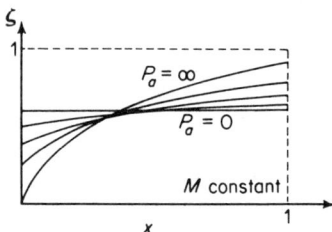

Fig. 9.24 Conversion profiles for various Peclet numbers.

Figure 9.24 shows the solution to the equation for a fixed M and various P_a; in this figure $\zeta = 1 - c(x)/c_0$. This shows the way in which the conversion profile changes from the case of no diffusion (P_a infinite) to the completely mixed reactor ($P_a = 0$).

Sec. 9.10 Axial Dispersion in Tubular Reactors

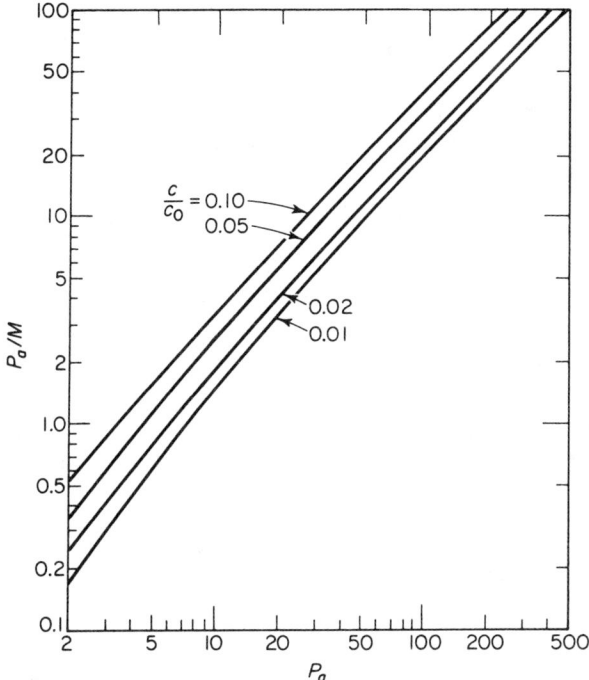

Fig. 9.25 Contours of conversion in the plane of P_a and P_a/M for a first order reaction.

To estimate the effect of longitudinal diffusion, the contours of constant conversion are drawn in the plane of P_a and P_a/M in Fig. 9.25. If all the parameters are known the values of P_a and $P_a/M = v^2/kE_a$ can be calculated and the contour will give the conversion. If a certain conversion is required and all parameters except the length are given, this may be found from the value of P_a.

Exercise 9.10.1. Obtain the exit concentrations from Eq. (9.10.11) for the limiting cases of $P_a \longrightarrow \infty$ and $P_a \longrightarrow 0$, and show that these agree with the usual formulae for the plug flow reactor and mixed vessel.

Exercise 9.10.2. A reactor consists in a bed of $\frac{1}{4}$-in. diameter particles catalyzing the first order decomposition of a gaseous reactant. It is observed that the rate of reaction is 10^{-2} mole/sec cc of reactor volume when the partial pressure of the reactant is 1 atm. If the gas flow rate is 20 ft/sec, estimate the length of reactor required for 99% conversion.

Exercise 9.10.3. If the same catalyst were made up in $\frac{1}{2}$-in. diameter particles, what bed length would be needed for 99% conversion?

REFERENCES

9.1. For a recent survey see:

G. F. Froment, "Fixed-bed Catalytic Reactors—Current Design Status," *Ind. Eng. Chem.*, **59,** 18 (Feb. 1967).

9.2. For the use of secondary flow see:

R. J. Adler and J. A. Koutsky, "Minimization of Axial Dispersion by the Use of Secondary Flow in Helical Tubes," *Can. J. Chem. Eng.*, **42,** 239 (1964).

The straight tube with a homogeneous first order reaction has been discussed by:

R. H. Wilhelm and F. A. Cleland, "Diffusion and Reaction in Viscous-Flow Tubular Reactors," *A.I.Ch.E. Journal*, **1,** 489 (1956).

The higher order reactions are treated in:

J. P. Vignes and P. J. Trambouze, "Diffusion et Réaction Chimique dans un Réacteur Tubulaire en Régime Laminaire," *Chem. Eng. Sci.*, **17,** 73 (1962).

For reactions catalyzed on a tube wall, see:

S. Katz, "Chemical Reactions Catalyzed on a Tube Wall," *Chem. Eng. Sci.*, **10,** 202 (1959).

J. S. Dranoff, "An Eigenvalue Problem Arising in Mass and Heat Transfer Studies," *Math. Comp.*, **15,** 403 (1961).

E. H. Wissler and R. S. Schechter, "Turbulent Flow of Gas Through a Circular Tube with Chemical Reaction at the Wall," *Chem. Eng. Sci.*, **17,** 937 (1962).

R. L. Solomon and J. L. Hudson, "Heterogeneous and Homogeneous Reactions in a Tubular Reactor," *A. I. Ch. E. Journal*, **13,** 545 (1967).

See also;

J. Kjaer, *Measurement and Calculation of Temperature and Conversion in Fixed-bed Catalytic Reactors.* Copenhagen: Gjellerups Forlag, 1958.

H. Hinrichs and J. Niedetzky, "A New Type of Converter for Ammonia Synthesis," *Angew. Chem. (Intern. Ed.)*, **1,** 206 (1962).

See also the article by Beek on packed reactors referred to in the References of Chap. 1 for almost all points touched in this chapter and for a good presentation of the full partial differential equations. All the standard texts have something to say on this topic; for a treatment going beyond what we have been able to accomplish here the reader is referred to Petersen's book in particular.

9.3. For packed bed pressure drop see:

S. Ergun, "Fluid Flow Through Packed Columns," *Chem. Eng. Prog.*, **48**, 89 (1952).

For the empty tube, correlations and a discussion of bends etc. are given in:

J. H. Perry (ed.), *Chemical Engineering Handbook*, 4th ed. New York: McGraw-Hill Book Co., 1963.

The way in which the pressure drop and the transfer properties of the packed bed are related to the behavior of an individual particle is excellently presented in:

W. E. Ranz, "Friction and Transfer Coefficients for Single Particles and Packed Beds," *Chem. Eng. Progr.*, **48**, 247 (1952).

9.5. Uniformity of the packed bed is discussed in:

J. W. Tierney, R. M. Baird, and L. H. S. Roblee, "Radial Porosity Variations in Packed Beds," *A. I. Ch. E. Journal*, **4**, 460 (1958).

The velocity variation has been measured and reported in:

E. J. Cairns and J. M. Prausnitz, "Velocity Profiles in Packed and Fluidized Beds," *Ind. Eng. Chem.*, **51**, 1441 (1959).

Radial dispersion is discussed in the paper by W. E. Ranz mentioned above. See also:

T. Baron, "Generalized Graphical Method for the Design of Fixed Bed Catalytic Reactors," *Chem. Eng. Progr.*, **48**, 118 (1952).

For a valuable summary of much work on radial and longitudinal dispersion, and indeed, for a review of the state of the art of reactor design, see:

R. H. Wilhelm, "Progress Towards the a priori Design of Chemical Reactors," *Pure Appl. Chem.*, **5**, 403 (1962).

The computational problems of the full partial differential equations are discussed by Beek, *loc. cit.* The detailed equations for mass and heat transfer in a cylindrical bed of spheres with reaction represented by linear terms have been solved in:

N. R. Amundson, "Solid-fluid Interactions in Fixed and Moving Beds," *Ind. Eng. Chem.*, **48**, 26 and 35 (1956).

A pseudo-homogeneous cylindrical model with a linearized reaction rate expression is considered by:

G. F. Froment, "Design of Fixed-Bed Catalytic Reactors Based on Effective Transport Models," *Chem. Eng. Sci.*, **17**, 849 (1962).

A model of the fixed bed in terms of interconnected cells is given in:

> L. Lapidus and H. A. Deans, "A Computational Model for Predicting and Correlating the Behavior of Fixed-Bed Reactors," *A.I.Ch.E. Journal*, **6,** 656 and 663 (1960).

The structure of a computer program for fixed bed design with the one-dimensional model is outlined in:

> J. Kjaer, *A General Calculation Method for Fixed Bed Catalytic Reactors on a Digital Computer*. Copenhagen: Akademisk Forlag, 1963.

9.6. The fact that the temperature should always be so chosen to maximize the reaction rate was observed by Denbigh (*loc. cit.*, Chap. 7, References for Sec. 7.7). The case of consecutive reactions was first considered by:

> N. R. Amundson and O. Bilous, "Optimum Temperature Gradients in Tubular Reactors," *Chem. Eng. Sci.*, **5,** 81 and 115 (1956).

See also:

> F. Horn, "Optimale Temperature- und Konzentrationsverlaufe," *Chem. Eng. Sci.*, **14,** 77 (1961).

> F. Horn and U. Troltenier, "Der maximale Umsatz bei einigen einfachen Reaktionssystemen," *Chem. Ing. Tech.*, **33,** 413 (1961).

> R. Jackson and I. Coward, "Optimum Temperature Profiles in Tubular Reactors," *Chem. Eng. Sci.*, **20,** 911 (1965).

The effect of diffusion is discussed by:

> J. Adler and D. Vortmeyer, "The Effect of Axial Diffusion Processes on the Optimum Yield of Tubular Reactors," *Chem. Eng. Sci.*, **18,** 99 (1963).

9.7. See references to Kjaer and the texts for counter- and cocurrent cooled reactors. For a discussion of the ammonia synthesis reactor see:

> R. F. Baddour, P. L. T. Brian, B. A. Longeais, and J. P. Eymery, "Steady-State Simulation of an Ammonia Synthesis Conveyter," *Chem. Eng. Sci.*, **20,** 281 (1965).

An earlier paper of interest is:

> D. Annable, "Application of the Temkin Kinetic Equation to Ammonia Synthesis in Large Scale Reactors, *Chem. Eng. Sci.*, **1,** 145 (1952).

The maximum temperature in the wall cooled reactor is discussed by Barkelew [*Chem. Eng. Prog Symp. Ser.*, **55,** 37 (1959)]. His analysis is repeated in Kramers and Westerterp. However, the use of the positive exponential approximation leaves the range of validity of these results open to some question.

The case of steady heat generation (i.e., a temperature-independent zeroth order reaction) where the maximum temperature can exceed the adiabatic temperature rise is treated by:

E. A. Grens, II and R. A. McKean, "Temperature Maxima in Countercurrent Heat Exchangers with Internal Heat Generation," *Chem. Eng. Sci.*, **18**, 291 (1963).

9.8. Chapter 4 in H. Kramers and K. R. Westerterp's *Elements of Chemical Reactor Design and Operation* (New York: Academic Press Inc., 1963) gives a good treatment of this subject. The notion of parametric sensitivity was first developed by:

N. R. Amundson and O. Bilous, "Chemical Reactor Stability and Sensitivity. II. Effect of Parameters on Sensitivity of Empty Tubular Reactors," *A.I.Ch.E. Journal*, **2**, 116 (1956).

A different approach to the subject is given by:

J. Coste, N. R. Amundson, and R. Aris, "Tubular Reactor Sensitivity," *A.I.Ch.E. Journal*, **7**, 124 (1961).

Autothermicity was introduced in:

C. van Heerden, "Autothermic Processes. Properties and Reactor Design," *Ind. Eng. Chem.*, **45**, 1242 (1953).

Some more extended results are in:

C. van Heerden, "The Character of the Stationary State of Exothermic Processes," *Chem. Eng. Sci.*, **14**, 1 (1958).

An approximate method of allowing for radial diffusion is given in:

H. E. Hoelscher, "Temperature Stability of Fixed-Bed Catalytic Converters," *Chem. Eng. Sci.*, **6**, 183 (1957).

The stability of the packed bed with reference to the fact that each particle may have multiple steady states has been considered in a series of papers by N. R. Amundson and Shean-Lin Liu.

"Stability of Adiabatic Packed Bed Reactors. An Elementary Treatment," *Ind. Eng. Chem. Fundamentals*, **1**, 200 (1962).

"Stability of Nonadiabatic Packed Bed Reactors, An Elementary Treatment, *ibid.*, **2**, 12 (1963).

"Stability of Adiabatic Packed Bed Reactors. Effect of Axial Mixing," *ibid.*, **2**, 183 (1963).

9.9. See references under 9.1 for the analysis of velocity profiles and diffusion. The entrance effect of a developing laminar profile has been considered by:

D. L. Ulrichson and R. A. Schmitz, "Chemical Reaction in the Entrance

Length of a Tubular Reactor," *Ind. Eng. Chem. Fundamentals*, **4**, 2 (1965).

9.10. The literature on axial dispersion is vast, and the following is only a sampling of the relevant papers. Levenspiel's text is particularly good on this subject.

Careful determinations of the axial Peclet number using a frequency response technique will be found in:

K. W. McHenry and R. H. Wilhelm, "Axial Mixing of Binary Gas Mixtures Flowing in a Random Bed of Spheres," *A.I.Ch.E. Journal*, **3**, 83 (1957).

A later study which brings together the results of several investigators and presents some later results is:

E. J. Cairns and J. M. Prausnitz, "Longitudinal Mixing in Packed Beds," *Chem. Eng. Sci.*, **12**, 20 (1960).

A collection of references and a review of the subject is to be found in:

H. Hofmann, "Der derzeitige Stand bei der Vorausberechnung der Verweilzeitverteilung in technischen Reaktoren," *Chem. Eng. Sci.*, **14**, 193 (1961).

A model explaining the axial Peclet number in terms of the hydrodynamics of the bed is given in:

N. R. Amundson and R. Aris, "Some Remarks on Longitudinal Mixing or Diffusion in Fixed Beds," *A.I.Ch.E. Journal*, **3**, 280 (1957).

Various models of dispersion are related to the diffusion model in:

K. B. Bischoff and O. Levenspiel, "Fluid Dispersion—Generalization and Comparison of Mathematical Models," *Chem. Eng. Sci.*, **17**, 245 (1962).

The classic paper on the influence of diffusion is:

G. Damköhler, "Einflusse der Strömung, Diffusion und des Warmeüberganges auf die Leistung von Reaktionsöfen," *Z. Elektrochem.*, **42**, 846 (1936).

For the effective longitudinal dispersion in an empty tube see:

G. I. Taylor, "Dispersion of Soluble Matter in Solvent Flowing Slowly Through a Tube," *Proc. Roy. Soc. (London), Ser. A*: **219**, 186 (1953). "The Dispersion of Matter in Turbulent Flow Through a Pipe," *ibid.*, **220**, 440 (1954).

R. Aris, "On the Dispersion of a Solute in a Fluid Flowing Through a Tube," *ibid.*, **235**, 67 (1956).

L. J. Tichacek, C. H. Barkelew, and T. Baron, "Axial Mixing in Pipes," *A.I.Ch.E. Journal*, **3**, 439 (1957).

R. D. Hawthorn, "Effect of Radial Temperature Variation of Axial Mixing in Pipes," *A.I.Ch.E. Journal*, **6**, 443 (1960).

The correct boundary condition was first given by Danckwerts in:

P. V. Danckwerts, "Continuous Flow Systems. Distribution of Residence Times," *Chem. Eng. Sci.*, **2**, 1 (1953).

It has been extensively discussed. See, for example:

J. F. Wehner and R. H. Wilhelm, "Boundary Conditions of a Flow Reactor," *Chem. Eng. Sci.*, **6**, 89 (1956).

J. R. A. Pearson, "A Note on the 'Danckwerts' Conditions for Continuous Flow Reactors," *Chem. Eng. Sci.*, **10**, 281 (1959).

K. B. Bischoff, "A Note on Boundary Conditions for Flow Reactors," *Chem. Eng. Sci.*, **16**, 731 (1961).

L. T. Fan and Y. K. Ahn, "Critical Evaluation of Boundary Conditions for Tubular Flow Reactors," *Ind. Eng. Chem. Process Design and Develop.*, **1**, 190 (1962).

The stirred tank sequence was used as a computational model by:

J. Coste, D. Rudd, and N. R. Amundson, "Taylor Diffusion in Tubular Reactors," *Can. J. Chem. Eng.*, **39**, 149 (1961).

Solutions for several orders of reaction are presented in:

L. T. Fan and T. C. Bailie, "Axial Diffusion in Isothermal Tubular Flow Reactors," *Chem Eng. Sci.*, **13**, 63 (1960).

Results for heterogeneous kinetics are given by:

H. Hofmann and H. J. Astheimer, "Der Einfluss der Vermischung bei heterogenen Reaktionen in kontinuierlich-betriebenen isotherm arbeitenden Rohrreaktoren," *Chem. Eng. Sci.*, **18**, 643 (1963).

A paper applying these concepts to a tubular reactor with sharp bends is:

K. B. Bischoff, "An Example of the Use of Combined Models: Mixing in a Tubular Reactor with Return Bends," *A.I.Ch.E. Journal*, **10**, 584 (1964).

See also:

J. M. Douglas and K. B. Bischoff, "Variable Density Effects and Axial Dispersion in Chemical Reactors," *Ind. Eng. Chem. Process Design and Develop.*, **3**, 130 (1964).

Radial as well as axial dispersion has been considered by using the cell model by Deans and Lapidus (see above).

NOTATION

A, A'	pre-exponential factors
A_j	jth chemical species
A_r	cross-sectional area of the reactor
a	tube radius
a_c	perimeter of heat transfer area in any cross section
C_p	mean heat capacity per unit volume
$\hat{C}_p$	mean heat capacity per unit mass
$\hat{C}_{pc}$	specific heat of coolant
c_j	concentration of A_j
D	diffusivity
d	tube diameter
d_p	particle diameter
E, E'	activation energies
E_a	axial eddy diffusivity
E_r	radial eddy diffusivity
f	friction factor
G	mass flow rate of reactants per unit area of reactor
G_c	mass flow rate of coolant per unit area of reactor
g_c	gravitational constant
g_j	mass fraction of A_j
h_j	partial molar enthalpy of A_j
J, J''	$(-\Delta H)/C_P$, $(-\Delta H)/\hat{C}_P$
k, k'	rate constants
k	thermal conductivity of the reactants in Prandtl number (Sec. 9.7 only)
L	length of reactor
L°	minimum length of reactor
M	kL/v
m_j	molar weight of A_j
$\bar{m}$	mean molar weight
$\hat{N}$	SL/P
$P(z)$	pressure
P_j	partial pressure of A_j
Pe_a, Pe_r	axial and radial Peclet numbers
P_a	Lv/E_a
Pr	Prandtl number
p	hydraulic radius, A_r/a_c, $p = \dfrac{E}{(-\Delta H)}$ in Sec 9.6 only
$p(t)$	residence time probability density
Q^*	heat transferred per unit volume of reactor

Notation

Q, Q''	$Q^*/\bar{C}_P, Q^*/\hat{C}_P$
R	gas constant
Re_p	Reynolds number $Gd_p/\mu = vd_p/\nu$
r	radial distance from tube axis (Sec. 9.9)
$r(\xi, T), r(\xi'', T)$	reaction rate
$r_m(\xi)$	maximum reaction rate
S	Stanton number, U/vC_P
S_0, S_w	overall and wall Stanton numbers
T	temperature
T_0	inlet temperature
T_f	feed temperature
T_c	coolant temperaure
$T°$	optimal temperature
$T_m(\xi)$	temperature for maximum reaction rate
T^*, T_*	upper and lower bounds on temperature
t	z/v
U	heat transfer coefficient
v	linear velocity based on total cross-sectional area of reactor
$v(r)$	linear velocity at radial distance r
X	fractional conversion
X_e	fractional conversion at equilibrium at given temperature
$X_n(\kappa_n)$	ratio of extent with velocity profile to extent with plug flow
x	z/L (Sec. 9.10)
Z	compressibility factor
z	distance from inlet
α_j	stoichiometric coefficient of A_j
$\bar{\alpha}$	$\sum \alpha_j$
γ	$GC_p/G_c\hat{C}_{pc}$
ΔH	heat of reaction
ΔT	$T_0 - T_f$
ϵ	voidage of bed
θ	holding time
$\theta(\rho)$	$L/V(\rho)$ holding time at radius $a\rho$
κ	$A'A^{-(E'/E)}$
κ_1	$k\theta$
κ_2	$kc_0\theta$
λ	dimensionless reactor length (Secs. 9.2 and 9.6)
μ	viscosity of reaction mixture
ν	kinematic viscosity, μ/ρ

The Batch Reactor 10

We do not need to discuss the batch reactor in quite the detail that we have used for the other types, for most of its features have been covered in passing. This chapter will therefore serve to correlate the relevant sections and interpret them in terms suitable to a batch reaction. Broadly speaking, the course of a batch reaction in time corresponds to the course of the reaction in space in the tubular reactor. With the swing to continuous processes, the batch reactor has fallen into a certain degree of neglect, though Walas points out in his text that it should not be ruled out of a process design without a thorough economic evaluation.

We shall recapitulate the governing equations in the next section and discuss the economic operation in the one following. The results on optimal control are essentially a reinterpretation of the optimal design for the tubular reactor. We shall not attempt a full derivation but hope that the qualitative description will be sufficiently convincing. The isothermal operation of a batch reactor is completely covered by the discussion in Chap. 5 of the integration of the rate equations at constant temperature. The simplest form of nonisothermal operation occurs when the reactor is insulated and the reaction follows an adiabatic path; the behavior of the reactor is then entirely similar to that discussed in Chap. 8.

The term *semibatch reactor* has sometimes been used for the continual operation of a reactor in the transient state. Such would be the case if a stirred reactor were started up and shut down on a repeated cycle, with a significant proportion of the production contributed from periods during

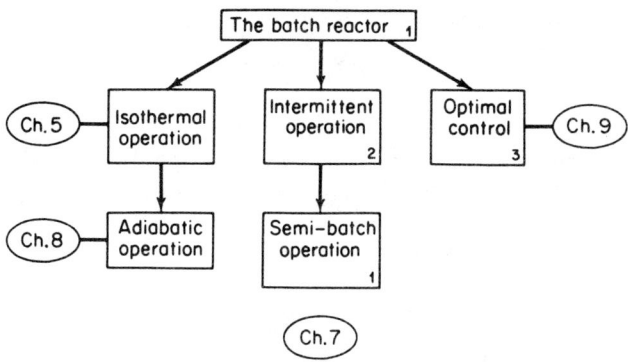

Fig. 10.1 The structure of the chapter.

which it was nowhere near a steady state. In another form, it might consist in partially charging a batch reactor and controlling the course of reaction by the addition of the remaining part of the charge. Figure 10.1 shows the structure of the chapter.

10.1 The Equations for the Batch Reactor

A batch reactor generally consists of a closed vessel provided with a means of stirring and with temperature control. It may be held at constant pressure or it can be entirely enclosed at a constant volume. If the R simultaneous reactions $\sum_{j=1}^{S} \alpha_{ij} A_j = 0$ are taking place and the volume is constant, we may work in concentrations and extents per unit volume and have, as in Sec. 2.9,

$$c_j = c_{j0} + \sum_{i=1}^{R} \alpha_{ij} \xi_i, \tag{10.1.1}$$

$$\frac{dc_j}{dt} = \sum_{i=1}^{R} \alpha_{ij} r_i(c_1, \ldots, c_S, T, P), \tag{10.1.2}$$

$$\frac{d\xi_i}{dt} = r_i(\xi_1, \ldots, \xi_R, T, P). \tag{10.1.3}$$

If the reaction takes place at constant pressure or if the volume is changed in some prescribed manner it is often preferable to work in terms of the total number of moles present or the gross extent of reaction, X. It will be recalled that the definition of the intensive rate of reaction $\sum \alpha_j A_j = 0$ is

$$r = \frac{1}{\alpha_j V} \frac{dN_j}{dt} = \frac{1}{V} \frac{dX}{dt},$$

and hence

$$\frac{dX}{dt} = Vr. \tag{10.1.4}$$

Now if r is expressed in terms of the concentrations c_j we must substitute

$$c_j = \frac{N_{j0} + \alpha_j X}{V}$$

and take account of the variation of V. In the case of a first order reaction, the volume would cancel out on the right-hand side, but for other orders some power of V will remain. If the reactants are ideal gases, the volume is

$$\begin{aligned} V &= \frac{RT}{P} \sum N_j = \frac{RT}{P} \sum (N_{j0} + \alpha_j X), \\ &= V_0 \frac{TP_0}{PT_0} \left\{ 1 + \bar{\alpha} \frac{X}{N_0} \right\}, \end{aligned} \tag{10.1.5}$$

where $N_0 = \sum N_{j0}$ and $\bar{\alpha} = \sum \alpha_j$.

In semibatch operation it may also be preferable to work with the total moles present instead of concentration, for the volume will be changing. In this case, we have

$$N_j = N_{j0} + \sum_{i=1}^{R} \alpha_{ij} X_i \tag{10.1.6}$$

for the change in number of moles by reaction only. The rate of change of the gross extent X_i is $R_i^* = V r_i$, but the amount of A_j present is also changing by the rate of addition of A_j, say n_j moles per unit time. Thus

$$\frac{dN_j}{dt} = V \sum_{i=1}^{R} \alpha_{ij} r_i + n_j, \tag{10.1.7}$$

$$V = \sum_{j=1}^{S} v_j N_j, \tag{10.1.8}$$

where v_j is the molar volume of A_j.

If the reaction is at constant pressure, we may take over the enthalpy balance we have commonly used and write

$$V C_p \frac{dT}{dt} = V \sum_{i=1}^{R} (-\Delta H)_i r_i - Q, \tag{10.1.9}$$

where Q is the rate of heat removal from the reactor. For liquid reactions, constant pressure and constant volume normally go together, but for gas reactions, the equations for constant volume should be formulated slightly differently, though in practice the difference may be small. We would express here the first law of thermodynamics as an internal energy balance and obtain by a similar argument (see Sec. 3.3)

$$V C_v \frac{dT}{dt} = V \sum_{i=1}^{R} (-\Delta U)_i r_i - Q, \tag{10.1.10}$$

where C_v = heat capacity per unit volume at constant volume, and

$$\Delta U = \Delta H - \left[P + \left(\frac{\partial U}{\partial V} \right)_T \right] \sum \alpha_j v_j, \text{ internal energy of reaction.}$$

In semibatch operation Eq. (10.1.7) must be modified to include the enthalpy brought in by the addition of a fresh reaction mixture, a term which can often be absorbed into Q. Q itself will be governed by the same sort of equations as were given for the stirred tank. Thus, for a coolant jacket of constant temperature T_{co}, we would have $Q = h(T - T_{co})$ and so on. This, of course, assumes that the rate of heat removal in the transient state would be instantaneously the same as in a steady state, which is not true. Here again, a more exact analysis with allowance for transient heat convection and conduction would be out of proportion to the problem. At best a lumped constant model with some time constants might be constructed for control purposes.

The first experiments on a process are commonly done in laboratory-scale batch reactors and the following illustrations are intended to show how the information gained there can be used for the design of a larger reactor.

Consider first a nitration. A constant volume batch reactor is used for the liquid phase nitration of benzenesulfonic acid (abbreviated to BSA; molecular weight 158) by nitric acid (molecular weight 63). In the presence of a strong dehydrating agent the reaction is second order (i.e., first order with respect to each reactant), virtually irreversible, and produces the ortho, meta, and para forms of nitrobenzenesulfonic acid (NBSA, molecular weight 203). In the batch experiment, the initial charge was in molar proportions of 2 moles of BSA to 1 mole of nitric acid and after 40 min, a third of the BSA had been consumed producing NBSA in the ratios: 21% ortho, 72% meta, and 7% para. On the basis of this data, we would like to consider using a stirred tank reactor to produce 146 lb/hr of meta-NBSA from a feed of 316 lb/hr of BSA and 126 lb/hr of nitric acid. The feed will be such that the nitric acid concentration will be the same as its initial concentration in the batch.

Here it is easiest to work in the concentration of nitric acid for the reaction is BSA + HNO$_3$ ⟶ NBSA + H$_2$O, and if n_0 denotes the initial concentration of nitric acid in the batch then

$2n_0 =$ initial concentration of BSA,

$n_0 + n =$ concentration of BSA, when the concentration of the remaining nitric acid is n.

The three forms of the product are produced by parallel reaction, so that, if we denote by k the total rate constant, the equation for the nitric acid concentration is

$$\frac{dn}{dt} = -kn(n_0 + n).$$

This can be written as

$$\frac{dy}{dt} = -(kn_0)y(1+y)$$

by taking $n/n_0 = y$. Now when a third of the BSA is gone its concentration is $\frac{2}{3}(2n_0)$ and $n = \frac{1}{3}n_0$. This happens in 40 min ($= \frac{2}{3}$ hr) so that taking kn_0 to have units of hr^{-1} we have

$$\frac{2}{3}kn_0 = \int_{1/3}^{1} \frac{dy}{y(1+y)} = \left[\ln \frac{y}{1+y}\right]_{1/3}^{1} = \ln 2.$$

Hence

$$kn_0 = 1.042 \text{ hr}^{-1}.$$

The proposed feed to the stirred tank has the same concentration of nitric acid, n_0, but from the mass flow rates given it is evident that the molar proportions are changed. In fact, the feed of both BSA and nitric acid is 2 lb-mole/hr and hence their feed concentrations will both be n_0 and their concentrations in the tank will be the same. Denoting the latter by n, a balance on the nitric acid gives

$$n_0 - n - \theta k n^2 = 0$$

or

$$1 - y - \theta(kn_0)y^2 = 0.$$

In the absence of other information it is assumed that the three forms of NBSA are produced in the same ratio so that 146 lb/hr of meta-NBSA corresponds to a total production of $146/0.72 = 203$ lb/hr of all three forms. This is just 1 lb-mole/hr so that a 50% conversion is needed, $y = 0.5$. Then the required holding time is

$$\theta = \frac{1-y}{y^2(kn_0)} = \frac{0.5}{0.25 \times 1.042} = 1.92 \text{ hr}.$$

Without further information on the amount of dehydrating agent used we cannot find the actual volume of the reactor.

As a second example, consider a gas phase reaction $A \longrightarrow 2B + 3C$ carried out in a constant pressure batch reactor at 800°F and 6 atm pressure. A conversion of 70% of A to its products was obtained after an hour. The pilot plant is to be an ideal plug flow reactor and is required to handle 500 cu ft/hr of A at 800°F and a constant pressure of 20 atm. What volume of reactor is needed to give 95% conversion?

We make the assumption that the reaction is first order and irreversible, so that if N is the number of moles of A in the batch reactor at time t and V is its volume,

$$\frac{dN}{dt} = -Vk\frac{N}{V} = -kN.$$

Taking the units of k to be hr^{-1}, we are told that $N = 0.3N_0$ when $t = 1$. Since the solution to the equation is

$$N = N_0 e^{-kt},$$

this gives

$$k = -\ln 0.3 = 1.205.$$

Let A_r be the cross-sectional area of the reactor and v_0 the inlet linear velocity, so that $v_0 A_r = 500$. Let c_0 denote the inlet concentration at a distance z from the inlet. Since the pressure and temperature are constant, the total number of moles per unit volume is constant through the reactor and equal to the feed concentration c_0. Since 5 moles of product are formed for every mole of feed used, $\bar{\alpha} = 4$ and

$$\frac{c}{c_0} = \frac{1-x}{1+4x},$$

where x is the extent in dimensionless (mole fraction) form. Also the mass flow rate is constant so that the linear velocity at any point z is given by

$$\frac{v}{v_0} = 1 + 4x.$$

Then from the equation

$$\frac{d}{dz}(vc) = -kc$$

we have, on substituting,

$$\frac{d}{dz}\left[v_0 c_0 (1-x)\right] = -kc_0 \frac{1-x}{1+4x}$$

or

$$v_0 \frac{dx}{dz} = k \frac{1-x}{1+4x}.$$

Since 95% of the feed is to be converted we require $vc = 0.05 v_0 c_0$ at the exit or $x = 0.95$ there. If L is the length of the reactor, $A_r L$ is its volume and

$$A_r L = \frac{A_r v_0}{k} \int_0^{0.95} \frac{1+4x}{1-x} dx = \frac{500}{1.205}\left(5 \ln \frac{1}{1-0.95} - 4 \times 0.95\right),$$
$$= 4650 \text{ cu ft.}$$

Exercise 10.1.1. If the reaction in the second example were second order and irreversible what volume of tubular reactor would be required?

10.2 Intermittent Operation

Since the batch reactor is always working in a transient condition, it is important to ask how it should be scheduled; for if left to itself indefinitely, the reaction will approach equilibrium and remain there. Moreover, this scheduling is not merely a function of the reaction itself but also of what has to be done between reaction periods in the way of removing the product and recharging the reactor. If w_j is the value or cost of a mole of A_j, the value of the contents of the reactor at any time is $\sum_{j=1}^{S} w_j N_j(t) = V \sum_{j=1}^{S} w_j c_j(t)$. Thus, the net increase in value of the batch from the beginning of the reaction is

$$\begin{aligned} W(t) &= V \sum_{j=1}^{S} w_j [c_j(t) - c_{j0}], \\ &= V \sum_{j=1}^{S} \sum_{i=1}^{R} \alpha_{ij} \xi_i w_j, \\ &= V \sum_{i=1}^{R} (\Delta W)_i \xi_i, \end{aligned} \quad (10.2.1)$$

where

$$(\Delta W)_i = \sum_{j=1}^{S} \alpha_{ij} w_j \quad (10.2.2)$$

is the rate of change of W with the extent of the ith reaction.

The complete schedule of the reactor operation can be divided into four types of periods with an associated duration θ and cost per unit time c. They are:

(i) Preparation time, during which the reactor is made ready and charged (θ_P, c_P);

(ii) Reaction time, during which reaction proceeds (θ_R, c_R);

(iii) Recovery time, during which the product is discharged (θ_Q, c_Q);

(iv) Idle time, during which the reactor is not in use at all (θ_0, c_0).

If these four steps follow one another in sequence, the total cost of one cycle of operation is

$$C_T = \theta_0 c_0 + \theta_P c_P + \theta_Q c_Q + \theta_R c_R \quad (10.2.3)$$

and the value of the product is $W(\theta_R)$. The net profit per unit time is

$$\frac{W(\theta_R) - C_T}{\theta_T} \quad (10.2.4)$$

where the total time

$$\theta_T = \theta_0 + \theta_P + \theta_Q + \theta_R. \quad (10.2.5)$$

If all the costs are fixed and all the durations except θ_R are fixed, then the profit per unit time will be greatest if θ_R satisfies

$$\theta_T \frac{dW}{d\theta_R} = \theta_T V \sum (\Delta W)_i r_i = W + c_R \theta_T - C_T. \quad (10.2.6)$$

Sec. 10.2   Intermittent Operation

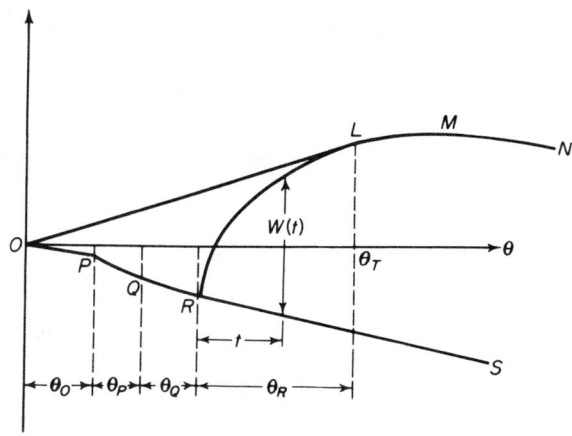

Fig. 10.2 Optimal reaction time.

The graphical solution of this equation is shown in Fig. 10.2. The parts $\theta_0 c_0$, $\theta_P c_P$, and $\theta_Q c_Q$ of the total cost are first laid off as shown by segments OP, PQ, and QR; then a line RS of slope $-c_R$ is drawn from point R. This is used as a base line from which curve $RLMN$ is drawn so that its height above RS is $W(t)$ at a distance t ahead of point R. The profit per unit time is greatest if θ_R is chosen so that θ_T is the abscissa of L, the point of contact of the tangent drawn through the origin. If the total profit rather than the profit per unit time is to be maximized, the highest point M would be chosen to determine θ_R. It is important to observe that $W(t)$ embodies the condition used for reactor control. In the isothermal case, $W(t)$ is calculated from the solution of the isothermal rate equations, and we shall see in the next section how $W(t)$ can be increased by proper temperature control. In the absence of kinetic equations, $W(t)$ might be calculated from empirical data.

Exercise 10.2.1. Show how Fig. 10.2 can be adapted to give the greatest value of θ_0, the idle time, that can be tolerated without absorbing all the profit of the reaction.

Exercise 10.2.2. A reaction between two reactants A and B in stoichiometric proportions is at present carried out in solution, batchwise to 90% conversion. A rate law of the form

$$-\frac{da}{dt} = \frac{k_1 ab}{1 + k_2 b}$$

may be assumed; it is known that if the present reaction time is halved, the conversion achieved is 73%. In order to step up production, it is proposed to cut reaction time to 65% of its present value. The interval necessary for emptying, cleaning, and refilling between batches will remain the same and is equal to the present reaction time. Show that a production increase of about 9 % may be expected. Is this the best that can be done? (C.U.)

Exercise 10.2.3. A compound B is prepared from A by a batch process. The reaction is reversible and first order in both directions. At 100°C the conversion of A at different times is as follows:

% of equilibrium conversion	10	30	50	70
time, min		42	145	280	482.

It takes 20 min to empty and refill the reactor. What percentage conversion (expressed as above) will lead to the maximum output per day? What further information is necessary in order to predict the effect upon output of an increased reaction temperature? (C.U.)

Exercise 10.2.4. Draw a careful figure for the extent of an exothermic reaction as a function of isothermal batch time for various temperatures. Then show how to find the optimal batch time and temperature in the kind of batch operations considered in this section.

10.3 Optimal Control

In the preceding chapter, we saw that for a reversible, exothermic reaction there was an optimal temperature profile along the tubular reactor which would achieve the greatest extent of reaction in a given time. Unfortunately, it is scarcely possible to control the temperature at every point along the length of a tubular reactor, but the same results carry over for the batch reactor, and here it is much more reasonable to assume a continuously varying temperature control. In particular, let us assume that we have a single exothermic reaction for which the equations would be

$$\frac{d\xi}{dt} = r(\xi, T), \quad \xi(0) = 0 \tag{10.3.1}$$

$$\frac{dT}{dt} = Jr(\xi, T) - q(t), \quad T(0) = T_0, \tag{10.3.2}$$

where $J = -\Delta H/C_p$ and $q(t) = Q/VC_p$ is the rate at which the temperature can be changed by the cooling system at time t. If we decide from a study of the first equation that a certain relation $T°(\xi)$ would be optimal for the course of the reaction, then the required heat removal program would be given by

$$q = \left(J - \frac{dT°}{d\xi}\right) r[\xi, T°(\xi)]; \tag{10.3.3}$$

this expresses the rate at which heat would have to be removed when the

Sec. 10.3 Optimal Control

extent of reaction is ξ. It could then be found as a function of t by solving

$$\frac{d\xi}{dt} = r[\xi, T^\circ(\xi)]. \tag{10.3.4}$$

If we set as our objective the maximization of the profit during the reaction $W(\theta_R)$ we wish to maximize the extent $\xi(t)$ achieved at time $t = \theta_R$. Now

$$\xi(\theta_R) = \int_0^{\theta_R} r[\xi, T(\xi)] \, dt, \tag{10.3.5}$$

and this is to be maximized by choosing $T(\xi)$ optimally. Since T may be freely chosen for each ξ, it is almost obvious that it should be chosen so as to maximize the integrand at all values of ξ—almost obvious, because we have seen that it is dangerous to generalize in the case of more than one reaction. This is a consequence either of considering the batch reactor as the limit of an infinite number of infinitesimal stirred tanks of total holding time θ_R or of adapting the proof given in Sec. 9.6 for the tubular reactor. Indeed, we know that if T is not restricted, then $T^\circ(\xi)$ lies on the curve Γ_m in the ξ, T plane (see Fig. 10.3), and $r[\xi, T^\circ(\xi)] = r_m(\xi)$. For small ξ, $T^\circ(\xi)$, $r_m(\xi)$ and $-dT^\circ/d\xi$ all become very large, so that one or another of the obvious physical bounds will be violated and it will be impossible to stay on Γ_m for small ξ. Let us suppose that T^*, the upper bound in temperature, is sufficiently high to impose no restriction, but that the value of q exceeds the maximum heat cooling capacity, q^*, at a point L on Γ_m. This implies that it is possible to stay on the optimal path above L but not below it, and

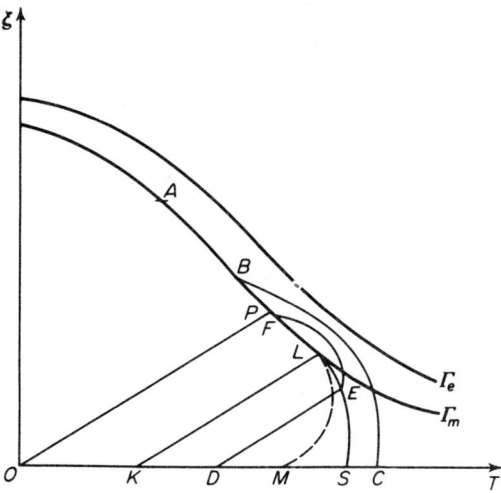

Fig. 10.3 Optimal control paths in the ξ, T plane.

we have to show what the optimal policy is for smaller values of the extent. If A is the point at which

$$r_m(\xi) = \frac{c_R}{(\Delta W)V}, \qquad (10.3.6)$$

it is clear that we shall never want to go beyond this point (it corresponds to M in Fig. 10.2), for then $W(\theta_R) - c_R\theta_R$ would begin to decrease.

A full analysis of this question requires some rather advanced control theory, but we may describe the results in terms that make good intuitive sense. Suppose that the reactor is not equipped with a heater but can only operate between the adiabatic, $q(t) = 0$, and the full cooling, $q(t) = q^*$, regimes. (Strictly speaking, we should make q^* a function of T but the results for this more realistic assumption are entirely similar.) Since the tangent to Γ_m at L has a slope corresponding to $q = q^*$, there is a solution of Eqs. (10.3.1) and (10.3.2) with $q = q^*$ that is just tangent to Γ_m at L. This starts at a temperature T_0 corresponding to M, and it might be thought that the fastest reaction path would be MLA. This indeed was reported erroneously by Aris and Blakemore and needs to be corrected by the more careful analysis of Aris and Siebenthal.

There is, in fact, a curve ALS in the ξ, T plane known as the *switching curve*. For a reactor state represented by a point (ξ, T) to the left of this, the adiabatic reaction is optimal, while for a state to the right, full cooling should be used. Thus, starting from a temperature T_0 corresponding to point O, the reaction is carried out adiabatically until point P on the switching curve is reached. It is then possible to stay on the optimal path PA since the rate of cooling required to do so is less than q^*. By contrast, a high initial temperature T_0, corresponding to point C, requires full cooling to be carried out immediately, and the reactor follows a path CB to the switching curve where, again, it is possible to continue on it to A. From an intermediate initial temperature, point D, the adiabatic path meets the switching surface at E, where the rate of cooling required to remain on it would exceed q^*. Hence a switch is made from adiabatic to full cooling operation, and the state of the reaction follows EF. When it meets the switching curve again at F it is now possible to remain on FA with a physically realizable cooling rate $q < q^*$. The only optimal approach to the critical point L is by the adiabatic path KL. The time course of heat removal rate is shown in Fig. 10.4, where the curves have been shifted in time so that the final parts on the optimal path Γ_m are coincident.

It is at once evident that though this optimal control condition satisfied the requirement $0 \leq q \leq q^*$, it calls for immediate changes of q or infinite values of dq/dt. This is physically unreasonable, and we should impose a further restriction

$$-p_* \leq \frac{dq}{dt} \leq p^*, \qquad (10.3.7)$$

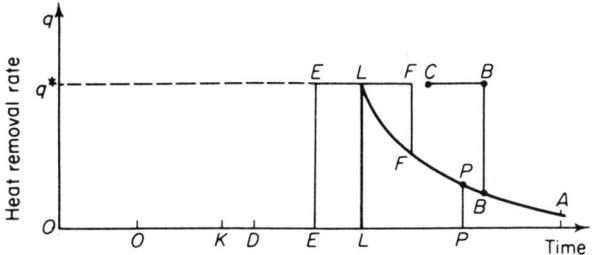

Fig. 10.4 Optimal heat removal programs.

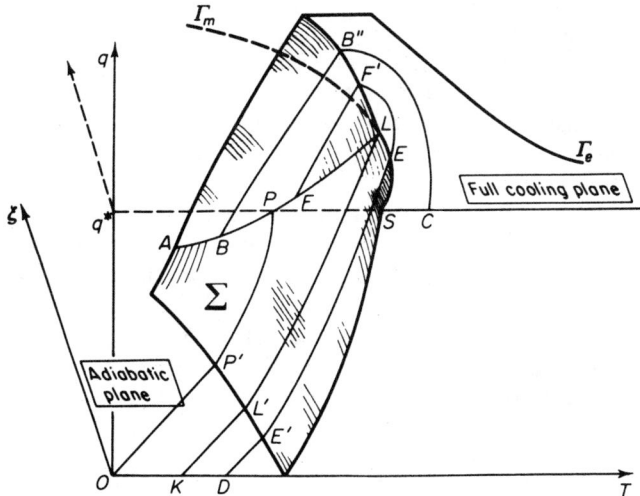

Fig. 10.5 Optimal control policy in ξ, T, q space.

where p_* and p^* are two positive numbers. It turns out that in this case the attainable part of Γ_m is still the optimal path but that the switching curve must be anticipated a little, so that it may be reached by increasing or decreasing q as fast as possible. The switching curve thus becomes a switching surface in [3] (ξ, T, q) space as shown in Fig. 10.5. To the left of the switching surface the cooling rate is decreased as rapidly as possible until the adiabatic plane $q = 0$ or the switching surface Σ is reached. To its right, the rate of cooling is increased as rapidly as possible until Σ or the full cooling plane $q = q^*$ is attained. Similarly, transitions like FF in Fig. 10.4 are no longer sudden but are anticipated and carried out as fast as possible like $F'F$ in Fig. 10.5. In this last figure, all paths corresponding to those in Fig. 10.3 are shown.

Many other variants of this type of control can be found in which the rate of cooling is governed by a coolant flow rate or by the addition of more reactant. These are too difficult to discuss here but may be found in the references given. Another interesting form of control suggests itself for the batch reactor. Suppose the equilibrium properties of the reaction $\sum \alpha_j A_j$ are known but not the kinetic constants. In this case we know Γ_e but not Γ_m. However, if we assume that the reaction begins at a low enough temperature we know that it should start adiabatically. We then ask whether it is possible to take measurements during the adiabatic course which could be analyzed to provide the missing data, and so to determine the subsequent optimal control. Studies of this adaptive scheme show that it is fraught with many practical difficulties, but simpler alternative schemes show that more realistic measurement errors can be tolerated.

The ideas given in the brief discussion of the optimal temperature profile for multiple reaction in a tubular reactor (Sec. 9.6) can also be applied to the programming of the temperature for a batch reactor. Thus for the consecutive reactions $A \longrightarrow B \longrightarrow C$, in which C is a waste product, we would expect a decreasing temperature to be beneficial if the activation energy of the second reaction were greater than that of the first. Such a program of temperature change could be calculated if the kinetics were thoroughly known, and Fig. 9.8 can be interpreted as such a program. Katz and Millman have recently made an interesting suggestion for a simple form of nearly optimal control. Since we know that the optimal temperature decreases as the reaction goes on and B is formed, we might try adjusting the temperature so that the rate constant of the first reaction, $k(T)$, decreases as b, the concentration of B, increases. In particular, they consider a linear relation

$$k(T) = k_0 - k_1 b$$

and determine the best values of k_0 and k_1. If a further term in the derivative of b is added, the approximation to the true optimum is quite remarkable.

REFERENCES

10.3. One of the earliest considerations of optimal temperature is:

K. G. Denbigh, "Velocity and Yield in Continuous Reaction Systems," *Trans. Faraday Soc.*, **40,** 352 (1944).

See also:

F. Horn, "Optimale Temperatur- und Konzentrationsverlaufe," *Chem. Eng. Sci.*, **14,** 77 (1961).

S. Katz, "Best Temperature Profiles in Plug Flow Reactors: Methods of the Calculus of Variations," *Annals N. Y. Acad. Sci.*, **84,** 441 (1960).

Optimal batch reactor control is discussed in:

N. Blakemore and R. Aris, "The Bang-bang Control of a Batch Reactor," *Chem. Eng. Sci.*, **17**, 591 (1962).

C. D. Siebenthal and R. Aris, "The Application of Pontryagin's Methods to the Control of Tubular and Batch Reactors," *Chem. Eng. Sci.*, **19**, 729, 747 (1964).

Simple ways of approximating optimal control are given in:

M. C. Millman and S. Katz, "Linear Temperature Control in Batch Reactors," *Ind. Eng. Chem. Process Design and Development*, **6**, 447 (1967).

NOTATION

A_j	jth chemical species
C_p	heat capacity per unit volume at constant pressure
C_v	heat capacity per unit volume at constant volume
C_T	total cost
c_j	concentration of A_j
c_0	cost per unit time when reactor is idle
c_P	cost per unit time when reactor is being prepared
c_Q	cost per unit time when reactor is being discharged
c_R	cost per unit time when reaction is going on
J	$-\Delta H/C_p$
N_j	number of moles of A_j present
N_{j0}	number of moles of A_j charged to reactor
n_j	rate of addition of A_j
P	pressure
Q	rate of heat removal
q	Q/VC_p
q^*	upper bound on q
R_i^*	gross rate of ith reaction
$r(\xi, T)$	rate of single reaction
r_i	R_i^*/V, rate of ith reaction
$r_m(\xi)$	maximum rate of reaction at extent
T	temperature
T_0	initial temperature
$T^\circ(\xi)$	optimal temperature
t	time from beginning of reaction
V	volume of reaction mixture
v_j	molar volume of A_j
W	increase in value of reaction mixture; $V \sum (\Delta W_i)\xi_i$
w_j	value of one mole of A_j

X_i	gross extent of ith reaction
α_j, α_{ij}	stoichiometric coefficient of A_j, in ith reaction
$\Delta H, (\Delta H)_i$	heat of reaction, of ith reaction
$\Delta U, (\Delta U)_i$	internal energy of reaction, of ith reaction
$\Delta W, (\Delta W)_i$	$\sum \alpha_j w_j, \sum \alpha_{ij} w_j$
θ_0	duration of idle period
θ_P	duration of preparation period
θ_Q	duration of discharge period
θ_R	duration of reaction
θ_T	total duration of batch cycle
ξ, ξ_i	extent, of ith reaction

Appendix: Some Less Well-known Functions of Use in Reactor Analysis

It is assumed that the exponential and logarithmic functions are thoroughly familiar to the reader as also are their relatives the circular and hyperbolic functions. However, the Bessel functions are introduced to the student much later and have less claim to familiarity. The exponential integral is also a function which occurs in several places and is worthy of some explanation. This appendix is therefore intended to provide a little background on the applications of these functions.

Modified Bessel Functions

The familiar hyperbolic cosine, cosh x, could be defined as the solution of the equation

$$\frac{d^2y}{dx^2} = y, \qquad (A.1)$$

for which $y = 1$ and $dy/dx = 0$ at $x = 0$. From the elementary theory of differential equations we know that the solution must be $A_1 e^{m_1 x} + A_2 e^{m_2 x}$, where m_1 and m_2 are the roots of $m^2 = 1$, i.e., $m_1 = 1$, $m_2 = -1$. The boundary conditions immediately give $y = \frac{1}{2}(e^x + e^{-x})$. It is also easy to see that the series

$$y = 1 + \frac{x^2}{2!} + \frac{x^4}{4!} + \cdots + \frac{x^{2n}}{(2n)!} + \cdots \tag{A.2}$$

satisfies the equation and the boundary conditions. Its convergence is not difficult to prove so that we have

$$y = \cosh x = \frac{1}{2}(e^x + e^{-x}) = \sum_{n=0}^{\infty} \frac{x^{2n}}{(2n)!} \tag{A.3}$$

Let the derivative of $y = \cosh x$ be denoted by y'. Then, by differentiating Eq. (A.1), we see that it satisfies the same differential equation

$$\frac{d^2 y'}{dx^2} = y',$$

but at $x = 0$ and $y' = dy/dx = 0$, while $dy'/dx = d^2y/dx^2 = y = 1$. Again, the methods of solution are elementary and we have

$$y' = \sinh x = \frac{1}{2}(e^x - e^{-x}) = \sum_{n=0}^{\infty} \frac{x^{2n+1}}{(2n+1)!} \tag{A.4}$$

An entirely analogous circumstance attends the solution of the equation

$$\frac{1}{r} \frac{d}{dr} \left(r \frac{dz}{dr} \right) = z, \tag{A.5}$$

subject to $z = 1$ and $dz/dx = 0$ at $r = 0$. This is the analog of Eq. (A.1) when d^2/dx^2, the Laplacian operator for one dimension and plane geometry, is replaced by

$$\frac{1}{r} \frac{d}{dr} \left(r \frac{d}{dr} \right),$$

the Laplacian operator for cylindrical symmetry. The method of trying solutions of the form e^{mr} is no longer applicable, so that the solution cannot be expressed in terms of elementary functions, but the method of solution in series gives

$$z = I_0(r) = 1 + \frac{r^2}{4} + \frac{r^4}{4^2(2!)^2} + \cdots + \frac{r^{2n}}{2^{2n}(n!)^2} + \cdots. \tag{A.6}$$

The convergence of this series is assured for all r and the function so defined is called the *modified Bessel function of zeroth order*. It is tabulated just as is the hyperbolic cosine: the two are shown in Fig. A.1.

If we write z' for dz/dr, Eq. (A.5) becomes

$$\frac{d}{dr}(rz') = z' + r\frac{dz'}{dr} = rz.$$

Appendix   Modified Bessel Functions

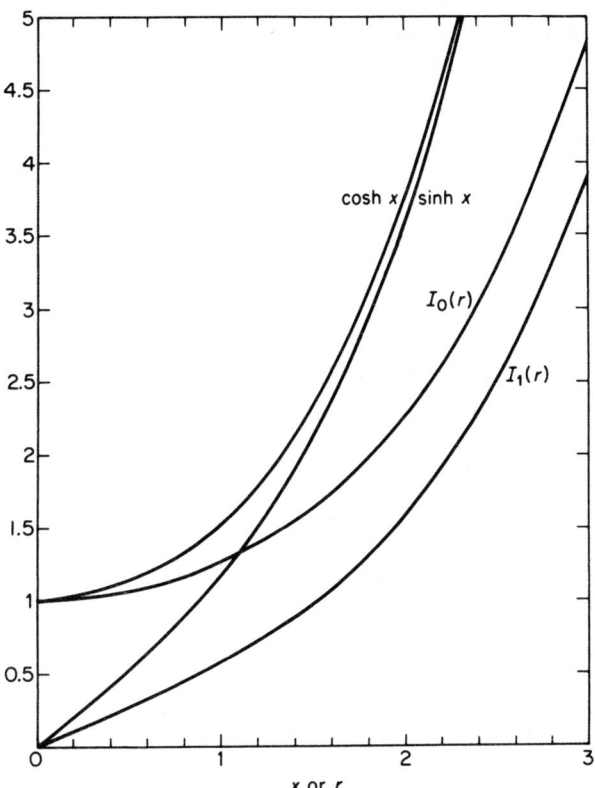

Fig. A.1 Modified Bessel functions and hyperbolic functions.

Differentiating again with respect to r and dividing by r gives

$$\frac{1}{r}\frac{d}{dr}\left(r\frac{dz'}{dr}\right) = \frac{1}{r}\frac{d}{dr}(rz - z'),$$

$$= z' + \frac{1}{r}z - \frac{1}{r}\frac{dz'}{dr}, \qquad (A.7)$$

$$= z'\left(1 + \frac{1}{r^2}\right)$$

a slightly more complicated equation than Eq. (A.5). Differentiating the series (A.6) gives

$$z' = I_1(r) = \frac{r}{2} + \frac{r^3}{8.2!} + \cdots + \frac{r^{2n-1}}{2^{2n-1}n!(n-1)!} + \cdots, \qquad (A.8)$$

and both the convergence of this and the fact that it satisfies Eq. (A.7) can be readily established. It is a modified Bessel function of the first order and has a behavior analogous to the hyperbolic sine. Both are shown in Fig. A.1.

Two further analogies are useful. First, as $x \longrightarrow \infty$, the exponential e^{-x} becomes negligible in comparison with e^x; hence

$$\cosh x \sim \tfrac{1}{2} e^x, \qquad \sinh x \sim \tfrac{1}{2} e^x.$$

The corresponding asymptotic formulae for I_0 and I_1 are

$$I_0(r) \sim I_1(r) \sim (2\pi r)^{-1/2} e^r. \tag{A.9}$$

Also we notice that by integrating both sides of Eq. (A.1) from zero to x we have

$$\int_0^x y \, dx = \int_0^x \frac{d^2 y}{dx^2} \, dx = \left[\frac{dy}{dx}\right]_0^x = y'.$$

Similarly, mutliplying both sides of Eq. (A.5) by r and integrating from zero to r we have

$$\int_0^r rz \, dr = \int_0^r \frac{d}{dr}\left(r \frac{dz}{dr}\right) dr = rz'.$$

The first of these gives the well-known relation

$$\int_0^x \cosh x' \, dx' = \sinh x;$$

the second gives

$$\int_0^r r' I_0(r') \, dr' = r I_1(r). \tag{A.10}$$

In Sec. 6.4 we saw that a first order irreversible reaction with diffusion in a plane slab was governed by the equation

$$\frac{d^2 a}{dx^2} = \lambda^2 a,$$

with $a = a_s$ at $x = \ell$ and $da/dx = 0$ at $x = 0$. In the case of a cylinder with sealed ends in which the concentration $a(r)$ would be a function only of radial distance from the axis, a balance over the annular region between radii r and $r + dr$ would give

$$\frac{1}{r} \frac{d}{dr}\left(r \frac{da}{dr}\right) = \lambda^2 a,$$

with $da/dr = 0$ at $r = 0$ and $a = a_s$ at $r = \ell$, the outer radius of the cylinder.

By replacing r by λr in Eq. (8.5) we can see immediately that the solution to this equation is

$$a(r) = A I_0(\lambda r).$$

This satisfies the condition at $r = 0$ and A is evaluated to suit $a(\ell) = a_s$. Thus

$$a(r) = \frac{a_s I_0(\lambda r)}{I_0(h)},$$

Appendix The Exponential Integral

where $h = \lambda \ell$. The effectiveness factor is the ratio of the total reaction rate to what it would be if $a(r)$ were everywhere equal to a_s. Thus

$$\eta = \frac{\int_0^\ell 2\pi r k \rho_b S_g a(r)\, dr}{\pi \ell^2 k \rho_b S_g a_s} = \frac{2}{\ell^2 I_0(h)} \int_0^\ell r I_0(\lambda r)\, dr.$$

By replacing r' in Eq. (A.10) by λr and r by $h = \lambda \ell$ we see that

$$\eta = \frac{2 I_1(h)}{h I_0(h)},$$

as given in Eq. (6.4.18).

When $h \longrightarrow 0$, Eqs. (A.6) and (A.8) show that $\eta \longrightarrow 1$, whereas when $h \longrightarrow \infty$, Eq. (A.9) shows that $\eta \sim 2/h$.

The Exponential Integral

The commonly given definition of the exponential integral is

$$Ei(x) = \int_{-\infty}^{x} e^t \frac{dt}{t}, \qquad (A.11)$$

which converges so long as $x < 0$. The integrand looks very much like t^{-1} in the neighborhood of the origin and so the integral is improper if $x > 0$. However, the negative area in an interval $-\epsilon \leq x < 0$ to the left of the origin is almost the same as the positive area in the interval $0 < x \leq \epsilon$ to the right, so that the principal value of the integral, namely

$$\int_{-\infty}^{x} \frac{e^t\, dt}{t} = \lim_{\epsilon \to 0} \left(\int_{-\infty}^{-\epsilon} \frac{e^t\, dt}{t} + \int_{\epsilon}^{x} \frac{e^t\, dt}{t} \right), \qquad (A.12)$$

does exist. It is this which is tabulated for positive x.

The integral $\int_a^b e^{kt}\, dt/(c + t)$ can be expressed in terms of the exponential integral by substituting $s = k(t + c)$; then it is evident that the integral is

$$e^{-kc} \{ Ei[k(b + c)] - Ei[k(a + c)] \}. \qquad (A.13)$$

When x is small and positive, $Ei(x)$ is large and negative, for

$$Ei(x) = \gamma + \ln x + \sum_{n=1}^{\infty} \frac{x^n}{n \cdot n!} \quad (x > 0), \qquad (A.14)$$

where $\gamma = 0.57721$ is Euler's constant.

A related integral is

$$E_1(z) = \int_z^\infty e^{-t} \frac{dt}{t} = -Ei(-z), \qquad (A.15)$$

and there are further members of the family,

$$E_n(z) = z^{n-1} \int_z^\infty e^{-t} \frac{dt}{t^n} = \int_1^\infty e^{-zt} \frac{dt}{t^n}. \qquad (A.16)$$

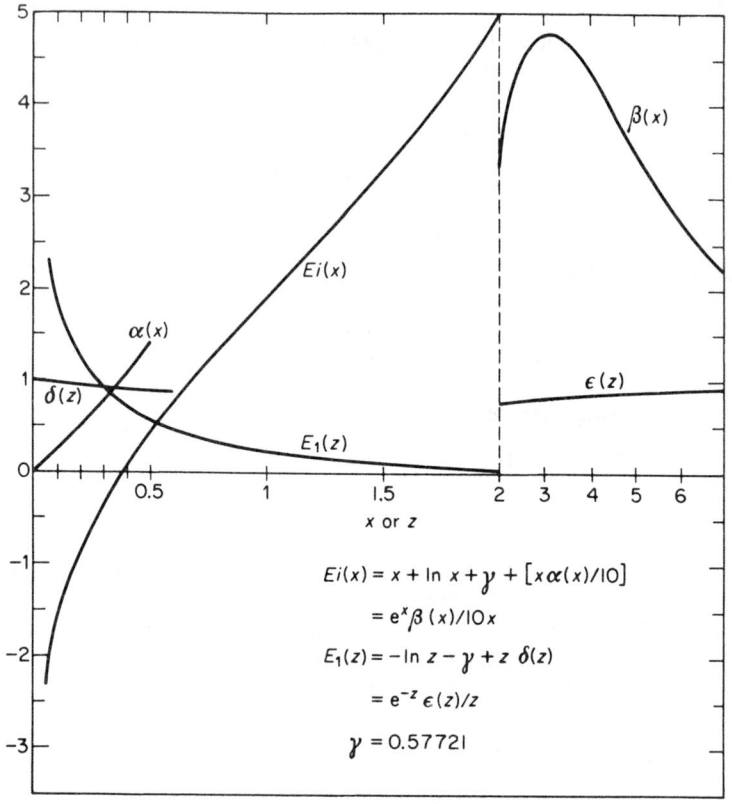

Fig. A.2 The exponential integral.

All these are tabulated in many places, the most convenient being the U. S. Department of Commerce's *Handbook of Mathematical Functions*, edited by M. Abramowitz and T. A. Stegun. (Figure A.2 shows the form of these functions.)

These functions usually turn up in trying to evaluate integrals such as

$$\int_a^b \frac{e^{-kt}}{(c+t)^n} dt = e^{kc} \left\{ \frac{E_n[k(a+c)]}{(a+c)^{n-1}} - \frac{E_n[k(b+c)]}{(b+c)^{n-1}} \right\}. \quad (A.17)$$

The functions $E_n(z)$ are related to one another by the recurrence relation

$$nE_{n+1}(z) = e^{-z} - zE_n(z). \quad (A.18)$$

As $z \longrightarrow 0$, $E_1(z)$ becomes infinite as $\ln(1/z) - \gamma$, but $E_n(0) = (n-1)^{-1}$ for $n < 1$. As $z \longrightarrow \infty$

$$E_n(z) \sim \frac{e^{-z}}{z} \left[1 - \frac{n}{z} + \frac{n(n+1)}{z^2} - \cdots \right]. \quad (A.19)$$

REFERENCES

The most convenient compilation of formulae and tables is:

M. Abramowitz and I. A. Stegun (eds.), *Handbook of Mathematical Tables*, National Bureau of Standards Applied Math. Series 55. Washington: U. S. Department of Commerce (1964).

It is also published in paperback by Dover Publications, Inc. (1965).

The classic reference for tables of the less familiar functions is:

E. Jahnke and F. Emde, *Tables of Functions*. New York: Dover Publications, Inc. (1945).

It has recently had a sixth edition with the added co-authorship of F. Loesch [New York: McGraw-Hill Book Co. (1960)].

For an introduction to Bessel functions see:

N. W. McLachlan, *Bessel Function for Engineers* (2nd ed.). Oxford: Oxford University Press (1955).

Index

A

Abramowitz, M., 342, 343
Acetic anhydride, hydrolysis of, 163–64
Acetone, cracking of, 288–89
Acetylcholine, 77
Acrivos, A., 82
Activation energy, 58
Adiabaticity (see Reactor, adiabatic)
Adler, J., 316
Adler, R. J., 226, 314
Adsorption, 117–25
  rate of, 118
Advection, 8
Ahn, Y. K., 319
Ammonia synthesis, equilibrium of, 47
  sensitivity of production rate, 298
  of blow-out temperature, 300
  of peak temperature, 301
  stoichiometry of, 22, 23
  Temkin equation for kinetics of, 301, 316
  in TVA reactor, 295–301
Amundson, N. R., 190, 197, 223, 224, 256, 302, 315–19
Analytical solutions, adiabatic reactors, 242, 255
Anderson, J. B., 153
Annable, D., 316
Aris, R., 82, 152, 153, 197, 223–25, 255, 256, 317, 318, 335
Arrhenius, S., 58
Arrhenius rate constant, definition, 58–60
  variation with temperature, 59
Asbjørnsen, O. A., 226
Astheimer, H. J., 319
Asymptotic behavior, of effectiveness factor, 131, 135
  of hyperbolic and Bessel functions, 340
Autothermicity of operation, 303
Avogadro's hypothesis, 15
Axial mixing, in tubular reactor, 274, 309–13

B

Backmixing in tubular reactor (see Axial mixing)
Backmix reactor (see Reactor, stirred tank)
Baddour, R. F., 295–301, 303, 316
Bailie, T. C., 319
Baird, R. M., 315
Bakemeier, H., 256
Barkelew, C. H., 316, 319

Baron, T., 315, 319
Barrow, G. M., 51
Batch reactor (see Reactor, batch)
Beek, J., 7, 285, 314, 315
Bellman, R., 212
Benson, S. W., 82
Benzene, 47
Benzene sulfonic acid, nitration of, 325–26
Bessel functions, modified, 134, 337–41
Bilous, O., 196, 223, 302, 316, 317
Bird, R. B., 132, 152
Bischoff, K. B., 7, 148, 149, 152, 153, 225, 226, 318, 319
Blair, J. E., 112
Blakemore, N., 335
Blanchet, J., 112, 218, 226, 256
Blow-out temperature, for stirred tank reactor, 175–76
  for tubular reactor, 299–300
Boudart, M., 3, 7, 82
Boundary conditions for tubular reactor with diffusion, 311
Bowen, J. R., 82
Brian, P. L. T., 316
Brötz, W., 6, 224
Brunauer, S., 151
Butt, J. B., 153

Colburn equation, 285
Cold-shot cooling, 244
Collatz, L., 112
Complete segregation, 219–20
Composition, change by reaction, 16, 21–25
  compatibility of feed and initial for stirred tank, 162
  rate of change by reaction, 26
Concentration, change by reaction, 21–24
  different measures of, 19–20
Control, of stirred tank reactor, 198–201
Cooling, methods of, for stirred tank reactor, 169
  for tubular reactor, 291
Coste, J., 317, 319
Costs, of construction of stirred tank reactor, 183–85
  of operating a batch reactor, 328
Coulson, J. M., 223
Coward, I., 281, 316
Cracking reaction, 268
  of acetone, 288
Criteria for stability of stirred tank reactor, 193
  physical interpretation of, 194–95
Crowell, A. D., 151

C

Cairns, E. J., 315, 318
Calcium carbonate, dissociation of, 50
Carberry, J. J., 151, 218, 223, 256
Catalyst particle, 113–50 pass., 253
  diffusion within, 123, 128–40, 141–46
  physical properties of, 116, 130–33
  shape effects, 134–36
  thermal conductivity of, 142
  transport to outside of, 140–41
Catalytic unit, height of, 141
Chapman, F. S., 223
Chemical equilibrium (see Equilibrium)
Chemical kinetics, relation between pure and applied, 3
Chemical potential, 37
Chermin, H. A. G., 111
Chollette, A., 218, 226, 256
Cleland, F. A., 314
Cloutier, L., 218, 226
Coddington, E. A., 111

D

Damköhler, G., 130, 151, 224, 318
Danckwerts, P. V., 219, 226, 319
Daniels, F., 112
Davidson, J. F., 7
De Acetis, J., 151
Deans, H. A., 153, 316
De Boer, J., 151
Degree of advancement, 16
Denbigh, K. G., 6, 51, 111, 203, 224, 225, 334
Denbigh's reaction, 104, 181
Design problems:
  basic for adiabatic reactors, 231–33, 244
  basic for stirred tank reactor, 168–81
  basic for tubular reactor, 271–74
  detailed considerations, 181–88
Desorption, 117–25
  rate of, 119
Dialer, K., 6

Differential equations:
  numerical methods for, 107–10
  revision of, 85–88
  singular perturbation of, 76
Diffusion:
  combined with external mass transfer, 140–41
  effective coefficient in porous solid, 132–33
  Fick's law, 133
  Knudsen, in small pores, 133
  longitudinal diffusion in reactor, 311–13
  multicomponent, 132–33
  within catalyst pellet, 122, 128–47
Dissociation energy, 35
Dissociation pressure, 50
Dittus-Boelter relation, 284
Douglas, J. M., 224, 225, 226, 241, 255, 256, 319
Dranoff, J. S., 224, 314

Equilibrium (cont.)
  reaction rate near, 72–73
  uniqueness of, 41, 45
  variation with temperature and pressure, 43
Equilibrium temperature, 45, 65
Ergun, S., 269, 315
Ethyl acetate, hydrolysis of, 61
Ethylene, hydration of, 47
  hydrogenation of, 121, 122
Euler's constant, 223, 341
Exponential integral, 103, 221, 242–43, 341–43, 308
Extent of intensive reaction, forms of, 21–22
  for simultaneous reactions, 23–24
Extent of molar reaction, definition, 16
  for simultaneous reactions, 17
  use in batch calculations, 323
Eymery, J. P., 316

E

Eagleton, L. C., 241, 255
Effectiveness factor, 117, 130
  cylinder, 134
  general reaction, 138
  nonisothermal case, 144–46
  slab, 134
  sphere, 135
Eldridge, J. W., 163, 164, 204
Emde, F., 343
Energy balance, stirred tank reactor, 166–68
  tubular reactor, 269–71
Enthalpy, 32
Entropy, 34, 38
Environments within stirred tank reactor, 219
Enzyme, 75
Equilibrium, chemical, 37–47 pass.
  calculation of, adiabatic conditions, 45
  heterogeneous reactions, 50
  homogeneous reactions, 40
  simultaneous reactions, 47–49
  consistency with rate expression, 58
  ideal mixtures, 37
  nonideal mixtures, 38
  portrait of second order reaction, 43–44

F

Fan, L. T., 319
Fangel, P., 224
Fanning friction factor, 269
Feeling for form, importance of, 3
Flow profile, effect of, in tubular reactor, 306–09
  interaction with diffusion, 310–11
Foss, A. S., 224
Frequency factor, 58
Froment, G. F., 314, 315
Frost, A. A., 82
Fugacity, 38

G

Gaitonde, N. Y., 225
Generation of heat by reaction, 36
Gibbs' free energy, 34, 37, 49
Gillespie, B., 256
Graphical constructions, adiabatic reactors, 241
  optimal sequence of stirred tanks, 214
  sequence of equal stirred tanks, 203
Grens, E. A., 317
Grunwald, E., 82

H

Hall, N. A., 51
Harrison, D., 7
Hawthorn, R. D., 319
Hayward, D. O., 151
HCU, 141
Heat balance, dimensionless form for first order reaction in stirred tank, 180
co- and countercurrent cooled reactors, 291–93
the stirred tank reactor, 172
total or overall, for tubular reactor, 271
Heat generation in stirred tank reactor, for various reactions, 177
variation with parameters, 172
Heat of formation, definition, 30
standard conditions, 30
Heat of reaction:
definitions, at constant pressure, 31
at constant volume, 32
relation to heats of formation, 33
relation to molar enthalpy, 32
variation with pressure and temperature, 35–36
Heat transfer, to catalyst particle, 142
to wall of tubular reactor, 284–85
Height, of transfer unit, 141
of catalytic unit, 141
Heineken, F. G., 82
Hexane, 47
Hicks, J. S., 144, 153
Hicks, M., 224
Hinrichs, H., 314
Hirschfelder, J. O., 112
Hoelscher, H. E., 317
Hoffmann, H., 224, 318, 319
Hoftyzer, P. J., 225
Holding time, of stirred tank reactor, 158
Holland, F. A., 223
Horn, F. J. M., 6, 225, 255, 282, 316, 334
Hot-spot temperature, 286, 300–301
Hougen, O. A., 6, 21, 51, 151
HTU, 141
Hudgins, R. R., 153
Hudson, J. L., 314
Hydraulic radius, 284
Hydrogen bromide, mechanism of formation, 13
Hydrogen iodide, 97
Hydrolysis of acetic anhydride, 163–64

Hyperbolic functions, comparison with Bessel functions, 337–40
Hysteresis of steady state, of catalyst pellet, 146
of stirred tank reactor, 173–76

I

Ibele, W. E., 51
Ignition, 146, 175–77
Ince, E. L., 111
Independence, linear, definition, 12
test for, 12–14
Internal energy of reaction, 324
Irving, J. P., 153

J

Jackson, R., 281, 316
Jahnke, E., 343
j_D factor, 127–28
Jeffreys, G. V., 288–89
j_H factor, 142
Jungers, J. C., 6

K

Katz, S., 314, 334–35
Kenney, M. E., 51
Ketene, 288
King, R. P., 256
Kjaer, J., 314, 316
Knudsen diffusion, 133
Konowalow, D. D., 112
Koros, R. M., 141, 152
Koutsky, J. A., 314
Krabetz, R., 256
Kramers, H., 6, 316, 317
Kuchler, L., 6, 235
Kunii, D., 7

L

Lapidus, L., 224, 316
Lassila, J. D., 51
Lead oxide, 31
Lead sulfate, 31–32
LeChatelier's principle, 43

Index 349

Lee, K. Y., 255
Leffler, J. E., 82
Levenspiel, O., 4, 6, 7, 111, 225, 318
Lightfoot, E. N., 152
Limit cycle, 199–200
Litte, R. D., 51
Liu, S. L., 256, 317
Loesch, F., 343
Long, W. M., 226
Longeais, B. A., 316

M

McHenry, K. W., 318
McKean, R. A., 317
McLachlan, N. W., 343
Malenge, J. P., 256
Mass balance, for stirred tank, 158–65
  reduction of number of equations, 163, 263
  for tubular reactor, 262–67
Mass transfer to catalyst pellet, 122, 124, 127–28
  combined with internal diffusion, 140
Maximum mixedness, 219
Methanol synthesis, 11
Michaelis constant, 76
Microscopic reversibility, principle of, 105
Miller, F. W., 153
Millman, M. C., 335
Mixing, in stirred tank reactor, 215–21
  effects of imperfect mixing, 220–21
  longitudinal, in tubular reactor, 309–13
Mole, definition, 15
Molecularity of reaction, 79
Molecular weight, mean, 265

N

Neutralization of acid in stirred tank, 164
Ng, D. Y. C., 226
Niedetsky, J., 314
Nowak, E. J., 141, 152
Nusselt number, 284

O

Olivier, J. P., 151
Oppelt, W., 224

Oppenheim, A. K., 82
Optimal design, sequence of adiabatic
  reactors, 234–37, 246–49
  sequence of stirred tanks, 210
  tubular reactor, 275
Optimal temperature (see Temperature)
Order of reaction, 57, 79

P

Packed bed, friction factor, 269
  heat transfer to walls, 285
  uniformity of voidage, 273
Page, F. M., 224
Pearson, J. R. A., 319
Pearson, R. G., 82
Peclet number, 273
  in packed bed, axial, 309–12
  radial, 274
Perry, J. H., 315
Petersen, E. E., 6, 142, 148, 150, 151, 152, 153
Phase plane, 195–201
Phosgene, 61, 74
Phosphine, 243
Piret, E. L., 163, 164, 204, 225
Platt, A. E., 61
Pollard, W. C., 152
Pore size distribution, 131
Prandtl number, 142, 284
Prater, C. D., 112, 143, 148, 153
Prausnitz, J. M., 315, 318
Pre-exponential factor, 58
Prerequisites, general, 5
  thermodynamical, 29
Present, R. D., 152
Pressure, decomposition, 50
  drop through reactor, 268–69
  partial, 21
Principle of optimality, 212

Q

Quenching, 146, 175–77

R

Ragatz, R. A., 51
Ranz, W. E., 315

Rate determining step, 74, 118–26
Rate of reaction, at catalytic site, 116
  definition, 25
  expressed in various concentration measures, 78
  general expression, 56
  heterogeneous, 60, 77
  homogeneous form, 54–61
  importance of feel for, 53
  influence of physical processes, 113–53 pass.
  maximum at given composition, 63–65, 206, 279
  near equilibrium, 72
  orders of magnitude, 80–82
  pseudo-homogeneous expression, 114–18
  variation with, extent, 62
  temperature, 63–68
Raymond, L., 256
Reactants, 10, 54
Reaction, autocatalytic, 61
  rate of, 68
  course of, 97
Reaction, back, 10, 55
  chain, 80
  endothermic, 32, 63–65
  entire, 9
  exothermic, 32, 64–67
  first order, adsorption desorption and, 118–21
  course of, 88–91
  diffusion and, 128–30, 134–36, 142–45
  dimensionless form in stirred tank, 180
  forward, 10, 55
  general homogeneous, course of, 94–96
  general irreversible, course of, 92–93
  invariants of, 18
  irreversible, 63, 79
  kinetic vs. stoichiometric description of, 8–9
  mechanism of, 9, 13, 74–77
  molecularity of, 79
  order of, 57, 79
  reversible, 79
  second order, dimensionless form in stirred tank, 181
  equilibrium portrait, 43–44
  rate portrait, 69–71
Reaction (cont.)
  time of, optimal in batch operation, 328–29
Reactions, concurrent, course of, 98–100
  consecutive, course of, 100–104
  optimal temperature profile for, 280
  in stirred tank reactor, 165, 178–80
  first order systems of, 104
  independence of, 11
  test for, 12–14
Reactor, adiabatic (see under type of reactor below):
  batch, adiabatic, 235–43
    energy balance, 324
    intermittent operation, 327–29
    mass balance, 323
    with volume change, 324
    optimal reaction time, 329
    optimal temperature control, 330–34
  packed bed, stability of, 253
  recycle, 218
  semibatch, 322
  stirred tank:
    adiabatic, 230–33
    stability of, 189, 252
    control of, 198–201
    general description, 158
    mixing in, 215–21
    multistage (see Reactors, sequences of)
    steady state of, effect of feed temperature on, 174
    effect of flow rate on, 175
    transient behavior near, 192
  tubular:
    adiabatic, 238–43
    stability, 253
    axial diffusion or mixing in, 309–13
    boundary conditions for, 311
    co- and counter-current cooled, 289–300
    multiple steady states of, 304–05
    cooling by constant wall temperature, 283–89
    energy balance, 269–71
    flow profile and its effect, 306–09
    general discussion of types, 259–61
    as limit of stirred tanks, 217–18, 274
    mass balances, 262–67
    with density variation, 263–65
    optimal isothermal temperature, 275–78

Index 351

Reactor (cont.)
 optimal temperature profile, 279–81
 types of cooling, 291
 types of, 4
Reactors, adiabatic, general types, 229
 combination of different types, 219
 adiabatic, 249–51
 convention for referring to different, 156n
 sequences of, stirred tank, adiabatic, 230–33
 isothermal, 201–05
 optimal isothermal, 207
 optimal nonisothermal, 208–15
 sequences of, tubular, adiabatic, 244–49
 optimal adiabatic, 246–49
Regenass, W., 224
Residence time distribution, definition, 215
 for laminar flow, 306–07
 mean of, 216–17
 in sequence of tanks, 217
 in stirred tank, 158, 215–17
 variance of, 216–17
Resnick, W., 218–26
Reynolds number, 127, 128, 142, 269, 273, 284
Richardson, J. F., 223
Rippin, D. W. T., 221, 224, 226
Roblee, L. H. S., 315
Rodiguin, N. M., 112
Rodiguina, E. N., 112
Rooze, J., 226
Rosenbrock, H. H., 112
Ross, S., 151
Rothfeld, L. B., 131, 152
Rudd, D. F., 319
Runge-Kutta method, 109

S

Sabo, D. S., 224
Satterfield, C. N., 132, 133, 150, 151
Schechter, R. S., 314
Schmidt number, 127, 128
Schmitz, R. A., 317
Schoenemann, K., 225
Sensitivity, coefficient, 298
 parametric, 301–05
Sherwood, T. K., 132, 133, 150, 151
Sherwood number, 127, 128

Siebenthal, C. D., 335
Smith, J. M., 6
Solomon, R. L., 314
Space velocity, 263
Stability, general discussion, 188–91
 global, 195
 of packed bed reactor, 253–54
 of stirred tank, adiabatic, 252
 criteria for, 193
 dynamic analysis, 191–94
 steady-state analysis, 188–90
 of tubular, adiabatic, 253
Stanton number, 284–97 pass.
Steady state hypothesis, 76
Stegun, T. A., 342, 343
Stewart, W. E., 152
Stirling's approximation, 222
Stoichiometric coefficient, 10
Stoichiometry, meaning of, 8
Storey, C. E., 112
Sugar, inversion of, 309
Sulfur dioxide, oxidation of, 266–67

T

Taylor, G. I., 318
Temkin equation for ammonia synthesis, 301, 316
Temperature, influence on rate constant, 58–59
 optimal for isothermal tubular reactor, 275–78
 optimal for stirred tank, 206
Thermochemistry, 29
Thiele, E. W., 151, 152
Thiele modulus, 130, 134, 143, 147
 normalized for kinetics, 138
 shape, 135–36
Thodos, G., 151
Thomas, W. J., 151
Tichacek, L. J., 319
Trambouze, P. J., 225, 314
Transfer units, height of, 141
 number of, 303–05
Trapnell, B. M. W., 151
Troltenier, U., 316
Trypsinogen, 97
Tsuchiya, H. M., 82
Tube–particle diameter ratio, effect on pressure drop, 269
 effect on radial Peclet number, 273

Turner, J. C. R., 226
T.V.A. reactor, 295

U

Ulrichson, D. L., 317

V

Van de Vusse, J. G., 219, 226
Van Heerden, C., 190, 223, 224, 256, 317
Van Krevelen, D. W., 111, 225
Variance of residence times, 217
Vignes, J. P., 314
Villermaux, J., 256
Voidage, in packed bed, 116
  within catalyst particle, 131–33
Volume change with reaction, balance for tubular reactor, 263–65
Vortmeyer, D., 316

W

Walas, S. M., 6
Walles, W. E., 61

Watson, K. M., 6, 21, 51, 131
Wehner, J. F., 319
Wei, J., 112
Weinstein, H., 226
Weisz, P. B., 144, 148, 153
Westbrook, G. T., 185, 223
Westerterp, K. R., 6, 223, 316, 317
Wheeler, A., 152
Wicke, E., 152, 224
Wilhelm, R. H., 274, 314, 315, 318, 319
Wissler, E. H., 314
Wolf, D., 218, 226

Y

Yield, of intermediate in consecutive reactions, 178–81
  maximum, 165
Young, D. M., 151

Z

Zweitering, T. N., 219, 221, 226

A CATALOG OF SELECTED
DOVER BOOKS
IN SCIENCE AND MATHEMATICS

A CATALOG OF SELECTED

DOVER BOOKS

IN SCIENCE AND MATHEMATICS

QUALITATIVE THEORY OF DIFFERENTIAL EQUATIONS, V.V. Nemytskii and V.V. Stepanov. Classic graduate-level text by two prominent Soviet mathematicians covers classical differential equations as well as topological dynamics and ergodic theory. Bibliographies. 523pp. 5⅜ x 8½. 65954-2 Pa. $14.95

MATRICES AND LINEAR ALGEBRA, Hans Schneider and George Phillip Barker. Basic textbook covers theory of matrices and its applications to systems of linear equations and related topics such as determinants, eigenvalues and differential equations. Numerous exercises. 432pp. 5⅜ x 8½. 66014-1 Pa. $12.95

QUANTUM THEORY, David Bohm. This advanced undergraduate-level text presents the quantum theory in terms of qualitative and imaginative concepts, followed by specific applications worked out in mathematical detail. Preface. Index. 655pp. 5⅜ x 8½. 65969-0 Pa. $15.95

ATOMIC PHYSICS (8th edition), Max Born. Nobel laureate's lucid treatment of kinetic theory of gases, elementary particles, nuclear atom, wave-corpuscles, atomic structure and spectral lines, much more. Over 40 appendices, bibliography. 495pp. 5⅜ x 8½. 65984-4 Pa. $13.95

ELECTRONIC STRUCTURE AND THE PROPERTIES OF SOLIDS: The Physics of the Chemical Bond, Walter A. Harrison. Innovative text offers basic understanding of the electronic structure of covalent and ionic solids, simple metals, transition metals and their compounds. Problems. 1980 edition. 582pp. 6⅛ x 9¼. 66021-4 Pa. $19.95

BOUNDARY VALUE PROBLEMS OF HEAT CONDUCTION, M. Necati Özisik. Systematic, comprehensive treatment of modern mathematical methods of solving problems in heat conduction and diffusion. Numerous examples and problems. Selected references. Appendices. 505pp. 5⅜ x 8½. 65990-9 Pa. $12.95

A SHORT HISTORY OF CHEMISTRY (3rd edition), J.R. Partington. Classic exposition explores origins of chemistry, alchemy, early medical chemistry, nature of atmosphere, theory of valency, laws and structure of atomic theory, much more. 428pp. 5⅜ x 8½. (Available in U.S. only) 65977-1 Pa. $12.95

A HISTORY OF ASTRONOMY, A. Pannekoek. Well-balanced, carefully reasoned study covers such topics as Ptolemaic theory, work of Copernicus, Kepler, Newton, Eddington's work on stars, much more. Illustrated. References. 521pp. 5⅜ x 8½. 65994-1 Pa. $15.95

PRINCIPLES OF METEOROLOGICAL ANALYSIS, Walter J. Saucier. Highly respected, abundantly illustrated classic reviews atmospheric variables, hydrostatics, static stability, various analyses (scalar, cross-section, isobaric, isentropic, more). For intermediate meteorology students. 454pp. 6⅛ x 9¼. 65979-8 Pa. $14.95

CATALOG OF DOVER BOOKS

RELATIVITY, THERMODYNAMICS AND COSMOLOGY, Richard C. Tolman. Landmark study extends thermodynamics to special, general relativity; also applications of relativistic mechanics, thermodynamics to cosmological models. 501pp. 5⅜ x 8½. 65383-8 Pa. $15.95

APPLIED ANALYSIS, Cornelius Lanczos. Classic work on analysis and design of finite processes for approximating solution of analytical problems. Algebraic equations, matrices, harmonic analysis, quadrature methods, much more. 559pp. 5⅜ x 8½. 65656-X Pa. $16.95

INTRODUCTION TO ANALYSIS, Maxwell Rosenlicht. Unusually clear, accessible coverage of set theory, real number system, metric spaces, continuous functions, Riemann integration, multiple integrals, more. Wide range of problems. Undergraduate level. Bibliography. 254pp. 5⅜ x 8½. 65038-3 Pa. $9.95

INTRODUCTION TO QUANTUM MECHANICS With Applications to Chemistry, Linus Pauling & E. Bright Wilson, Jr. Classic undergraduate text by Nobel Prize winner applies quantum mechanics to chemical and physical problems. Numerous tables and figures enhance the text. Chapter bibliographies. Appendices. Index. 468pp. 5⅜ x 8½. 64871-0 Pa. $12.95

ASYMPTOTIC EXPANSIONS OF INTEGRALS, Norman Bleistein & Richard A. Handelsman. Best introduction to important field with applications in a variety of scientific disciplines. New preface. Problems. Diagrams. Tables. Bibliography. Index. 448pp. 5⅜ x 8½. 65082-0 Pa. $13.95

MATHEMATICS APPLIED TO CONTINUUM MECHANICS, Lee A. Segel. Analyzes models of fluid flow and solid deformation. For upper-level math, science and engineering students. 608pp. 5⅜ x 8½. 65369-2 Pa. $14.95

ELEMENTS OF REAL ANALYSIS, David A. Sprecher. Classic text covers fundamental concepts, real number system, point sets, functions of a real variable, Fourier series, much more. Over 500 exercises. 352pp. 5⅜ x 8½. 65385-4 Pa. $11.95

PHYSICAL PRINCIPLES OF THE QUANTUM THEORY, Werner Heisenberg. Nobel Laureate discusses quantum theory, uncertainty, wave mechanics, work of Dirac, Schroedinger, Compton, Wilson, Einstein, etc. 184pp. 5⅜ x 8½. 60113-7 Pa. $8.95

INTRODUCTORY REAL ANALYSIS, A.N. Kolmogorov, S.V. Fomin. Translated by Richard A. Silverman. Self-contained, evenly paced introduction to real and functional analysis. Some 350 problems. 403pp. 5⅜ x 8½. 61226-0 Pa. $11.95

PROBLEMS AND SOLUTIONS IN QUANTUM CHEMISTRY AND PHYSICS, Charles S. Johnson, Jr. and Lee G. Pedersen. Unusually varied problems, detailed solutions in coverage of quantum mechanics, wave mechanics, angular momentum, molecular spectroscopy, scattering theory, more. 280 problems plus 139 supplementary exercises. 430pp. 6½ x 9¼. 65236-X Pa. $14.95

ASYMPTOTIC METHODS IN ANALYSIS, N.G. de Bruijn. An inexpensive, comprehensive guide to asymptotic methods–the pioneering work that teaches by explaining worked examples in detail. Index. 224pp. $5\frac{3}{8} \times 8\frac{1}{2}$. 64221-6 Pa. $7.95

OPTICAL RESONANCE AND TWO-LEVEL ATOMS, L. Allen and J. H. Eberly. Clear, comprehensive introduction to basic principles behind all quantum optical resonance phenomena. 53 illustrations. Preface. Index. 256pp. $5\frac{3}{8} \times 8\frac{1}{2}$. 65533-4 Pa. $10.95

COMPLEX VARIABLES, Francis J. Flanigan. Unusual approach, delaying complex algebra till harmonic functions have been analyzed from real variable viewpoint. Includes problems with answers. 364pp. $5\frac{3}{8} \times 8\frac{1}{2}$. 61388-7 Pa. $10.95

ATOMIC SPECTRA AND ATOMIC STRUCTURE, Gerhard Herzberg. One of best introductions; especially for specialist in other fields. Treatment is physical rather than mathematical. 80 illustrations. 257pp. $5\frac{3}{8} \times 8\frac{1}{2}$. 60115-3 Pa. $7.95

APPLIED COMPLEX VARIABLES, John W. Dettman. Step-by-step coverage of fundamentals of analytic function theory–plus lucid exposition of five important applications: Potential Theory; Ordinary Differential Equations; Fourier Transforms; Laplace Transforms; Asymptotic Expansions. 66 figures. Exercises at chapter ends. 512pp. $5\frac{3}{8} \times 8\frac{1}{2}$. 64670-X Pa. $14.95

ULTRASONIC ABSORPTION: An Introduction to the Theory of Sound Absorption and Dispersion in Gases, Liquids and Solids, A.B. Bhatia. Standard reference in the field provides a clear, systematically organized introductory review of fundamental concepts for advanced graduate students, research workers. Numerous diagrams. Bibliography. 440pp. $5\frac{3}{8} \times 8\frac{1}{2}$. 64917-2 Pa. $11.95

UNBOUNDED LINEAR OPERATORS: Theory and Applications, Seymour Goldberg. Classic presents systematic treatment of the theory of unbounded linear operators in normed linear spaces with applications to differential equations. Bibliography. I99pp. $5\frac{3}{8} \times 8\frac{1}{2}$. 64830-3 Pa. $7.95

LIGHT SCATTERING BY SMALL PARTICLES, H.C. van de Hulst. Comprehensive treatment including full range of useful approximation methods for researchers in chemistry, meteorology and astronomy. 44 illustrations. 470pp. $5\frac{3}{8} \times 8\frac{1}{2}$. 64228-3 Pa. $12.95

CONFORMAL MAPPING ON RIEMANN SURFACES, Harvey Cohn. Lucid, insightful book presents ideal coverage of subject. 334 exercises make book perfect for self-study. 55 figures. 352pp. $5\frac{5}{8} \times 8\frac{1}{4}$. 64025-6 Pa. $11.95

OPTICKS, Sir Isaac Newton. Newton's own experiments with spectroscopy, colors, lenses, reflection, refraction, etc., in language the layman can follow. Foreword by Albert Einstein. 532pp. $5\frac{3}{8} \times 8\frac{1}{2}$. 60205-2 Pa. $13.95

GENERALIZED INTEGRAL TRANSFORMATIONS, A.H. Zemanian. Graduate-level study of recent generalizations of the Laplace, Mellin, Hankel, K. Weierstrass, convolution and other simple transformations. Bibliography. 320pp. $5\frac{3}{8} \times 8\frac{1}{2}$. 65375-7 Pa. $8.95

THE ELECTROMAGNETIC FIELD, Albert Shadowitz. Comprehensive undergraduate text covers basics of electric and magnetic fields, builds up to electromagnetic theory. Also related topics, including relativity. Over 900 problems. 768pp. 5⅝ x 8¼. 65660-8 Pa. $19.95

FOURIER SERIES, Georgi P. Tolstov. Translated by Richard A. Silverman. A valuable addition to the literature on the subject, moving clearly from subject to subject and theorem to theorem. 107 problems, answers. 336pp. 5⅜ x 8½. 63317-9 Pa. $11.95

THEORY OF ELECTROMAGNETIC WAVE PROPAGATION, Charles Herach Papas. Graduate-level study discusses the Maxwell field equations, radiation from wire antennas, the Doppler effect and more. xiii + 244pp. 5⅜ x 8½. 65678-0 Pa. $9.95

DISTRIBUTION THEORY AND TRANSFORM ANALYSIS: An Introduction to Generalized Functions, with Applications, A.H. Zemanian. Provides basics of distribution theory, describes generalized Fourier and Laplace transformations. Numerous problems. 384pp. 5⅜ x 8½. 65479-6 Pa. $13.95

THE PHYSICS OF WAVES, William C. Elmore and Mark A. Heald. Unique overview of classical wave theory. Acoustics, optics, electromagnetic radiation, more. Ideal as classroom text or for self-study. Problems. 477pp. 5⅜ x 8½. 64926-1 Pa. $14.95

CALCULUS OF VARIATIONS WITH APPLICATIONS, George M. Ewing. Applications-oriented introduction to variational theory develops insight and promotes understanding of specialized books, research papers. Suitable for advanced undergraduate/graduate students as primary, supplementary text. 352pp. 5⅜ x 8½. 64856-7 Pa. $9.95

A TREATISE ON ELECTRICITY AND MAGNETISM, James Clerk Maxwell. Important foundation work of modern physics. Brings to final form Maxwell's theory of electromagnetism and rigorously derives his general equations of field theory. 1,084pp. 5⅜ x 8½. 60636-8, 60637-6 Pa., Two-vol. set $27.90

AN INTRODUCTION TO THE CALCULUS OF VARIATIONS, Charles Fox. Graduate-level text covers variations of an integral, isoperimetrical problems, least action, special relativity, approximations, more. References. 279pp. 5⅜ x 8½. 65499-0 Pa. $8.95

HYDRODYNAMIC AND HYDROMAGNETIC STABILITY, S. Chandrasekhar. Lucid examination of the Rayleigh-Benard problem; clear coverage of the theory of instabilities causing convection. 704pp. 5⅝ x 8¼. 64071-X Pa. $17.95

CALCULUS OF VARIATIONS, Robert Weinstock. Basic introduction covering isoperimetric problems, theory of elasticity, quantum mechanics, electrostatics, etc. Exercises throughout. 326pp. 5⅜ x 8½. 63069-2 Pa. $9.95

DYNAMICS OF FLUIDS IN POROUS MEDIA, Jacob Bear. For advanced students of ground water hydrology, soil mechanics and physics, drainage and irrigation engineering and more. 335 illustrations. Exercises, with answers. 784pp. 6⅛ x 9¼. 65675-6 Pa. $19.95

CATALOG OF DOVER BOOKS

NUMERICAL METHODS FOR SCIENTISTS AND ENGINEERS, Richard Hamming. Classic text stresses frequency approach in coverage of algorithms, polynomial approximation, Fourier approximation, exponential approximation, other topics. Revised and enlarged 2nd edition. 721pp. 5⅜ x 8½. 65241-6 Pa. $16.95

THEORETICAL SOLID STATE PHYSICS, Vol. 1: Perfect Lattices in Equilibrium; Vol. II: Non-Equilibrium and Disorder, William Jones and Norman H. March. Monumental reference work covers fundamental theory of equilibrium properties of perfect crystalline solids, non-equilibrium properties, defects and disordered systems. Appendices. Problems. Preface. Diagrams. Index. Bibliography. Total of 1,301pp. 5⅜ x 8½. Two volumes. Vol. I: 65015-4 Pa. $16.95
Vol. II: 65016-2 Pa. $16.95

OPTIMIZATION THEORY WITH APPLICATIONS, Donald A. Pierre. Broad spectrum approach to important topic. Classical theory of minima and maxima, calculus of variations, simplex technique and linear programming, more. Many problems, examples. 640pp. 5⅜ x 8½. 65205-X Pa. $17.95

THE CONTINUUM: A Critical Examination of the Foundation of Analysis, Hermann Weyl. Classic of 20th-century foundational research deals with the conceptual problem posed by the continuum. 156pp. 5⅜ x 8½. 67982-9 Pa. $8.95

ESSAYS ON THE THEORY OF NUMBERS, Richard Dedekind. Two classic essays by great German mathematician: on the theory of irrational numbers; and on transfinite numbers and properties of natural numbers. 115pp. 5⅜ x 8½.
21010-3 Pa. $6.95

THE FUNCTIONS OF MATHEMATICAL PHYSICS, Harry Hochstadt. Comprehensive treatment of orthogonal polynomials, hypergeometric functions, Hill's equation, much more. Bibliography. Index. 322pp. 5⅜ x 8½. 65214-9 Pa. $12.95

NUMBER THEORY AND ITS HISTORY, Oystein Ore. Unusually clear, accessible introduction covers counting, properties of numbers, prime numbers, much more. Bibliography. 380pp. 5⅜ x 8½. 65620-9 Pa. $10.95

THE VARIATIONAL PRINCIPLES OF MECHANICS, Cornelius Lanczos. Graduate level coverage of calculus of variations, equations of motion, relativistic mechanics, more. First inexpensive paperbound edition of classic treatise. Index. Bibliography. 418pp. 5⅜ x 8½. 65067-7 Pa. $14.95

COMBINATORIAL TOPOLOGY, P. S. Alexandrov. Clearly written, well-organized, three-part text begins by dealing with certain classic problems without using the formal techniques of homology theory and advances to the central concept, the Betti groups. Numerous detailed examples. 654pp. 5⅜ x 8½. 40179-0 Pa. $18.95

THEORETICAL PHYSICS, Georg Joos, with Ira M. Freeman. Classic overview covers essential math, mechanics, electromagnetic theory, thermodynamics, quantum mechanics, nuclear physics, other topics. First paperback edition. xxiii + 885pp. 5⅜ x 8½. 65227-0 Pa. $21.95

CATALOG OF DOVER BOOKS

HANDBOOK OF MATHEMATICAL FUNCTIONS WITH FORMULAS, GRAPHS, AND MATHEMATICAL TABLES, edited by Milton Abramowitz and Irene A. Stegun. Vast compendium: 29 sets of tables, some to as high as 20 places. 1,046pp. 8 x 10½. 61272-4 Pa. $29.95

MATHEMATICAL METHODS IN PHYSICS AND ENGINEERING, John W. Dettman. Algebraically based approach to vectors, mapping, diffraction, other topics in applied math. Also generalized functions, analytic function theory, more. Exercises. 448pp. 5⅜ x 8¼. 65649-7 Pa. $12.95

A SURVEY OF NUMERICAL MATHEMATICS, David M. Young and Robert Todd Gregory. Broad self-contained coverage of computer-oriented numerical algorithms for solving various types of mathematical problems in linear algebra, ordinary and partial, differential equations, much more. Exercises. Total of 1,248pp. 5⅜ x 8½.
Two volumes. Vol. I: 65691-8 Pa. $16.95
Vol. II: 65692-6 Pa. $16.95

TENSOR ANALYSIS FOR PHYSICISTS, J.A. Schouten. Concise exposition of the mathematical basis of tensor analysis, integrated with well-chosen physical examples of the theory. Exercises. Index. Bibliography. 289pp. 5⅜ x 8½. 65582-2 Pa. $10.95

INTRODUCTION TO NUMERICAL ANALYSIS (2nd Edition), F.B. Hildebrand. Classic, fundamental treatment covers computation, approximation, interpolation, numerical differentiation and integration, other topics. 150 new problems. 669pp. 5⅜ x 8½. 65363-3 Pa. $16.95

INVESTIGATIONS ON THE THEORY OF THE BROWNIAN MOVEMENT, Albert Einstein. Five papers (1905-8) investigating dynamics of Brownian motion and evolving elementary theory. Notes by R. Fürth. 122pp. 5⅜ x 8½.
60304-0 Pa. $5.95

CATASTROPHE THEORY FOR SCIENTISTS AND ENGINEERS, Robert Gilmore. Advanced-level treatment describes mathematics of theory grounded in the work of Poincaré, R. Thom, other mathematicians. Also important applications to problems in mathematics, physics, chemistry and engineering. 1981 edition. References. 28 tables. 397 black-and-white illustrations. xvii + 666pp. 6⅛ x 9¼.
67539-4 Pa. $17.95

AN INTRODUCTION TO STATISTICAL THERMODYNAMICS, Terrell L. Hill. Excellent basic text offers wide-ranging coverage of quantum statistical mechanics, systems of interacting molecules, quantum statistics, more. 523pp. 5⅜ x 8½.
65242-4 Pa. $13.95

STATISTICAL PHYSICS, Gregory H. Wannier. Classic text combines thermodynamics, statistical mechanics and kinetic theory in one unified presentation of thermal physics. Problems with solutions. Bibliography. 532pp. 5⅜ x 8½.
65401-X Pa. $14.95

CATALOG OF DOVER BOOKS

ORDINARY DIFFERENTIAL EQUATIONS, Morris Tenenbaum and Harry Pollard. Exhaustive survey of ordinary differential equations for undergraduates in mathematics, engineering, science. Thorough analysis of theorems. Diagrams. Bibliography. Index. 818pp. 5⅜ x 8½. 64940-7 Pa. $19.95

STATISTICAL MECHANICS: Principles and Applications, Terrell L. Hill. Standard text covers fundamentals of statistical mechanics, applications to fluctuation theory, imperfect gases, distribution functions, more. 448pp. 5⅜ x 8½.
65390-0 Pa. $14.95

ORDINARY DIFFERENTIAL EQUATIONS AND STABILITY THEORY: An Introduction, David A. Sánchez. Brief, modern treatment. Linear equation, stability theory for autonomous and nonautonomous systems, etc. 164pp. 5⅜ x 8¼.
63828-6 Pa. $6.95

THIRTY YEARS THAT SHOOK PHYSICS: The Story of Quantum Theory, George Gamow. Lucid, accessible introduction to influential theory of energy and matter. Careful explanations of Dirac's anti-particles, Bohr's model of the atom, much more. 12 plates. Numerous drawings. 240pp. 5⅜ x 8½. 24895-X Pa. $7.95

THEORY OF MATRICES, Sam Perlis. Outstanding text covering rank, nonsingularity and inverses in connection with the development of canonical matrices under the relation of equivalence, and without the intervention of determinants. Includes exercises. 237pp. 5⅜ x 8½. 66810-X Pa. $8.95

GREAT EXPERIMENTS IN PHYSICS: Firsthand Accounts from Galileo to Einstein, edited by Morris H. Shamos. 25 crucial discoveries: Newton's laws of motion, Chadwick's study of the neutron, Hertz on electromagnetic waves, more. Original accounts clearly annotated. 370pp. 5⅜ x 8½. 25346-5 Pa. $11.95

INTRODUCTION TO PARTIAL DIFFERENTIAL EQUATIONS WITH APPLICATIONS, E.C. Zachmanoglou and Dale W. Thoe. Essentials of partial differential equations applied to common problems in engineering and the physical sciences. Problems and answers. 416pp. 5⅜ x 8½. 65251-3 Pa. $11.95

BURNHAM'S CELESTIAL HANDBOOK, Robert Burnham, Jr. Thorough guide to the stars beyond our solar system. Exhaustive treatment. Alphabetical by constellation: Andromeda to Cetus in Vol. 1; Chamaeleon to Orion in Vol. 2; and Pavo to Vulpecula in Vol. 3. Hundreds of illustrations. Index in Vol. 3. 2,000pp. 6⅛ x 9¼.
23567-X, 23568-8, 23673-0 Pa., Three-vol. set $46.85

CHEMICAL MAGIC, Leonard A. Ford. Second Edition, Revised by E. Winston Grundmeier. Over 100 unusual stunts demonstrating cold fire, dust explosions, much more. Text explains scientific principles and stresses safety precautions. 128pp. 5⅜ x 8½. 67628-5 Pa. $5.95

AMATEUR ASTRONOMER'S HANDBOOK, J.B. Sidgwick. Timeless, comprehensive coverage of telescopes, mirrors, lenses, mountings, telescope drives, micrometers, spectroscopes, more. 189 illustrations. 576pp. 5⅜ x 8¼. (Available in U.S. only) 24034-7 Pa. $13.95

CATALOG OF DOVER BOOKS

SPECIAL FUNCTIONS, N.N. Lebedev. Translated by Richard Silverman. Famous Russian work treating more important special functions, with applications to specific problems of physics and engineering. 38 figures. 308pp. 5⅜ x 8½. 60624-4 Pa. $9.95

THE EXTRATERRESTRIAL LIFE DEBATE, 1750–1900, Michael J. Crowe. First detailed, scholarly study in English of the many ideas that developed between 1750 and 1900 regarding the existence of intelligent extraterrestrial life. Examines ideas of Kant, Herschel, Voltaire, Percival Lowell, many other scientists and thinkers. 16 illustrations. 704pp. 5⅜ x 8½. 40675-X Pa. $19.95

INTEGRAL EQUATIONS, F.G. Tricomi. Authoritative, well-written treatment of extremely useful mathematical tool with wide applications. Volterra Equations, Fredholm Equations, much more. Advanced undergraduate to graduate level. Exercises. Bibliography. 238pp. 5⅜ x 8½. 64828-1 Pa. $8.95

POPULAR LECTURES ON MATHEMATICAL LOGIC, Hao Wang. Noted logician's lucid treatment of historical developments, set theory, model theory, recursion theory and constructivism, proof theory, more. 3 appendixes. Bibliography. 1981 edition. ix + 283pp. 5⅜ x 8½. 67632-3 Pa. $10.95

MODERN NONLINEAR EQUATIONS, Thomas L. Saaty. Emphasizes practical solution of problems; covers seven types of equations. "... a welcome contribution to the existing literature...."–*Math Reviews.* 490pp. 5⅜ x 8½. 64232-1 Pa. $13.95

FUNDAMENTALS OF ASTRODYNAMICS, Roger Bate et al. Modern approach developed by U.S. Air Force Academy. Designed as a first course. Problems, exercises. Numerous illustrations. 455pp. 5⅜ x 8½. 60061-0 Pa. $12.95

INTRODUCTION TO LINEAR ALGEBRA AND DIFFERENTIAL EQUATIONS, John W. Dettman. Excellent text covers complex numbers, determinants, orthonormal bases, Laplace transforms, much more. Exercises with solutions. Undergraduate level. 416pp. 5⅜ x 8½. 65191-6 Pa. $11.95

INCOMPRESSIBLE AERODYNAMICS, edited by Bryan Thwaites. Covers theoretical and experimental treatment of the uniform flow of air and viscous fluids past two-dimensional aerofoils and three-dimensional wings; many other topics. 654pp. 5⅜ x 8½. 65465-6 Pa. $16.95

INTRODUCTION TO DIFFERENCE EQUATIONS, Samuel Goldberg. Exceptionally clear exposition of important discipline with applications to sociology, psychology, economics. Many illustrative examples; over 250 problems. 260pp. 5⅜ x 8½. 65084-7 Pa. $10.95

THREE PEARLS OF NUMBER THEORY, A. Y. Khinchin. Three compelling puzzles require proof of a basic law governing the world of numbers. Challenges concern van der Waerden's theorem, the Landau-Schnirelmann hypothesis and Mann's theorem, and a solution to Waring's problem. Solutions included. 64pp. 5⅜ x 8½. 40026-3 Pa. $4.95

LECTURES ON CLASSICAL DIFFERENTIAL GEOMETRY, Second Edition, Dirk J. Struik. Excellent brief introduction covers curves, theory of surfaces, fundamental equations, geometry on a surface, conformal mapping, other topics. Problems. 240pp. 5⅜ x 8½. 65609-8 Pa. $9.95

CATALOG OF DOVER BOOKS

ROTARY-WING AERODYNAMICS, W.Z. Stepniewski. Clear, concise text covers aerodynamic phenomena of the rotor and offers guidelines for helicopter performance evaluation. Originally prepared for NASA. 537 figures. 640pp. 6⅛ x 9¼.
64647-5 Pa. $16.95

DIFFERENTIAL GEOMETRY, Heinrich W. Guggenheimer. Local differential geometry as an application of advanced calculus and linear algebra. Curvature, transformation groups, surfaces, more. Exercises. 62 figures. 378pp. 5⅜ x 8½.
63433-7 Pa. $11.95

INTRODUCTION TO SPACE DYNAMICS, William Tyrrell Thomson. Comprehensive, classic introduction to space-flight engineering for advanced undergraduate and graduate students. Includes vector algebra, kinematics, transformation of coordinates. Bibliography. Index. 352pp. 5⅜ x 8½. 65113-4 Pa. $10.95

THE THEORY OF GROUPS, Hans J. Zassenhaus. Well-written graduate-level text acquaints reader with group-theoretic methods and demonstrates their usefulness in mathematics. Axioms, the calculus of complexes, homomorphic mapping, p-group theory, more. Many proofs shorter and more transparent than older ones. 276pp. 5⅜ x 8½. 40922-8 Pa. $12.95

ANALYTICAL MECHANICS OF GEARS, Earle Buckingham. Indispensable reference for modern gear manufacture covers conjugate gear-tooth action, gear-tooth profiles of various gears, many other topics. 263 figures. 102 tables. 546pp. 5⅜ x 8½.
65712-4 Pa. $16.95

SET THEORY AND LOGIC, Robert R. Stoll. Lucid introduction to unified theory of mathematical concepts. Set theory and logic seen as tools for conceptual understanding of real number system. 496pp. 5⅜ x 8¼. 63829-4 Pa. $14.95

A HISTORY OF MECHANICS, René Dugas. Monumental study of mechanical principles from antiquity to quantum mechanics. Contributions of ancient Greeks, Galileo, Leonardo, Kepler, Lagrange, many others. 671pp. 5⅜ x 8½.
65632-2 Pa. $18.95

FAMOUS PROBLEMS OF GEOMETRY AND HOW TO SOLVE THEM, Benjamin Bold. Squaring the circle, trisecting the angle, duplicating the cube: learn their history, why they are impossible to solve, then solve them yourself. 128pp. 5⅜ x 8½. 24297-8 Pa. $5.95

MECHANICAL VIBRATIONS, J.P. Den Hartog. Classic textbook offers lucid explanations and illustrative models, applying theories of vibrations to a variety of practical industrial engineering problems. Numerous figures. 233 problems, solutions. Appendix. Index. Preface. 436pp. 5⅜ x 8½. 64785-4 Pa. $13.95

CURVATURE AND HOMOLOGY: Enlarged Edition, Samuel I. Goldberg. Revised edition examines topology of differentiable manifolds; curvature, homology of Riemannian manifolds; compact Lie groups; complex manifolds; curvature, homology of Kaehler manifolds. New Preface. Four new appendixes. 416pp. 5⅜ x 8½.
40207-X Pa. $14.95

HISTORY OF STRENGTH OF MATERIALS, Stephen P. Timoshenko. Excellent historical survey of the strength of materials with many references to the theories of elasticity and structure. 245 figures. 452pp. 5⅜ x 8½. 61187-6 Pa. $14.95

GEOMETRY OF COMPLEX NUMBERS, Hans Schwerdtfeger. Illuminating, widely praised book on analytic geometry of circles, the Moebius transformation, and two-dimensional non-Euclidean geometries. 200pp. 5⅝ x 8¼. 63830-8 Pa. $8.95

MECHANICS, J.P. Den Hartog. A classic introductory text or refresher. Hundreds of applications and design problems illuminate fundamentals of trusses, loaded beams and cables, etc. 334 answered problems. 462pp. 5⅜ x 8½. 60754-2 Pa. $12.95

TOPOLOGY, John G. Hocking and Gail S. Young. Superb one-year course in classical topology. Topological spaces and functions, point-set topology, much more. Examples and problems. Bibliography. Index. 384pp. 5⅝ x 8¼. 65676-4 Pa. $11.95

STRENGTH OF MATERIALS, J.P. Den Hartog. Full, clear treatment of basic material (tension, torsion, bending, etc.) plus advanced material on engineering methods, applications. 350 answered problems. 323pp. 5⅜ x 8½. 60755-0 Pa. $10.95

ELEMENTARY CONCEPTS OF TOPOLOGY, Paul Alexandroff. Elegant, intuitive approach to topology from set-theoretic topology to Betti groups; how concepts of topology are useful in math and physics. 25 figures. 57pp. 5⅜ x 8½. 60747-X Pa. $4.95

ADVANCED STRENGTH OF MATERIALS, J.P. Den Hartog. Superbly written advanced text covers torsion, rotating disks, membrane stresses in shells, much more. Many problems and answers. 388pp. 5⅜ x 8½. 65407-9 Pa. $11.95

COMPUTABILITY AND UNSOLVABILITY, Martin Davis. Classic graduate-level introduction to theory of computability, usually referred to as theory of recurrent functions. New preface and appendix. 288pp. 5⅜ x 8½. 61471-9 Pa. $8.95

GENERAL CHEMISTRY, Linus Pauling. Revised 3rd edition of classic first-year text by Nobel laureate. Atomic and molecular structure, quantum mechanics, statistical mechanics, thermodynamics correlated with descriptive chemistry. Problems. 992pp. 5⅜ x 8½. 65622-5 Pa. $19.95

AN INTRODUCTION TO MATRICES, SETS AND GROUPS FOR SCIENCE STUDENTS, G. Stephenson. Concise, readable text introduces sets, groups, and most importantly, matrices to undergraduate students of physics, chemistry, and engineering. Problems. 164pp. 5⅜ x 8½. 65077-4 Pa. $7.95

THE HISTORICAL BACKGROUND OF CHEMISTRY, Henry M. Leicester. Evolution of ideas, not individual biography. Concentrates on formulation of a coherent set of chemical laws. 260pp. 5⅜ x 8½. 61053-5 Pa. $8.95

THE PHILOSOPHY OF MATHEMATICS: An Introductory Essay, Stephan Körner. Surveys the views of Plato, Aristotle, Leibniz & Kant concerning propositions and theories of applied and pure mathematics. Introduction. Two appendices. Index. 198pp. 5⅜ x 8½. 25048-2 Pa. $8.95

THE DEVELOPMENT OF MODERN CHEMISTRY, Aaron J. Ihde. Authoritative history of chemistry from ancient Greek theory to 20th-century innovation. Covers major chemists and their discoveries. 209 illustrations. 14 tables. Bibliographies. Indices. Appendices. 851pp. 5⅜ x 8½. 64235-6 Pa. $18.95

CATALOG OF DOVER BOOKS

DE RE METALLICA, Georgius Agricola. The famous Hoover translation of greatest treatise on technological chemistry, engineering, geology, mining of early modern times (1556). All 289 original woodcuts. 638pp. 6¾ x 11. 60006-8 Pa. $21.95

SOME THEORY OF SAMPLING, William Edwards Deming. Analysis of the problems, theory and design of sampling techniques for social scientists, industrial managers and others who find statistics increasingly important in their work. 61 tables. 90 figures. xvii + 602pp. 5⅜ x 8½. 64684-X Pa. $16.95

THE VARIOUS AND INGENIOUS MACHINES OF AGOSTINO RAMELLI: A Classic Sixteenth-Century Illustrated Treatise on Technology, Agostino Ramelli. One of the most widely known and copied works on machinery in the 16th century. 194 detailed plates of water pumps, grain mills, cranes, more. 608pp. 9 x 12. 28180-9 Pa. $24.95

LINEAR PROGRAMMING AND ECONOMIC ANALYSIS, Robert Dorfman, Paul A. Samuelson and Robert M. Solow. First comprehensive treatment of linear programming in standard economic analysis. Game theory, modern welfare economics, Leontief input-output, more. 525pp. 5⅜ x 8½. 65491-5 Pa. $17.95

ELEMENTARY DECISION THEORY, Herman Chernoff and Lincoln E. Moses. Clear introduction to statistics and statistical theory covers data processing, probability and random variables, testing hypotheses, much more. Exercises. 364pp. 5⅜ x 8½. 65218-1 Pa. $10.95

THE COMPLEAT STRATEGYST: Being a Primer on the Theory of Games of Strategy, J.D. Williams. Highly entertaining classic describes, with many illustrated examples, how to select best strategies in conflict situations. Prefaces. Appendices. 268pp. 5⅜ x 8½. 25101-2 Pa. $8.95

CONSTRUCTIONS AND COMBINATORIAL PROBLEMS IN DESIGN OF EXPERIMENTS, Damaraju Raghavarao. In-depth reference work examines orthogonal Latin squares, incomplete block designs, tactical configuration, partial geometry, much more. Abundant explanations, examples. 416pp. 5⅜ x 8¼. 65685-3 Pa. $10.95

THE ABSOLUTE DIFFERENTIAL CALCULUS (CALCULUS OF TENSORS), Tullio Levi-Civita. Great 20th-century mathematician's classic work on material necessary for mathematical grasp of theory of relativity. 452pp. 5⅜ x 8½. 63401-9 Pa. $11.95

VECTOR AND TENSOR ANALYSIS WITH APPLICATIONS, A.I. Borisenko and I.E. Tarapov. Concise introduction. Worked-out problems, solutions, exercises. 257pp. 5⅜ x 8¼. 63833-2 Pa. $9.95

THE FOUR-COLOR PROBLEM: Assaults and Conquest, Thomas L. Saaty and Paul G. Kainen. Engrossing, comprehensive account of the century-old combinatorial topological problem, its history and solution. Bibliographies. Index. 110 figures. 228pp. 5⅜ x 8½. 65092-8 Pa. $7.95

CATALOG OF DOVER BOOKS

CATALYSIS IN CHEMISTRY AND ENZYMOLOGY, William P. Jencks. Exceptionally clear coverage of mechanisms for catalysis, forces in aqueous solution, carbonyl- and acyl-group reactions, practical kinetics, more. 864pp. 5⅜ x 8½.
65460-5 Pa. $19.95

PROBABILITY: An Introduction, Samuel Goldberg. Excellent basic text covers set theory, probability theory for finite sample spaces, binomial theorem, much more. 360 problems. Bibliographies. 322pp. 5⅜ x 8½. 65252-1 Pa. $10.95

LIGHTNING, Martin A. Uman. Revised, updated edition of classic work on the physics of lightning. Phenomena, terminology, measurement, photography, spectroscopy, thunder, more. Reviews recent research. Bibliography. Indices. 320pp. 5⅜ x 8¼. 64575-4 Pa. $8.95

PROBABILITY THEORY: A Concise Course, Y.A. Rozanov. Highly readable, self-contained introduction covers combination of events, dependent events, Bernoulli trials, etc. Translation by Richard Silverman. 148pp. 5⅜ x 8¼. 63544-9 Pa. $8.95

AN INTRODUCTION TO HAMILTONIAN OPTICS, H. A. Buchdahl. Detailed account of the Hamiltonian treatment of aberration theory in geometrical optics. Many classes of optical systems defined in terms of the symmetries they possess. Problems with detailed solutions. 1970 edition. xv + 360pp. 5⅜ x 8½.
67597-1 Pa. $10.95

STATISTICS MANUAL, Edwin L. Crow, et al. Comprehensive, practical collection of classical and modern methods prepared by U.S. Naval Ordnance Test Station. Stress on use. Basics of statistics assumed. 288pp. 5⅜ x 8½. 60599-X Pa. $8.95

DICTIONARY/OUTLINE OF BASIC STATISTICS, John E. Freund and Frank J. Williams. A clear concise dictionary of over 1,000 statistical terms and an outline of statistical formulas covering probability, nonparametric tests, much more. 208pp. 5⅜ x 8½. 66796-0 Pa. $7.95

STATISTICAL METHOD FROM THE VIEWPOINT OF QUALITY CONTROL, Walter A. Shewhart. Important text explains regulation of variables, uses of statistical control to achieve quality control in industry, agriculture, other areas. 192pp. 5⅜ x 8½. 65232-7 Pa. $8.95

METHODS OF THERMODYNAMICS, Howard Reiss. Outstanding text focuses on physical technique of thermodynamics, typical problem areas of understanding, and significance and use of thermodynamic potential. 1965 edition. 238pp. 5⅜ x 8½.
69445-3 Pa. $8.95

STATISTICAL ADJUSTMENT OF DATA, W. Edwards Deming. Introduction to basic concepts of statistics, curve fitting, least squares solution, conditions without parameter, conditions containing parameters. 26 exercises worked out. 271pp. 5⅜ x 8½.
64685-8 Pa. $9.95

TENSOR CALCULUS, J.L. Synge and A. Schild. Widely used introductory text covers spaces and tensors, basic operations in Riemannian space, non-Riemannian spaces, etc. 324pp. 5⅜ x 8¼. 63612-7 Pa. $11.95

CATALOG OF DOVER BOOKS

A CONCISE HISTORY OF MATHEMATICS, Dirk J. Struik. The best brief history of mathematics. Stresses origins and covers every major figure from ancient Near East to 19th century. 41 illustrations. 195pp. 5⅜ x 8½. 60255-9 Pa. $8.95

A SHORT ACCOUNT OF THE HISTORY OF MATHEMATICS, W.W. Rouse Ball. One of clearest, most authoritative surveys from the Egyptians and Phoenicians through 19th-century figures such as Grassman, Galois, Riemann. Fourth edition. 522pp. 5⅜ x 8½. 20630-0 Pa. $13.95

HISTORY OF MATHEMATICS, David E. Smith. Nontechnical survey from ancient Greece and Orient to late 19th century; evolution of arithmetic, geometry, trigonometry, calculating devices, algebra, the calculus. 362 illustrations. 1,355pp. 5⅜ x 8½. 20429-4, 20430-8 Pa., Two-vol. set $27.90

THE GEOMETRY OF RENÉ DESCARTES, René Descartes. The great work founded analytical geometry. Original French text, Descartes' own diagrams, together with definitive Smith-Latham translation. 244pp. 5⅜ x 8½. 60068-8 Pa. $8.95

GAMES, GODS & GAMBLING: A History of Probability and Statistical Ideas, F. N. David. Episodes from the lives of Galileo, Fermat, Pascal, and others illustrate this fascinating account of the roots of mathematics. Features thought-provoking references to classics, archaeology, biography, poetry. 1962 edition. 304pp. 5⅜ x 8½. (USO) 40023-9 Pa. $9.95

THE HISTORY OF THE CALCULUS AND ITS CONCEPTUAL DEVELOPMENT, Carl B. Boyer. Origins in antiquity, medieval contributions, work of Newton, Leibniz, rigorous formulation. Treatment is verbal. 346pp. 5⅜ x 8½. 60509-4 Pa. $9.95

THE THIRTEEN BOOKS OF EUCLID'S ELEMENTS, translated with introduction and commentary by Sir Thomas L. Heath. Definitive edition. Textual and linguistic notes, mathematical analysis. 2,500 years of critical commentary. Not abridged. 1,414pp. 5⅜ x 8½. 60088-2, 60089-0, 60090-4 Pa., Three-vol. set $34.85

GAMES AND DECISIONS: Introduction and Critical Survey, R. Duncan Luce and Howard Raiffa. Superb nontechnical introduction to game theory, primarily applied to social sciences. Utility theory, zero-sum games, n-person games, decision-making, much more. Bibliography. 509pp. 5⅜ x 8½. 65943-7 Pa. $14.95

THE HISTORICAL ROOTS OF ELEMENTARY MATHEMATICS, Lucas N.H. Bunt, Phillip S. Jones, and Jack D. Bedient. Fundamental underpinnings of modern arithmetic, algebra, geometry and number systems derived from ancient civilizations. 320pp. 5⅜ x 8½. 25563-8 Pa. $9.95

CALCULUS REFRESHER FOR TECHNICAL PEOPLE, A. Albert Klaf. Covers important aspects of integral and differential calculus via 756 questions. 566 problems, most answered. 431pp. 5⅜ x 8½. 20370-0 Pa. $9.95

CATALOG OF DOVER BOOKS

CHALLENGING MATHEMATICAL PROBLEMS WITH ELEMENTARY SOLUTIONS, A.M. Yaglom and I.M. Yaglom. Over 170 challenging problems on probability theory, combinatorial analysis, points and lines, topology, convex polygons, many other topics. Solutions. Total of 445pp. 5⅜ x 8½. Two-vol. set.
Vol. I: 65536-9 Pa. $8.95
Vol. II: 65537-7 Pa. $7.95

FIFTY CHALLENGING PROBLEMS IN PROBABILITY WITH SOLUTIONS, Frederick Mosteller. Remarkable puzzlers, graded in difficulty, illustrate elementary and advanced aspects of probability. Detailed solutions. 88pp. 5⅜ x 8½.
65355-2 Pa. $4.95

EXPERIMENTS IN TOPOLOGY, Stephen Barr. Classic, lively explanation of one of the byways of mathematics. Klein bottles, Moebius strips, projective planes, map coloring, problem of the Koenigsberg bridges, much more, described with clarity and wit. 43 figures. 210pp. 5⅜ x 8½. 25933-1 Pa. $8.95

RELATIVITY IN ILLUSTRATIONS, Jacob T. Schwartz. Clear nontechnical treatment makes relativity more accessible than ever before. Over 60 drawings illustrate concepts more clearly than text alone. Only high school geometry needed. Bibliography. 128pp. 6⅛ x 9¼. 25965-X Pa. $7.95

AN INTRODUCTION TO ORDINARY DIFFERENTIAL EQUATIONS, Earl A. Coddington. A thorough and systematic first course in elementary differential equations for undergraduates in mathematics and science, with many exercises and problems (with answers). Index. 304pp. 5⅜ x 8½. 65942-9 Pa. $9.95

FOURIER SERIES AND ORTHOGONAL FUNCTIONS, Harry F. Davis. An incisive text combining theory and practical example to introduce Fourier series, orthogonal functions and applications of the Fourier method to boundary-value problems. 570 exercises. Answers and notes. 416pp. 5⅜ x 8½. 65973-9 Pa. $13.95

AN INTRODUCTION TO ALGEBRAIC STRUCTURES, Joseph Landin. Superb self-contained text covers "abstract algebra": sets and numbers, theory of groups, theory of rings, much more. Numerous well-chosen examples, exercises. 247pp. 5⅜ x 8½.
65940-2 Pa. $8.95

STARS AND RELATIVITY, Ya. B. Zel'dovich and I. D. Novikov. Vol. 1 of *Relativistic Astrophysics* by famed Russian scientists. General relativity, properties of matter under astrophysical conditions, stars and stellar systems. Deep physical insights, clear presentation. 1971 edition. References. 544pp. 5⅜ x 8½.
69424-0 Pa. $14.95

Prices subject to change without notice.
Available at your book dealer or write for free Mathematics and Science Catalog to Dept. GI, Dover Publications, Inc., 31 East 2nd St., Mineola, N.Y. 11501. Dover publishes more than 250 books each year on science, elementary and advanced mathematics, biology, music, art, literature, history, social sciences and other areas.